Neural and Fuzzy Logic Control
of Drives and Power Systems

# Neural and Fuzzy Logic Control of Drives and Power Systems

M.N. Cirstea, A. Dinu, J.G. Khor,
M. McCormick

Newnes

OXFORD AMSTERDAM BOSTON LONDON NEW YORK PARIS
SAN DIEGO SAN FRANCISCO SINGAPORE SYDNEY TOKYO

Newnes
An imprint of Elsevier Science
Linacre House, Jordan Hill, Oxford OX2 8DP
225 Wildwood Avenue, Woburn, MA 01801-2041

First published 2002



**British Library Cataloguing in Publication Data**
A catalogue record for this book is available from the British Library

ISBN 0 7506 55585



Typeset at Replika Press Pvt Ltd, Delhi 110 040, India
Printed and bound in Great Britain

# Contents

# Preface

The idea of writing this book arose from the need to investigate the main principles of modern power electronic control strategies, using fuzzy logic and neural networks, for research and teaching. Primarily, the book aims to be a quick learning guide for postgraduate/undergraduate students or design engineers interested in learning the fundamentals of modern control of drives and power systems in conjunction with the powerful design methodology based on VHDL.

At the same time, the book is structured to address the more complex needs of professional designers, using VHDL for neural and fuzzy logic systems design, by including comprehensive design examples. This facilitates the understanding of hardware description language applications and provides a practical approach to the development of advanced controllers for power electronics.

The first section of the book contains a brief review of control strategies for electric drives/power systems and a summary description of neural networks, fuzzy logic, electronic design automation (EDA) techniques, ASICs/FPGAs and VHDL. The aspects covered allow a basic understanding of the main principles of modern control. The second section contains two comprehensive case studies. The first deals with neural current and speed control of induction motor drives, whereas the second presents the environmentally friendly fuzzy logic control of a diesel-driven stand-alone synchronous generator set. Both control strategies were implemented in Xilinx FPGAs and comprehensively tested by simulation and experimental measurements.

This book brings together the complex features of control strategies, EDA, neural networks, fuzzy logic, electric machines and drives, power systems and VHDL and forms a basic guide for the understanding of the fundamental principles of modern power electronic control systems design. To be expert in the design of advanced digital controllers for drives and power systems, extra reading is strongly recommended and comprehensive material is referenced in the bibliographical section. The book includes a number of recent research results from work carried out by the authors, who are members of the electronic control and drives research group at De Montfort University, Leicester, UK.

The facilities provided by the university and the support of NEWAGE AVK SEG, Stamford, UK, a major international manufacturer of electric generators, are gratefully acknowledged.

Dr Marcian N. Cirstea
Dr Andrei Dinu
Dr Jeen G. Khor
Prof. Malcolm McCormick

# 1

# Control systems

## 1.1 Control theory: historical review

The function of a control mechanism is to maintain certain essential properties of a system at a desired value under perturbations. Historical control systems which are simple but effective have been employed in water regulation and control of liquid level in wine vessels for centuries. Some of these concepts are still used today, for example the float system in the water tank of the toilet flush. However, modern control systems used in today's industry are much more complex and owe their beginnings to the development of control theory. The earliest significant work in modern automatic control can be traced to James Watt's design of the fly-ball governor (1788) for the speed control of a steam engine. In 1868, Maxwell [170] presented the first mathematical analysis of feedback control. It was during this time that systematic studies into control systems and feedback dynamics began. One significant development was the well-known Routh's stability criterion (1877) which won E.J. Routh the Adam's Prize.

The early twentieth century saw the beginning of what is now known as **classical control theory**. Minorsky's work (1922) on the determination of stability from the differential equation describing the system (characteristic equation) and Nyquist's development (1932) of a graphical procedure for determining stability (frequency response) substantially contributed to the study of control theory. In 1934, Hazen [111] introduced the term 'servomechanism' to describe position control systems in his attempt to develop a generalised theory of servomechanisms. Two years later, the development of the *proportional integral derivative (PID) controller* was described by Callender *et al.* (1936). Control theory, like many branches of engineering, underwent significant development during World War II. Based on Nyquist's work, H.W. Bode introduced a method for feedback amplifier design, now known as the *Bode plot* (1945). By 1948, the *root locus* method of design and stability analysis was developed by W.R. Evans [93]. With the introduction of digital computers in the 1960s, the use of frequency response and characteristic equations began to give way to ordinary differential equations (ODEs), which worked well with computers. This led to the birth of **modern control theory**.

While the term classical control theory is used to describe the design methods of Bode, Nyquist, Minorsky and similar workers, modern control theory relies on ODE design methods that are more suitable for computer aided engineering, for example the *state space approach*. Both these branches of control theory rely on mathematical representation of the control plant from which to derive its performance. To address the issues of non-linearities and time-variant parameters in plant models, control strategies that continuously adapt to the variations of plant characteristics have been introduced. Generally known as adaptive control systems, they include techniques such as self-tuning control, H-infinity control, model referencing adaptive control and sliding mode control, Studies also include the use of non-linear state observers that continuously estimate the parameters of the control plant [174]. They can be employed to tackle the issue of *non-observability*, that is the condition whereby not all of the required states are available for feedback. This may be the cheaper solution because it does not require as many sensors, such as in variable speed drives [59], or because it is physically difficult or even impossible to obtain the feedback states such as in a nuclear reactor.

In many instances, the mathematical model of the plant is simply unknown or ill-defined, leading to greater complexities in the design of the control system. It has been proposed that **intelligent control systems** give a better performance in such cases. Unlike conventional control techniques, intelligent controllers are based on *artificial intelligence* (AI) rather than on a plant model. They imitate the human decision-making process and can often be implemented in complex systems with more success than conventional control techniques. AI can be classified into expert systems, fuzzy logic, artificial neural networks and genetic algorithms. With the exception of expert systems, these techniques are based on *soft-computing* methods. The result is that they are capable of making approximations and 'intelligent guesses' where necessary, in order to come out with a 'good enough' result under a given set of constraints. Intelligent control systems may employ one or more AI techniques in their design.

## 1.2 Introduction to control systems

A system is a group of physical components assembled to perform a specific function. A system may be electrical, mechanical, hydraulic, pneumatic, thermal, biomedical, or a combination of any of these systems. An ideal control system is one in which an output is a direct function of input. However, in practice disturbances affect the output being controlled and cause it to deviate from the desired value. A control system may be defined in a variety of ways, but the most basic definition is:

> *A control system is a group of components assembled in such a way as to regulate an energy input to achieve the desired output.*

### 1.2.1 Classification

Control systems are classified based on the following characteristics:

(A) The type of operating techniques used in driving the output to a desired value:
- *Analogue control systems* – analogue techniques are used to process the input signal and control the output signal.
- *Digital control systems* – digital techniques are employed to control the output. Analogue, digital, or both analogue and digital techniques may be used to control a desired physical quantity, which can be any physical variable (temperature, pressure, electric voltage, mechanical position, etc.). At the beginning

of the control era, most control systems were analogue employing analogue techniques, but these systems were relatively bulky, complex and cumbersome, both to design and to maintain. However, with the development of digital technology the design of control systems became easier as well as more economical. Nowadays, digital control systems are used more and more due to their accuracy, precision, high speed of response, wide range of applications and, why not, elegance. The main difference between an analogue control system and a digital control system is that the first processes continuous signals while the second processes discrete signals, which are in fact periodically taken samples of continuous signals.

(B) The use of feedback:

- *Closed-loop systems* with either *positive* (regenerative) feedback or *negative* (degenerative) feedback. If an output or part of an output is fed back so that it can be compared with an input, the system is said to use feedback and the arrangement forms a closed loop. If the feedback signal aids an input signal – the feedback is positive; if the feedback signal opposes the input signal – the feedback is negative.
- *Open-loop systems* – systems that don't use a feedback. Advantages of open-loop control systems are that they are relatively simple, economical and easy to maintain. On the other hand, closed-loop systems are more accurate, stable and less sensitive to outside disturbances, although they are relatively expensive, complex and not easy to maintain.

(C) The nature of system behaviour:

- *Linear systems* – if the amplitude proportionality property (a) and the principle of superposition (b) are satisfied. (a) If the system output is $o(t)$ for a given input $i(t)$, then for an input $K_i(t)$ the output should be $K_o(t)$; $K$ is the proportionality constant. (b) According to the superposition principle if $i_1(t)$ and $i_2(t)$ are inputs and their corresponding outputs are $o_1(t)$ and $o_2(t)$, then the input $i_1(t) + i_2(t)$ must produce the output $o_1(t) + o_2(t)$. Example d.c. motor speed control system.
- *Non-linear systems* – these do not follow amplitude proportionality and the superposition principle.

(D) The application area:

- *Servomechanisms* – control systems in which the output or the controlled variable is a mechanical position or the rate of change of mechanical position (a motion). Example: d.c. motor speed control.
- *Sequential control systems* – systems in which a prescribed set of operations are performed. Example: automatic washing machine.
- *Numerical control systems* – they act on 'numerical information' (controlled variables as position, speed, direction – coded in the form of instructions) stored on a 'control medium' (simply a storage medium: punched cards, paper tape, magnetic tape, CD-ROM). The control medium contains all the instructions necessary to accomplish a desired manufacturing operation (milling, welding, drilling). The major advantage of a numerical control system is the flexibility of its control medium.
- *Process control systems* – the variables in a manufacturing process are controlled. Examples: temperature, pressure, conductivity. They can be either closed-loop or open-loop control systems.

(E) The method of generating the control pulses:
- Single-channel control systems.
- Multi-channel control systems.

(F) The synchronisation between the signals within the control system and input voltages:
- Synchronous control systems.
- Asynchronous control systems.

### 1.2.2 Characteristics of control systems

Although different systems are designed to perform different functions, all of them have to meet some common requirements. The major characteristics of a typical control system, which are often used as measures of performance to evaluate a system under consideration, are the following:

#### *1.2.2.1 Stability*

A system is said to be stable if its output attains a certain value in a finite time after the input is applied. When the output of a system remains constant and does not change as a function of time, the output is said to attain a steady-state value. On the contrary, an unstable system never attains a steady-state value. A practical system must be stable. An unstable system may be made stable by using certain techniques, of which the most common is the use of compensating networks. Often, an unstable system is made stable simply by using negative feedback.

#### *1.2.2.2 Accuracy*

The accuracy indicates deviation of the actual output from its desired value and it is a relative measure of system performance. Generally, the accuracy of a control system is improved by using control models such as integral or integral plus proportional.

#### *1.2.2.3 Speed of response*

The speed of response is a measure of how quickly an output attains a steady-state value after the input is applied. A practical system must have a finite response time.

#### *1.2.2.4 Sensitivity*

The sensitivity of a system is a measure of how sensitive the output is to changes in the values of physical components as well as environmental conditions. The dependence of output on disturbances can be minimised by using certain compensating networks.

#### *1.2.2.5 Representation*

The most common methods used to represent control systems in order to improve communication between design engineers and users are block diagrams and signal flow graphs. They help visualisation of the system under consideration at a glance. The block diagram of a system consists of blocks, directed line segments joining these blocks and the summing junctions or error detectors that are used to add the signals algebraically.

A signal flow graph is a diagram that indicates the manner in which the signal flows in a given system. It is a one-line diagram that uses directed segments.

This short overview on control systems and their general features aimed to familiarise the reader with basic characteristics of control systems. The next section focuses on some general aspects of control systems for electrical drives, especially for a.c. electrical drives.

## 1.3 Control systems for a.c. drives

A specific definition of a process control system may be: '*A control system is a combination of amplifiers, transducers, and actuators, which collectively act on a process to maintain some condition at a required value.*' The adjustable speed a.c. drive constitutes a multivariable control system and therefore, in principle, the general theories of multivariable control system should be applicable. Here, the voltages and the frequency are the control inputs and the outputs may be speed, position, torque, airgap flux, stator current or a combination of all of them. If the mathematical model of the system is considered precise and no extraneous disturbances are possible, then theoretically open loop control of the drive system should be satisfactory. This means that the control functions can be defined uniquely to give the specified performance of the drive system. The performance of the drive can be optimised by generating critical control functions using modern optimal control theories. Optimal control theory is extremely difficult to apply to a real life industrial drive system because of the laborious computational requirement and the inaccuracies of the system model.

### 1.3.1 The objects of control systems in a.c. drives

Before the advent of power semiconductor devices, a.c. machines were commonly accepted as fixed speed machines due to their connection to a fixed voltage and frequency supply. Similarly, d.c. motors were considered the workhorses in industry for variable speed applications. Although control principles and converter equipment are simple, the d.c. machine is expensive when compared to the simple and rugged cage type induction motor. In addition, the principal problem of a d.c. machine is that commutators and brushes make it unreliable, unsuitable to operate in dusty and explosive environments and it requires frequent maintenance. The a.c. machine is more rugged and reliable, as well as less expensive and more efficient, especially the cage type induction motor; however, the cost of the converter and the control is considerably higher, which makes the a.c. drive more expensive than the d.c. drive. In addition, the control of a.c. drives is very complex and requires intricate signal processing to obtain a performance comparable to the d.c. drive. Present technology aims to provide substantial cost reductions and performance improvements for a.c. drive systems to make them more universally used. Some of the expanding application areas are:

- Replacement of variable speed d.c. drives by appropriate a.c. drive systems.
- Application of adjustable speed a.c. drives to constant speed process control, thereby saving energy.

- Replacement of heat engines (which use petroleum-based energy), hydraulic and pneumatic controlled drive systems by electric a.c. drive systems (as in the electric car).

An electrical a.c. machine is a complex electromagnetic and mechanical structure that is designed for optimal conversion of electrical energy into mechanical energy, and vice versa. In a conventional multiphase machine, the time phase distribution of power supply and space phase distribution of stator windings produce a rotating airgap flux wave, and the speed of rotation correlates with the frequency of the power supply. The airgap flux reacts with the rotor magnetomotive force (MMF) wave to develop the electrical torque, the magnitude of which depends on the flux and MMF amplitudes and their phase displacement angle. The rotor MMF in a synchronous machine is created by a separate field winding that carries d.c. current, whereas in an induction motor it is produced by the stator induction effect. The speed to frequency relationship is unique in a synchronous machine, but for induction motors, the rotor must 'slip' from synchronous speed to induce rotor MMF, which results in the development of the torque.

In adjustable speed a.c. drive systems the static power converter constitutes an interface between the primary power supply and the machine. The converter generally converts and controls the 60 Hz, three-phase a.c. supply for the machine, which may be at variable-voltage-constant-frequency, constant-voltage-variable-frequency or variable-voltage-variable-frequency. A converter consists of a matrix of power semiconductor switching devices which may be thyristors, gate turn-off (GTO) devices, power transistors, or power MOS. This acts like a switch mode power amplifier between the control signals and the output, with inherently rich harmonics at the input and the output. The output harmonics cause machine heating and torque pulsation problems and the input harmonics cause line voltage distortion and electromagnetic interference (EMI) problems. Since generally no additional dynamics are involved in the converter circuit, the input and output powers match at any instant, and the output waveform may be constructed from input waves and the characteristic switching functions.

A well-designed drive system should carefully consider the interaction between the converter and the machine, and the various design trade-off considerations. As the converter operation and its mode of control severely affect the machine performance, the machine parameters similarly affect the converter performance. The power switching devices of a converter are delicate and very sensitive to voltage and current transients. While a machine may have large overload current capability, the semiconductor device overload capability is very limited because of the short transient thermal time constant. In addition, the commutation capability of a converter may soon reach the limiting condition due to overcurrent. Therefore, the converter is normally designed to match the peak power capability of the machine, which is an expensive proposition. Because of the possibility of overvoltage and overcurrent failures, a converter normally requires well-designed control and protection schemes.

### 1.3.2 Basic principle of microcomputer control

Traditional control systems are normally implemented using analogue and digital hardware. In its relatively short existence, digital computer technology has touched, and had a profound effect upon, many areas of life. Its enormous success is due largely to the flexibility and reliability that computer systems offer to potential users. This, coupled with the ability to handle and manipulate vast amounts of data quickly, efficiently and repeatedly, has made computers extremely useful in many varied applications. In control systems the digital computer acts as the controller and provides the enabling technology that allows the design and implementation of the overall system, so that satisfactory performance is obtained.

Digital control systems differ from continuous systems in that the computer acts only at instants of time rather than continuously. This is because a computer can execute only one operation at a time, and so the overall algorithm proceeds in a sequential manner. Hence, taking measurements from the system and processing them to compute an activating signal, which is then applied to the system, is a standard procedure in a typical control application. Having applied a control action, the computer collects the next set of measurements and repeats the complete iteration in an endless loop. The maximum frequency of control update is defined by the time taken to complete one cycle of the loop. This is obviously dependent upon the complexity of the control task and the capabilities of the hardware.

At first glance this appears to be a poorly matched situation, where a digital computer is attempting to control a continuous system by applying impulsive signals to it every now and then; from this viewpoint it seems unlikely that satisfactory results are possible. Fortunately, the setup is not as awkward as it first appears. If the cycle iteration speed of the computer and the dynamics of the system are taken into account, adequate performance can be expected when the former is much faster than the latter. Indeed, digital controllers have been used to give results as good as, or better than, analogue controllers in numerous situations, with the added feature that the control strategies can be varied by simply reprogramming the computer instead of having to change the hardware. In addition, analogue controllers are susceptible to ageing and drift, which in turn causes degradation in performance. These advantages have attracted many users to adopt digital technology in preference to conventional methods and made computer control applicable to many areas. Some of the current interest areas are: auto-pilots for aeroplanes/missiles, satellite altitude control, industrial and process control, robotics, navigational systems and radar and building energy management and control systems.

With advances in VLSI (very large scale integration) and denser packing capabilities, faster integrated circuits can be manufactured which result in quicker and more powerful computers. Therefore, application to control areas which a few years ago were considered to be impractical or impossible because of computer limitations, are now entering the realms of possibility.

Another recent advance in computer systems is in the area of parallel processing, where the computational task is shared out between several processors that can communicate with each other in an efficient manner. Individual processors can solve sub-problems, with the results brought together in some ordered way, to arrive at the solution to the overall problem. Since many processors can be incorporated to execute the computations, it is possible to solve large and complex problems quickly and efficiently

One of the problems in a computer control system is the interfacing between computers and continuous systems so that the analogue plant signals can first be read into the computer, and then digital control signals can be applied to the system. Analogue signals must be converted into digital form for analysis in the computer, and the digital signals from the computer have to be converted back to analogue form for application to the plant under control. This kind of converter can introduce significant conversion time delays into digital computer control system applications. These, together with other sequential processing delays, mean that when continuous analogue signals are to be converted into digital form, the conversions can only be performed at discrete instants, separated by finite intervals.

In computer control applications impulsive signals are inappropriate for controlling analogue systems, since these require an input signal to be present all the time. To overcome this difficulty, hold devices are inserted at the digital-to-analogue interfaces. The simplest device available is a zero-order-hold (ZOH), which holds the output constant at the value fed to it at the last sampling instant; hence a piecewise constant signal is generated. Higher order holds are also available, which use a number of previous sampling instant values to generate the signal over the current sampling interval.

Mainly, in a digital control loop, the following procedure must take place:

- Measure system output and compare with the desired value to give an error.
- Use the error, via a control law, to compute an actuating signal.
- Apply this corrective input to the system.
- Wait for the next sampling instant.
- Repeat this algorithm.

The functions that can be incorporated in microcomputer software are summarised as follows:

- Converter control, including firing pulse generation.
- Feedback control.
- Signal estimation for system control.
- Drive mode sequencing.
- Diagnostics.

The superiority of microcomputer control over conventional hardware-based control can be recognised as evident when dealing with complex drive control systems. The simplification of hardware saves control electronics cost and improves the system reliability. Digital control has inherently improved noise immunity, which is particularly important in drive systems because of large power switching transients in the converters. Additionally, the software control algorithms can easily be altered or improved in the future without changing the hardware. Another important feature is that the structure and parameters of the control system can be altered in real time, making the control adaptive to the plant characteristics. The complex computation and decision-taking capabilities of microcomputers enables the application of the modern optimal and adaptive control theories to optimise the drive system performance. In addition, powerful diagnoses can be written in the software. Microcomputer technology is moving at such a fast rate that the use of efficient high level language with large hardware integration and VLSI implementation of the controller is easily possible.

Unlike dedicated hardware control, a microcomputer executes control in serial fashion, i.e. multitasking operations are performed in a time multiplexed method. As a result, a slow computation capability may pose serious problems in executing the fast control loops. However, the problem can be solved by multi-microprocessor control, where judicious partitioning of tasks can significantly enhance the execution speed. The different stages necessary in microcomputer control development of a drive system are:

- Develop control strategy.
- Make simplified system study and determine control parameters.
- Translate into digital control algorithm.
- Simulate drive system on hybrid/digital computer-iterate control.
- Develop hardware and software.
- Design and build breadboard test.

The foregoing outlines some basic aspects of microcomputer/microprocessor control. Presently, many digital control systems are microprocessor-based, primarily because of the availability of control integrated circuits (ICs), cheaper memories and tremendous advancements in data handling capabilities. A big step forward in control is the use of application specific integrated circuits (ASICs), which have successfully replaced microprocessors due to their ease of design using modern computer-aided design (CAD)/ electronic design automation (EDA) techniques.

# 2

# Modern control systems design using CAD techniques

## 2.1 Electronic design automation (EDA)

Following the traditional design route, the engineer begins with the idea, then normally proceeds to the paper circuit design stage. The design then continues through to the prototype stage, using any of the many traditional construction methods. The prototype design is then tested and verified against the specification. At this point if any conceptual fault is found, a redesign is carried out and the process is repeated.

The use and simulation of mathematical models for electrical systems design has been employed for some considerable time, but the functional models derived must then be translated into hardware and it is at this stage that the technology-based design rules and delays are taken into account. Electronic design automation (EDA) enables this transition to take place with a higher degree of confidence than was previously possible.

EDA tools are well suited to providing low level, high speed hardware, to implement the control functions in power electronic systems. Computer-aided design (CAD) software enables the design and evaluation of these complex digital circuits within the PC/workstation environment, without the requirement for physical hardware at this stage. For the successful development of the specialised microelectronics hardware needed, a knowledge of available technologies and EDA techniques for design, simulation, layout, PCB production and verification is required. The design cycle can be considerably reduced by removing three parts of the design cycle before the design is verified, by a technique known as the modelling and simulation method. This allows a product to be produced for the market in a much shorter time than using traditional methods. The method is illustrated in the block diagram in Fig. 2.1.

The method allows the development of the design using the CAD system, whereby verification is carried out by simulating the circuit design using software models. At this point any design faults should be identified and rectified without going through the costly step of prototype construction for verification. The modelling and simulation method allows the design to be about 98 per cent certain of working correctly first time [186].

The work of multidisciplinary teams is facilitated by the large variety of software integrated into the EDA environment which improves the efficiency of the design process by integrating the expertise of the specialists into an enabling environment. Further development of the methodology leads to a *concurrent engineering* approach to the design process. The basic concept of concurrent engineering is that all parts of the design, production, manufacture, marketing, financing and managing of a product are

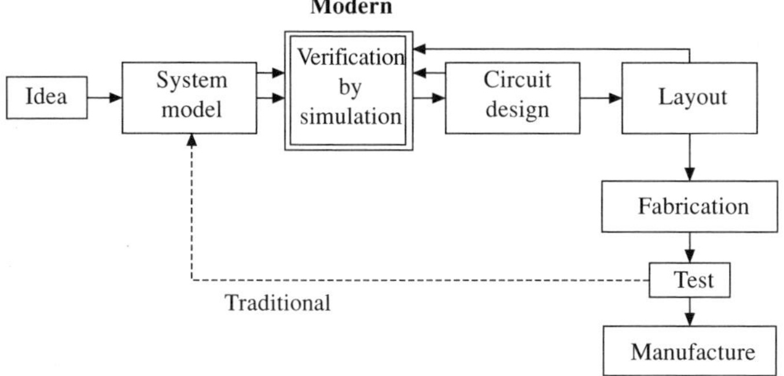


**Fig. 2.1** Modern modelling and simulation design methodology versus traditional approach

carried out in a computer and workstation environment. This allows access to a common database where any modification to a product is updated to all members of the design and support team, but only key personnel are allowed to alter data [51].

The basic forces of change that affect product development are: technology, tools, tasks, talent and time. These forces are at work in disturbing or stabilising a specific company setting the product development environment. This environment includes people, concepts and technologies necessary to design a product, manufacture it and market it. According to Carter and Sullivan [52], change forces not only exist in parallel, but also are fully integrated vertically and horizontally in the product development environment.

With the increasingly competitive nature of the electronics industry, the development time for new products is rapidly decreasing. Engineers are constantly expected to develop new products for the market within a short time. The introduction of electronic design automation in the late 1970s and early 1980s has allowed the development time of electronic designs to be shortened considerably. EDA is a design methodology in which dedicated tools, primarily software products, are used to assist in the development of integrated circuits, printed circuit boards (PCBs) and electronic systems. In the early days, EDA tools were nothing more than a set of incoherent design tools that aided a specific stage in the development cycle, providing what are called '*islands of automation*'. Where the different tools need to share data, user-written data translators were sometimes used. EDA tools have since evolved into an integration of design tool-sets that conform to a standard data management protocol, thus eliminating the need for data translators. Some of the advantages of EDA include [40]:

- Enabling more thorough verification of design using simulation tools. This allows the design to be verified before being implemented into hardware, thus design faults can be detected in the early stages of the design process.
- Exploring alternative designs using the synthesis and implementation tools. The designer can create a few alternative designs before selecting the best design for the implementation.
- Automating some of the design steps, thus allowing the designer to concentrate on more important activities.
- Ease in design data management.
- Enabling the designer to operate at higher levels of abstraction, i.e. 'top-down' design method.

Using hardware description languages such as VHDL and Verilog HDL, top-down design is realisable. The designs are first described at register transfer level (RTL) where the design functions are addressed, with no reference to the hardware required for implementation. RTL descriptions can then be automatically translated into gate level using logic synthesis tools. This design methodology is similar to software programming, where the programme is written in a high level language before being converted into machine language.

The popularity of EDA tools has increased rapidly with the widespread use of application specific integrated circuits (ASICs) and field programmable gate arrays (FPGAs) in the 1980s. In ASIC technology, the cost of correcting a design flaw late in the design process can be very high. The need for 'right-first-time' designs led to demands for reliable EDA tools. With increasing use of ASICs and FPGAs in power electronic control systems, EDA techniques are increasingly being employed [60], [186], [187]. This has led to the development of a new design approach that relies more on verification by simulation, allowing new products to be developed and produced for the market in a shorter time.

## 2.2 Application specific integrated circuit (ASIC) basics

For many years the designers of electronic circuits and systems have been totally dependent upon the semiconductor manufacturers for the type of integrated circuit from which their circuits and systems may be built. In areas where very large volumes are required, such as calculators, televisions, radios and washing machines, the semiconductor manufacturers have produced full custom designs. The high cost of this process has prevented the exploitation of the size, speed, weight and reliability benefits of silicon design for all but the mass production market or certain military products.

The introduction of computer-aided design (CAD) in the 1980s brought silicon design costs within the bounds of possibility for an increased number of products. In most cases, if the total production of a few thousand pieces is anticipated, then it is likely that a semi-custom integrated circuit will prove viable. The uniqueness of a design in silicon is also an important commercial consideration. It will take a competitor much longer to copy the key features of a silicon chip than it would for him to produce a comparable printed circuit board. Due to the availability of CAD systems, circuit and system designers now have the ability to produce the design to be implemented in silicon and no longer have to use SSI/MSI devices supplied by semiconductor manufacturers. A designer can now consider what type of integration to use for the fabrication of his application specific integrated circuit (ASIC) design.

*Application specific integrated circuits* (ASICs) is a generic term used to designate any integrated circuit designed and built specifically for a particular application. The ASIC concept has been introduced with the advances of VLSI technology which permits the user to tailor his design during the development stages of an IC to suit his needs. The advancement of the large-scale integration process has resulted in two major ASIC technologies, CMOS and BiCMOS, that have attained feature sizes of 0.18 μm and smaller. With the CMOS process, it is possible to manufacture ASIC devices with

10 000 000 gates or higher (one gate is generally defined as a single NAND gate). On the other hand, BiCMOS gate arrays (containing bipolar and CMOS devices) will offer greater operating speed at the expense of a more complex process and lower densities. The frequency of BiCMOS devices is relatively high (100 MHz), because of the drive capacity of bipolar transistors. However, the density is lower. With 0.18 μm BiCMOS technology, it is possible to obtain ICs having up to 5 000 000 gates.

Mixed-signal ASICs (containing both digital and analogue components on the same chip) are recently offered by several chip suppliers providing more possibilities for integration of complex systems. These chip level systems can implement combined analogue/digital designs that formerly required board-level solutions. Analogue cells include operational amplifiers, comparators, D/A and A/D converters, sample-and-hold, voltage references, and RC active filters. Logic cells include gates, counters, registers, microsequencer, PLA (programmable logic array), RAM and ROM. Interface cells include 8- and 16-bit parallel I/O ports as well as synchronous serial ports and UARTs (universal asynchronous receiver–transmitters).

RISC and DSP cores are now offered as megacells by several chip suppliers permitting the design of customised advanced processors using an ASIC design methodology. Building blocks such as DSP cores, RISC cores, memory and logic modules can be integrated on a single chip by the user using advanced CAD (computer-aided design) tools. As an example, Texas Instruments Inc. offers DSP cores in the C1x, C2x, C3x and C5x families as ASIC core cells. Each core is a library cell including a schematic symbol, a timing simulation model for the simulation engine, chip layout files, and a set of test patterns.

The design process of an ASIC consists of three main stages:

- Logic design and simulation.
- Placement, routing layout.
- Prototype production.

The end-user can enter the design process following the semi-standard, semi-custom and full-custom paths, depending on the specific requirements of his application.

With semi-standard ASICs, cost is highly negotiable if predicted volume is sufficient and trustworthy, and the IC manufacturer might retain some rights to resell the chip or parts of its design to others.

In the semi-custom design path, the design engineer (end-user) establishes the specifications, performs the logic design (schematic capture and design verification) and simulation using CAD tools usually provided by the ASIC supplier. A CAD netlist (a list of simulated network connections) and the performance specifications are then submitted. The chip supplier performs the placement, routing, connectivity check and mask layout merging precharacterised physical blocks into a mosaic with its own unique customised metallisation and builds the prototype chip.

In the full-custom design path, in addition to the semi-custom design stages, the end-user also goes through a placement, routing and connectivity check of the design. The chip supplier takes responsibility only for mask layout and prototype production. The design of semi-custom ASICs can be performed using gate arrays or standard cells technologies. A gate array is a CMOS LSI chip consisting of p devices, n devices and tunnels in a repetitive, ordered structure on either a silicon or a sapphire substrate. All device nodes (gates, drains and sources) are accessible. Gate arrays are available for both single-layer and multilayer metallisation. To design an ASIC using a gate array, the end-user defines the connections of the individual devices to achieve the desired functions. At the fabrication stage, only metallisation layers are deposited on the silicon. Signal routing over the gates makes the gates beneath unusable. In this approach, gate utilisation factor is usually about 70–90 per cent. Macros such as RAM and ROM are very inefficient for implementation. However, lower cost and quicker production times are expected for this technology.

In the cell-based approach, no fixed positions for gates and routing channels are predefined. The integrated circuit is designed using libraries of building blocks with specific logic functions. The chip supplier generally provides extensive libraries of well-characterised and verified standard cells, supercells and megacells. To design the ASIC, the end-user combines the library cells into the configuration that performs the functions required by his specific application. The fabrication process involves the etching of the required gates as well as the deposition metallisation of layers. Standard-cell technology offers a better utilisation factor for silicon. Dedicated macros for RAM and ROM ensure reduced gates count and minimum silicon area. A longer fabrication time is expected since more steps are required.

The design of ASICs is performed usually in CAD systems. The stages are: schematic capture, simulation, logic optimisation and synthesis, placement and routing, layout versus schematic design rule check, and functions compiler. The design of a high performance mixed-signal IC is inherently more difficult than the design of a logic IC. The variety of analogue and digital functions requires a cell-based approach. Thorough simulation and layout verification is necessary to ensure the functionality of the prototype ASIC. Redesign of large ASICs typically uses a high level design language (HDL = hardware description language) to help designers to document designs and to simulate large systems. The most common hardware description languages are Verilog and VHDL (the latter conforms to IEEE Standard 1076).

Programmable logic devices (PLDs) are uncommitted arrays of AND and OR logic gates that can be organised to perform dedicated functions by selectively making the interconnections between the gates. Recent PLDs have additional elements (output logic macro cell, clock, security fuse, tri-state output buffers and programmable output feedback) that make them more adaptable for digital implementations. The most popular PLDs are PALs (programmable array logics), PLAs (programmable logic arrays) and EPROMs. Programming of PLDs can be done by blowing fuses (in PALs) or by EEPROM or SRAM technologies which provide reprogrammability. The main advantages of PLDs compared to FPGAs are the speed and ease of use without non-recurring engineering cost. The size of PLDs is, on the other hand, smaller than that of FPGAs. Current PLDs offer complexity equivalent to hundreds of thousands of gates and speed of the order of hundreds of MHz.

## 2.3 Field programmable gate arrays (FPGAs)

Field programmable gate arrays (FPGAs) are a special class of ASICs which differ from mask-programmed gate arrays in that their programming is done by end-users at their site with no IC masking steps. An FPGA consists of an array of logic blocks that can be programmed and connected to achieve different designs. Current commercial FPGAs utilise logic blocks that are based on one of the following: transistor pairs, basic small gates (two-input NANDs and exclusive-ORs), multiplexers, look-up tables, and wide fan-in AND–OR structures. Reprogramming of FPGAs is via electrically programmable switches that are implemented by one of three main technologies: static RAM (SRAM), antifuse and floating gate. Static RAM technology: the switch is a pass transistor that is controlled by the state of a static RAM bit. A SRAM-based FPGA is programmed by writing data in the static RAM. Antifuse technology: an antifuse is a two-terminal device that irreversibly changes from a high resistance to a low resistance link when electrically programmed by a high voltage. Floating-gate technology: the switch is a floating-gate transistor that can be turned off by injecting a charge on the floating gate. The charge can be removed by exposing the floating gate to ultraviolet (UV) light (EPROM technology) or by using an electric voltage (EEPROM technology). The design process of an FPGA consists of three main stages:

- Logic design and simulation.
- Placement, routing and connectivity check.
- Programming.

The process is the same as that used for a semi-custom ASIC gate array, except for the last stage, and uses mostly the same software tools. Current FPGAs offer complexity equivalent to a million gate conventional gate array and typical system clock speeds of hundreds of MHz. The size is much smaller than mask-programmed gate arrays but large enough to implement relatively complex functions on a single chip. The main advantage of FPGAs over mask-programmed ASICs is the fast turnaround that can significantly reduce design risk because a design error can be quickly and inexpensively corrected by reprogramming the FPGA.

The Foundation Series is an EDA software by Xilinx Inc. for designing and implementing programmable hardware such as field programmable gate arrays (FPGAs) and programmable logic devices (PLDs). The main component of the software is the Foundation Project Manager, an application that manages the EDA tools in the software and maintains a unified environment for the user. It comprises five groups: Design Entry, Simulation, Implementation, Verification and Programming. There are three Design Entries: HDL Editor, FSM (Finite State Machine) Editor and Schematic Editor. They allow the project design to be described either as an HDL program, a state machine description or as a schematic design. The designs presented as examples in this book use all three methods. After the Design Entry stage, the design can be *synthesised*, a process that converts the design, whether it is an HDL program or a schematic, into a netlist format. The netlists contain the structural description of the design and are used for functional simulation. At this stage, it is not yet specific to any technology.

In order to download the design into hardware, the target technology has to be specified. The netlist is compiled into a format that is compatible to the targeted device in a process that is called *implementation*. This is followed by accurate timing simulation. It is important to note that the targeted device has to be confirmed at the start of the implementation procedure. In the applications presented in the second part of this book, the Xilinx XC4010XL-PC84 FPGA device was used. Further information on each implementation segment as well as on the Foundation Series in general can be found in [14], [80]. For the present discussion, it is sufficient to point out that the final product of this procedure is a bitstream file, which can be directly downloaded into the targeted device via the serial or parallel interfaces of a PC.

## 2.4 ASICs for power systems and drives

The development of a traditional microprocessor-based motion control system is a complex task consisting of several stages usually completed by several engineers. It involves the design of both hardware and software components and their integration considering various factors such as system performance specifications, processor computing capacities, hardware availability, software development and debugging tools, and system cost. This development can follow the same guidelines as that adopted for any real-time control system. However, the motion control designer has to pay particular attention to the constraints imposed by the control configuration and strategy since the final design can be greatly affected.

In motion control systems, ASIC technology permits the design engineer to tailor the processor and the peripheral devices to obtain the desired specifications for his application. Using ASIC methodology, a motion control engineer can design a control system on one or several chips using building blocks such as DSP or RISC cores, memory, analogue and logic modules. Optimised integration level and performance can thus be achieved. The high integration level results in a reduced chips count that can lower significantly the fabrication cost and improve the system reliability. A disadvantage of ASICs in motion control systems is the lack of flexibility to modify or to adapt the design to different types of motor drives, once the chip is built. To change the design, even in small detail, it is necessary to go back to the initial design stages. The high development and fabrication cost for an ASIC can thus only be justified in large volume production. In small-volume production and in prototyping stages, FPGAs offer a realistic alternative to full gate arrays design to implement specific motion control functions of high complexity requiring up to a million gates.

Chip manufacturers are now offering a number of standard ASICs that perform complex functions in drive control systems such as coordinates conversion (abc/dq conversion), pulse width modulation, PID controllers, fuzzy controllers, neural networks, etc. Such devices can be used with advantage in motion control designs allowing reduction of processor computing load and increase of the sampling rate. In the following, some examples of commercial ASICs designed for motion control are presented.

The Analogue Devices AD2SIO0/AD2S110 a.c. vector controller performs the Clark and Park transformations, usually required for implementing field-oriented control of a.c. motors. The Clark transform converts a three-phase parameter (abc coordinates) into an equivalent two-phase parameter ($\alpha$-$\beta$ coordinates). The Park transform rotates the resulting vector into another one, represented in a new rectangular set of coordinates, normally linked to the rotor ($\alpha$-$\beta$ to d-q coordinates).

The Hewlett-Packard HCTL-1000 is a general-purpose digital motion control IC which provides position and velocity control for d.c., d.c. brushless and stepper motors. The HCTL-1000 executes any one of four control algorithms selected by the user: position control, proportional velocity control, trapezoidal profile control for point-to-point moves and integral velocity control.

The Signetics HEF4752V a.c. motor control circuit is an ASIC designed for the control of three-phase pulse width modulated (PWM) inverters in a.c. motor speed control systems. A pure digital waveform generation is used for synthesising three 120° out of phase signals, the average voltage of which varies sinusoidally with time in the frequency range 0 to 200 Hz.

The American Neuralogix NLX230 fuzzy microcontroller is a fully configurable fuzzy logic engine containing a 1-of-8 input selector, 16 fuzzifiers, a minimum comparator, a maximum comparator and a rule memory. Up to 64 rules can be stored in the on-chip, 24-bit-wide rule memory. The NLX230 can perform 30 million rules per second.

The Intel 80170X ETANN (Electrically Trainable Analogue Neural Network) simulates the data processing functions of 64 neurones, each of which is influenced by up to 128 weighted synapse inputs. The chip has 64 analogue inputs and outputs. Its control functions for setting and reading synapse weights are digital. The 80170X is capable of 2 billion multiply–accumulate operations (connections) per second.

The few dedicated circuit examples given above, together with the general modern trend towards 'systems-on-a-chip' integration in electronics, illustrate the need for further complex ASIC/FPGA designs for drives and power systems.

# 3

# Electric motors and power systems

## 3.1 Electric motors

Electric motors are major users of electricity in industrial plants and commercial premises. Motive power accounts for almost half of the total electrical energy used in the UK and nearly two-thirds of industrial electricity use. It is estimated that over ten million motors, with a total capacity of 70 GW, are installed in UK industry alone [11]. Although many motor types are currently in use (synchronous motors, PM synchronous motors, d.c. motors, d.c.-brushless motors, switched reluctance motors, stepping motors), most of the industrial drives are powered by three-phase induction motors. The majority of them are rated up to 300 kW and can be classified as illustrated by Fig. 3.1.

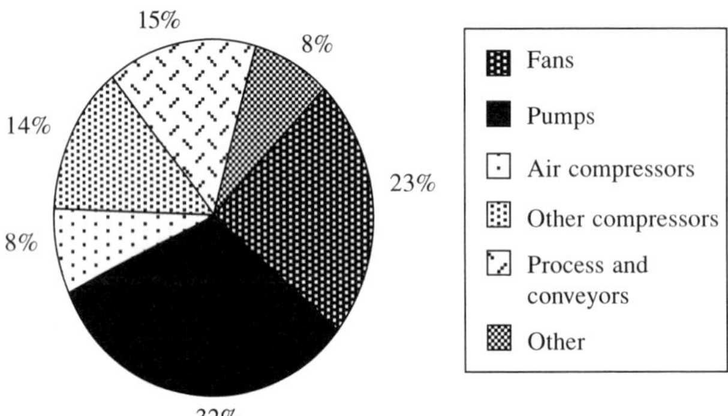


**Fig. 3.1** Energy consumption by induction motors up to 300 kW in industry

The large industrial use of induction motors has been stimulated over the years by their low prices and reliability. The low price of buying such a motor can, however, be deceptive. A modest-sized 11 kW induction motor costs as little as £300 to buy, but it could accumulate running costs of over £30 000 in ten years. The electricity bill for a motor for just a month can be more than its purchase price [11]. Therefore, even small efficiency improvements may produce impressive cost savings.

The most efficient and flexible solutions to the energy saving problem are based on variable speed drives (VSDs). Using VSDs the motor speed can be readily adapted to the requirements of particular applications. For instance, VSDs replace the old solution of using adjustable nozzles in applications involving fans or pumps. An adjustable nozzle can ensure a variable flow of fluid, but at the cost of decreasing the motor efficiency. A VSD is capable of performing the same task while maintaining the motor efficiency at high levels. In addition to the huge potential for saving energy, the use of induction-motor-based VSDs has other important benefits including:

- improved process control and hence enhanced productivity;
- soft starting, soft stopping and regenerative braking;
- unity power factor;
- wide range of speed, torque and power;
- good dynamic response (comparable with d.c. drives).

Previously, d.c. motors were extensively used in complex speed and position control applications, such as industrial robots and numerically controlled machinery, because their flux and torque can be easily controlled. However, d.c. motors have the disadvantage of using a commutator, which increases the motor size, the maintenance cost and reduces the motor life. Advances in digital technology and power electronics have made the induction motor control a cost-effective solution. Therefore, d.c. motors are currently being replaced by induction motors in many industrial plants. A large proportion of the induction-motor VSD cost is still due to the price of the sensors and digital controllers that are needed. However, the prices of the digital electronic circuits have decreased sharply during the last few years. This makes the sensor cost an important consideration in the total price of the VSD.

The speed and/or position sensors ensure high operation accuracy for the closed-loop systems. In some practical situations, however, there are strong reasons to eliminate the speed sensor due to both economical and technical reasons. For example, the pumps used in oilrigs to pump out the oil have to work under the surface of the sea, sometimes at depths of 50 metres. Obtaining the speed measurement data up to the surface means extra cables, which is extremely expensive, therefore reducing the number of sensors and measurement cables provides a major cost reduction [13]. Recently, it has been shown that speed can be calculated from the current and voltage across the a.c. motor thereby eliminating the need for speed sensors. There have been many alternative proposals addressing the problem of speed sensorless induction motor control. These methods are mathematically intensive as they imply the on-line calculation of the space-vector motor model. Therefore, they are implemented using fast state-of-the-art digital circuits (ASICs and DSPs). An example of modern sensorless neural control of an induction motor is presented in the second part of this book.

## 3.2 Power systems

The discovery of electromagnetism by Michael Faraday in 1831 led to the rapid development of electromagnet machines for converting mechanical energy into electricity. Within a few months of Faraday's announcement, an Italian scientist, Signor Salvator dal Negro, invented an electric generator in which a permanent magnet was pushed and pulled to provide the necessary motion. The first of the rotating electromagnet generators as we know today was invented by Hypolite Pixii in Paris. It was made public at a meeting of Académie des Science in 1832. Later that year, Pixii added a commutator to his machine to obtain direct current (d.c.) from the alternating current (a.c.) produced.

Early electric generators, or *dynamos* as they are known, produced d.c. electric current on a small scale. They were used mainly for supplying electroplating baths and later for providing power to arc lamps in lighthouses.

The invention of light bulbs and steam-engine-driven generators in America by Thomas A. Edison led to the commercial expansion of electric generation for lighting purposes towards the end of the nineteenth century. In the early days direct current was the preference, but when long distance transmission become necessary alternating current was found to be more suitable. Power transmission at high voltages is more economical and the voltage level of alternating current can be easily changed using transformers. By the second half of the twentieth century, alternating current became almost universal, leading to the widespread use of a.c. generators. Among the various types of a.c. generators, the polyphase synchronous generator is the largest single-unit electrical machine in production today, with power ratings of up to several hundred MVAs being common. They are widely used in large power stations as well as in industrial, marine, telecommunication and other standby or continuous power applications. Recent work in synchronous generators is mainly aimed at improving the efficiency of the machine, quality of the output power and the stability of the system. Synchronous generators are responsible for the bulk of the electrical power generated in the world. They are mainly used in power stations and are predominantly driven either by steam or hydraulic turbines. These generators are usually connected to an *infinite bus* where the terminal voltages are held at a constant value irrespective of loading due to the capacity ('momentum') of all the other generators also connected to it. Another common application of synchronous generators is their use in stand-alone or isolated power generation systems. The prime mover in such applications is usually a diesel engine.

Although a massive proportion of synchronous generators are electromagnetic, the use of permanent magnet synchronous machines as stand-alone generators has been studied for more than half a century. Permanent magnet synchronous generators (PMSGs) are more difficult to regulate and it is only with the recent developments in power electronics that they are seriously being considered for various applications [39], [191], [17]. One of the main advantages of the control system proposed in the examples section of this book is its ability to regulate stand-alone PMSGs as well as electromagnet generators. This functionality is duly demonstrated by the experiments presented, in which a PMSG is used. It has to be mentioned that synchronous machines are by no means the only type of electrical machine used for stand-alone power generation. Studies have been conducted into the use of induction generators [76], [77], [78], [54], reluctance generators [18] and other types of machines that might prove to be more suitable in certain applications.

Since the invention of electrical machines in the nineteenth century, there has been a need to convert electrical power for various applications such as electrical machine drives, voltage regulation, welding, heating, etc. Initially, rotating machines were predominantly used to control and convert electrical power. It was the introduction of the glass bulb mercury arc rectifier (1900) which led to the beginning of the power electronics era. Power electronics is the branch of engineering concerned with the application of electronics in the control and conversion of electrical power. Early power electronic devices such as thyratrons and ignitrons were crude and unreliable. The introduction of selenium rectifiers during World War II was particularly welcome due to their reliability.

In 1948, the invention of the p-n junction transistor by Bardeen, Brattian and Shockley from Bell Laboratories was seen as a revolutionary advancement in the field of electronics. This laid the foundation for the development of the p-n-p-n transistor switch by J.L. Moll *et al.* (1956), a device which later became known as the thyristor, or silicon controlled rectifier (SCR). By 1957, the first commercial thyristor was made available by General Electric Company. This marked the beginning of the modern power electronics era. This three-terminal device had a continuous current rating of 25 A and a blocking voltage of up to 300 V. Since then, the thyristor has become one of the most popular devices in power electronics. Circuit design engineers have constantly worked on improving the operating performance of the thyristor, resulting in the creation of a range of different types of thyristors optimised for different applications. They can generally be grouped into six categories, namely [16]:

- Phase control thyristor.
- Inverter thyristor.
- Asymmetrical thyristor.
- Reverse conducting thyristor (RCT).
- Gate-assisted turn-off thyristor (GATT).
- Light-triggered thyristor.

Another class of power electronic device subsequently developed were the *controllable power switches*. Thyristors, while being able to be latched on by a control signal, can only be turned off by the power circuit, which is a great drawback. However, controllable switches can be turned on as well as turned off by the control signals. Although controllable switches like the transistor have been around since 1948, designing them to possess high power handling capabilities was not achieved until much later. Compared to thyristors, controllable power switches offer greater flexibility in power applications, including the possibility of controlling d.c. circuits without complicated commutation circuitry. Thus, they are particularly attractive in inverter applications. Examples of devices in this category are gate turn-off thyristors (GTOs), power transistors, power MOSFETs, integrated gate-commutated thyristors (IGCTs) and insulated gate bipolar transistors (IGBTs). The GTO is a thyristor-like latching device but can be turned off by a negative gate current. Power transistors and power MOSFETs were developed from small-signal designs to later versions which are capable of handling higher voltage applications in the order of hundreds of volts. In the early 1980s, the IGBT was developed [26], which combines the low on-state conduction losses of the bipolar junction transistor (BJT) and the high switching frequency of the MOSFET. The IGBT has since gained widespread popularity in power electronic applications. Commercial IGBTs are currently available up to 3.3 kV. These components can be used in a range of power applications. The development of such power devices is expected to grow as the use of new materials such as monocrystalline silicon carbide (SiC) increases their voltage ratings and reduces thermal resistance [198], [196].

Generally, a power electronic system comprises two separate sets of circuits: the logic level control circuitry and the high power circuits. Recent developments in electronics made it possible to combine these two components into a single integrated circuit, the power integrated circuit (PIC). A PIC is defined by Thomas [217] as an integrated circuit which combines the logic level control and/or protection circuitry with power handling capability of supplying 1 A and withstanding at least 100 V. With the current trend towards integrated solutions, this technology is receiving a substantial amount of attention. Integrated power electronic devices are seen as the solution for smaller and lower cost power electronic systems in the future.

## 3.3 Pulse width modulation

Pulse width modulation (PWM) is currently the most widely used technique of inverter control and has received considerable attention in the last two decades. The PWM switching scheme essentially involves the strategic variation of the ON and OFF timing periods of each pair of switches in the inverter. This produces a waveform that contains a series of pulses which have the same voltage level but different widths, as illustrated in Fig. 3.2.

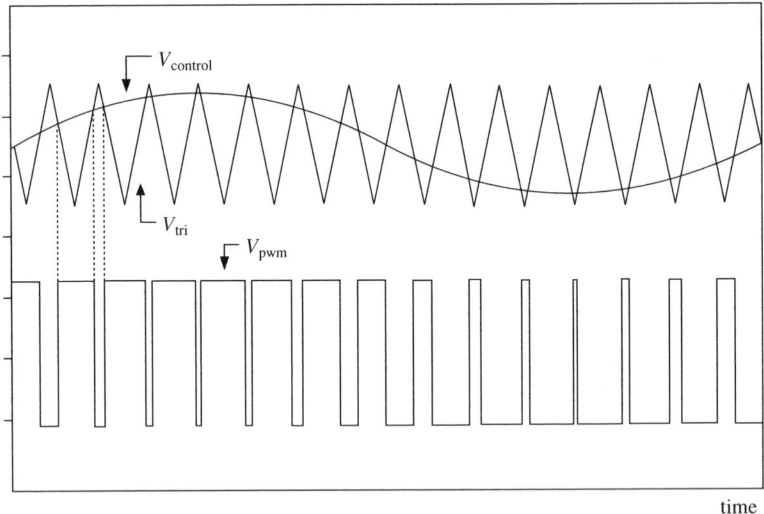


**Fig. 3.2** Pulse width modulation

The fundamental component of the PWM switching pattern $V_{PWM}$ in Fig. 3.2 is a sine wave, which is the required output voltage. To obtain the switching pattern, a sinusoidal signal $V_{control}$ is compared with a high frequency triangular carrier wave $V_{tri}$. This form of PWM control is sometimes called *sinusoidal-PWM* in order to explicitly differentiate it from other forms of PWM control schemes. Table 3.1 illustrates how $V_{control}$ and $V_{tri}$ can be used to determine the switching pattern in a single phase inverter. The two devices on the same branch (T1 and T2; T3 and T4) must not be ON at the same time, otherwise a short circuit will occur.

**Table 3.1** PWM control

| | T1 | T2 | T3 | T4 |
|---|---|---|---|---|
| $V_{control} \geq V_{tri}$ | ON | OFF | OFF | ON |
| $V_{control} < V_{tri}$ | OFF | ON | ON | OFF |

In sinusoidal-PWM control schemes, there are two characteristic ratios which are important factors in the design of the controllers. The amplitude modulation ratio $m_a$ is defined as the ratio of the peak amplitude of the control signal to the peak amplitude of the carrier signal,

$$m_a = \frac{\hat{V}_{\text{control}}}{\hat{V}_{\text{tri}}}$$

The frequency modulation ratio $m_f$ is defined as the ratio of the carrier frequency to the ratio of the control signal frequency.

$$m_f = \frac{f_{\text{control}}}{f_{\text{tri}}}$$

The standard sinusoidal-PWM technique suffers from the major drawback that the a.c. term gain ($G_{\text{ac}}$), which is the ratio of the amplitude of the output voltage to the amplitude of the PWM waveform, is limited to a maximum value of 0.866 ($G_{\text{ac}} \leq 0.866$). Several improved PWM techniques have been introduced to tackle this problem but they each have their own disadvantages. In general, improved techniques have higher a.c. gains but suffer from more harmonic distortions and require more complicated hardware for implementation. Further information of the improved techniques can be found in [44]. They include techniques such as sine + 3rd harmonic PWM, harmonic injection and programmed harmonic elimination. Other PWM techniques include random PWM schemes and sliding mode control. Random PWM schemes [124], [106] are based on the use of random number generation. They offer a more evenly spread harmonic spectrum and are found to have reduced radio interference, noise and vibration effects. Sliding mode control, on the other hand, is described by Jung and Tzou [137] to be especially suitable for closed-loop control of power converting systems under load variations.

However, improved PWM techniques require a more complex hardware implementation. For the present work, the standard PWM technique is found to be suitable for the application while being easier to implement in hardware when compared to the other techniques.

There are various design solutions to implement a PWM controller. The following section describes a traditional circuit implementation method: a C++ program is used to generate the switching pattern. A fairly straightforward method is to use an erasable programmable read only memory (EPROM) to store the PWM pattern. During the operation, this information is sequentially retrieved and fed into a driver circuit board, which will switch the IGBTs accordingly. Figure 3.3 shows a schematic of the circuit design. It comprises a voltage controlled oscillator NE566 (IC1), a counter (IC2), an EPROM (IC3) and some AND gates (IC4) to act as output buffers.

The information for producing one cycle of the power waveform, i.e. one period of the sinusoidal reference signal, is broken down into 4096 slices and stored in the EPROM memory locations. Each momory location corresponds to an address ranging from 0 to 4095 and each bit of information in a memory location controls one power switch in the inverter. For a single phase inverter which has four power switches, $4096 \times 4$ bits (16 kb) of memory are required while a three-phase inverter with six switches requires $4096 \times 6$ bits (24 kb) of memory. IC2 is a CMOS4040 12-bit counter, designed to count from 0 to 4095 in a repetitive cycle. This is used as the address input to retrieve information

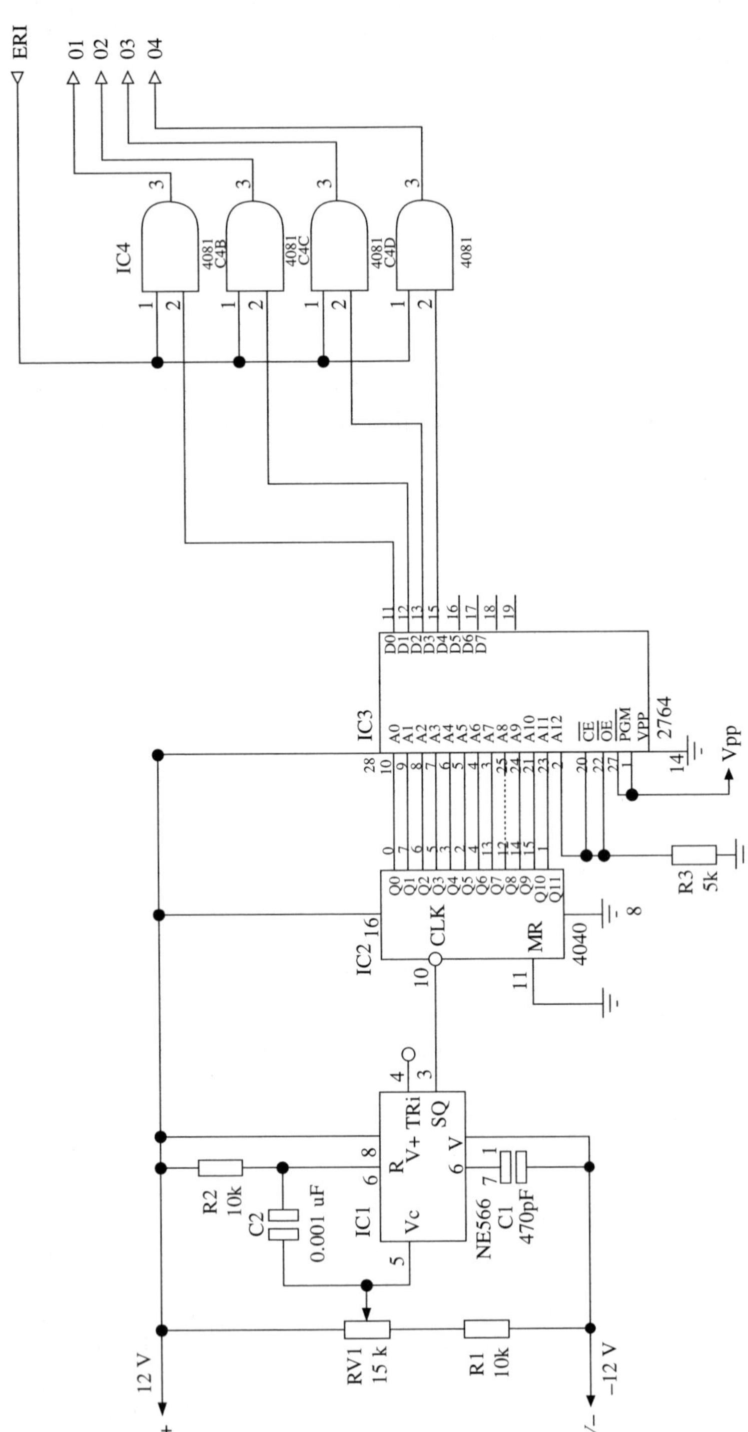


**Fig. 3.3** Circuit diagram of the EPROM-based PWM generator

from the EPROM. To obtain an output frequency of 50 Hz, the counter (IC2) has to complete 50 cycles in one second. Therefore, the sampling frequency must be:

$$f_s = 50 \times 4096 = 204.8 \text{ kHz}$$

The advantage of using a voltage controlled oscillator instead of a fixed frequency oscillator is that a voltage signal can be used to control the oscillator frequency and hence the sampling frequency of the inverter.

This makes it possible to control the inverter frequency with a closed-loop control circuit. Due to immediate availability during the implementation stage, a 64 kb EPROM is used in the circuit although 16 kb ($2^{12} \times 4$) of information is sufficient for single-phase operation (24 kb for three phase). The period of the triangular carrier wave is chosen to contain ten sampling units. Each sampling unit corresponds to one clock cycle hence the actual sampling time will be the inverse of the clock frequency.

In the C program, a comparison between the reference power waveform and the carrier waveform is made at every sampling point. The output is 1 if the reference power value is larger than the carrier value and 0 if vice versa. The necessary switching signal is generated from this comparison as shown in Fig. 3.4. However, as a result of introducing discrete sampling points, a certain amount of error is inevitable. The errors are labelled as $\pm\varepsilon_n$ in the diagram. The maximum value for each error is just under the length of one sampling unit which, in this case, is 10 per cent of the period of the switching signal (because one cycle of the switching signal consists of ten sampling units). The effects of these errors can be reduced by increasing the number of sampling points in each cycle of the switching signal. This can be done either by maintaining the frequency modulation ratio $m_f$ and increasing the total number of sampling points in the power cycle or by maintaining the number of sampling points in one power cycle and reducing $m_f$.

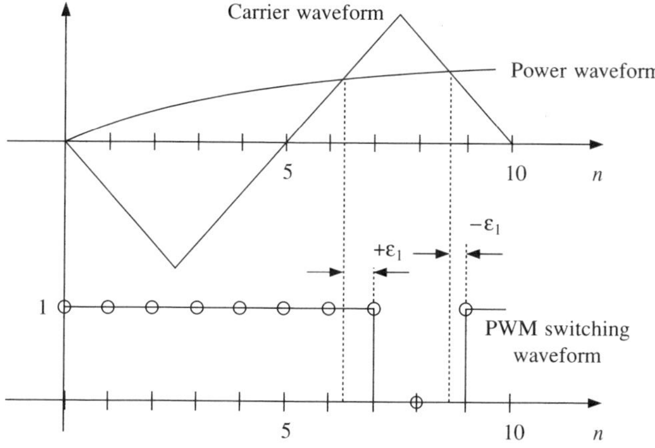


**Fig. 3.4** PWM waveform generation

The frequency of the triangular carrier waveform, also known as the switching frequency is given by:

$$f_{\text{tri}} = 1/(NT_s) \text{ Hz}$$

where $T_s$ is the sampling period which is determined by the desired power frequency and

$N$ is the number of sampling points in one cycle of the carrier signal. For a 50 Hz power frequency, the sampling frequency $T_s$ is:

$$T_s = \frac{1}{50} \cdot \frac{1}{4096} = 4.88\ \mu\text{s}$$

Therefore, the switching frequency is

$$f_{\text{tri}} = \frac{1}{10 \times 4.88 \times 10^{-6}}\ \text{Hz} = 20.48\ \text{kHz}$$

The frequency modulation factor is given by:

$$m_f = \frac{f_{\text{tri}}}{f_1} = \frac{(1/4096\ T_s)}{(1/10\ T_s)} = 409.6$$

where:
$f_{\text{tri}}$ is the frequency of the triangular carrier waveform (switching frequency)
$f_1$ is the frequency of the fundamental harmonic (sinusoidal power frequency)

A three-phase PWM waveform generator was also constructed by simply changing the contents of the EPROM with a new C program which is written to generate the three-phase switching data. In the program, the triangular carrier waveform is compared with three different sinusoidal power waveforms, each phase shifted from one another by 120°. The result of each comparison determines the switching signal of the IGBTs in each branch of the inverter. Instead of four outputs, the three-phase PWM controller has six outputs, as there are six IGBTs in a three-phase inverter. Therefore, two additional data outputs from the EPROM are used. The control circuit was successfully implemented and used in the experiments of the second case study presented in this book.

## 3.4 The space vector in electrical systems

The space vector concept originated in the study of Y-connected induction motors but it can be extended to describe all three-phase electric systems regardless of their exact nature: electrical generators, electrical motors, transformers, etc. The basic principle is to transform the scalar electromagnetic quantities describing the system (currents, voltages and magnetic fluxes) into two-dimensional vectors named space vectors. One space vector replaces a set of three scalar quantities of the same type, thereby generating a more compact notation for the mathematical equations. Therefore, space vectors are largely used to analyse the operation of three-phase electrical machines [159], [183], [227], [229].

If '$A$' is an electromagnetic quantity then $A_a$, $A_b$ and $A_c$ are the three values corresponding to the three system phases. They are initially associated with two-dimensional vectors situated on three directions 120° apart in a plane: $\vec{A}_a$, $\vec{A}_b$, and $\vec{A}_c$ as illustrated in Fig. 3.5. Adding the three vectors together, a single two-dimensional vector is obtained according to equation (3.1). $\vec{A}$ is the space vector associated with scalar quantities $A_a$, $A_b$ and $A_c$. The vector components on the real axis (axis '$d$') and on the imaginary axis (axis '$q$') are given in (3.2).

$$\vec{A} = \vec{A}_a + \vec{A}_b + \vec{A}_c \tag{3.1}$$

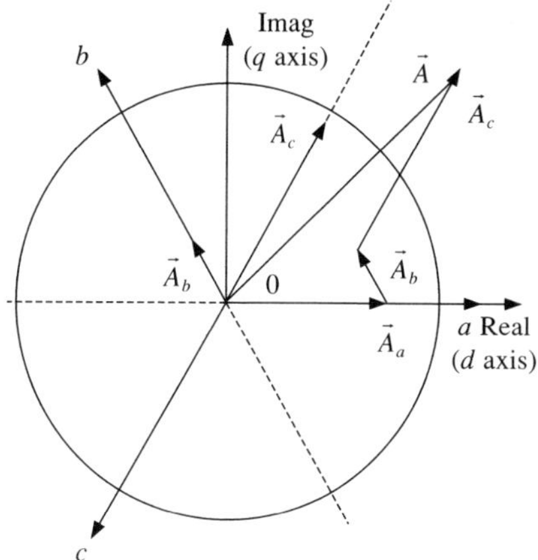


**Fig. 3.5** The relation between phase quantities and the corresponding space vector

$$\begin{cases} A_d = A_a \cdot \cos(0) + A_b \cdot \cos\left(\dfrac{2\pi}{3}\right) + A_c \cdot \cos\left(\dfrac{4\pi}{3}\right) = A_a - \dfrac{1}{2}A_b - \dfrac{1}{2}A_c \\ A_q = A_a \cdot \sin(0) + A_b \cdot \sin\left(\dfrac{2\pi}{3}\right) + A_c \cdot \sin\left(\dfrac{4\pi}{3}\right) = \dfrac{\sqrt{3}}{2}A_b - \dfrac{\sqrt{3}}{2}A_c \end{cases} \tag{3.2}$$

In practical calculations, the space vectors are represented either by $2 \times 1$ matrices or by complex quantities. Using matrix notation, equation (3.2) becomes (3.3) while (3.4) describes the complex number approach to space vector calculation (3.1). Two-dimensional vectors like the one in (3.1) are distinguished from the equivalent complex numbers by means of notation. Underlined symbols stand for complex values while vectors are represented by symbols placed under an arrow. Thus, $\underline{A}$ is a complex number while $\vec{A}$ is a vector.

$$\begin{pmatrix} A_d \\ A_q \end{pmatrix} = \begin{pmatrix} 1 & -\dfrac{1}{2} & -\dfrac{1}{2} \\ 0 & \dfrac{\sqrt{3}}{2} & -\dfrac{\sqrt{3}}{2} \end{pmatrix} \cdot \begin{pmatrix} A_a \\ A_b \\ A_c \end{pmatrix} \tag{3.3}$$

$$\begin{cases} \underline{A} = A_a + \varepsilon \cdot A_b + \varepsilon^2 \cdot A_c \\ \varepsilon = \cos\left(\dfrac{2\pi}{3}\right) + j \cdot \sin\left(\dfrac{2\pi}{3}\right) \end{cases} \tag{3.4}$$

The transformation of the set of three scalar variables into a space vector is equivalent to a transformation from a three-phase system into a two-phase system. The inverse transformation can be calculated based on the property that the algebraic sum of the three scalar values is always null. This property is shared by all electromagnetic quantities related to individual phases (currents, voltages and magnetic fluxes) if the power supply generates symmetric voltages and the load is symmetric and Y-connected.

$$A_a + A_b + A_c = 0 \tag{3.5}$$

Combining (3.5) with equation (3.2), the system (3.6) is generated from which (3.7) is derived. The system (3.7) describes the inverse transformation of a space vector into the corresponding set of three scalar phase quantities.

$$\begin{cases} A_d = A_a - \frac{1}{2} A_b - \frac{1}{2} A_c \\ A_q = \frac{\sqrt{3}}{2} A_b - \frac{\sqrt{3}}{2} A_c \\ A_a + A_b + A_c = 0 \end{cases} \tag{3.6}$$

$$\begin{cases} A_a = \frac{2}{3} \cdot A_d \\ A_b = -\frac{1}{3} \cdot A_d + \frac{1}{\sqrt{3}} \cdot A_q \\ A_c = -\frac{1}{3} \cdot A_d - \frac{1}{\sqrt{3}} \cdot A_q \end{cases} \tag{3.7}$$

## 3.5 Induction motor control

### 3.5.1 Space vector model of three-phase induction motor

The mathematical models of the electrical machines are classified as lumped-parameter circuit models and distributed-parameter models. The latter are more complex but more accurate than the former. The distributed-parameter models are used for very precise calculations necessary for optimal machine design. They allow an exact calculation of the electromagnetic field and heat distribution inside the machine. The lumped-parameter models can be obtained as a simplification of the distributed-parameter models. They are used for control system design where only global quantities like currents, torque and speed are important. Their internal distribution inside the machine is not relevant when designing controllers to govern the evolution of speed, torque and power consumption according to the particular application requirements.

Furthermore, the lumped-parameter circuit model is simpler and therefore more convenient to use in the study of electric drives. The space vector model of the induction motor is the lumped-parameter model with the largest use in the study and design of electrical drive applications. It is common to consider as a first approximation that the rotor windings and the stator windings have a sinusoidal distribution inside the motor and no magnetic saturation is present [43], [159]. Therefore, the magnetomotive force (MMF) space harmonics and slot harmonics are neglected. Although saturation is not taken into account, the model is considered to yield acceptable results for the study of common electric drive applications [159], [227].

The induction motor space vector model is derived from the basic electrical equations describing each of the stator windings and each of the rotor windings. The stator windings equations are given in (3.8) where $u_{as}$, $u_{bs}$ and $u_{cs}$ are the phase voltages, $i_{as}$, $i_{bs}$ and $i_{cs}$ are the phase currents, while $\Psi_{as}$, $\Psi_{bs}$ and $\Psi_{cs}$ are the phase magnetic fluxes.

$$\begin{cases} u_{as} = R_s i_{as} + \dfrac{d\Psi_{as}}{dt} \\ u_{bs} = R_s i_{bs} + \dfrac{d\Psi_{bs}}{dt} \\ u_{cs} = R_s i_{cs} + \dfrac{d\Psi_{cs}}{dt} \end{cases} \tag{3.8}$$

The associated space vectors (expressed as complex numbers) are obtained by multiplying the second equation in (3.8) with $\varepsilon$ and the third with $\varepsilon^2$, after which all the three equations are added together. The conversion of the three scalar equations into one space vector equation is illustrated by (3.9) and (3.10).

$$\begin{cases} u_{as} = R_s i_{as} + \dfrac{d\Psi_{as}}{dt} \\ \varepsilon u_{bs} = R_s \cdot \varepsilon i_{bs} + \varepsilon \dfrac{d\Psi_{bs}}{dt} \\ \varepsilon^2 u_{cs} = R_s \cdot \varepsilon^2 i_{cs} + \varepsilon^2 \dfrac{d\Psi_{cs}}{dt} \end{cases} \tag{3.9}$$

$$\begin{cases} \underline{u}_s^s = u_{as} + \varepsilon \cdot u_{bs} + \varepsilon^2 \cdot u_{cs} \\ \underline{i}_s^s = i_{as} + \varepsilon \cdot i_{bs} + \varepsilon^2 \cdot i_{cs} \\ \underline{\Psi}_s^s = \Psi_{as} + \varepsilon \cdot \Psi_{bs} + \varepsilon^2 \cdot \Psi_{cs} \end{cases} \Rightarrow \underline{u}_s^s = R_s \underline{i}_s^s + \frac{d\underline{\Psi}_s^s}{dt} \tag{3.10}$$

Different reference frames (still or rotating) can be used to calculate the coordinates of the electromagnetic space vectors [43]. Equations (3.10) are written in the stator reference frame.

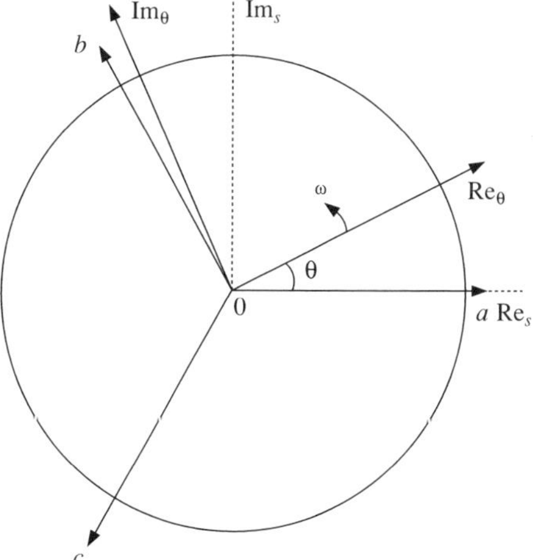


**Fig. 3.6** The fixed stator reference frame and the general mobile θ reference frame

Any rotating reference frame is defined by the electrical angle function $\theta(t)$ that indicates the relative position to the still reference frame. Alternatively, it can be defined by the electrical rotation speed $\omega_e(t)$ and the initial electrical angle $\theta(0)$. For a general rotating frame the equations (3.10) are transformed into (3.11). The fourth equation in (3.11) can be rewritten as (3.12). Equation (3.13) is eventually obtained by dividing (3.12) with $e^{j\theta}$.

$$\begin{cases} \underline{u}_s^\theta = \underline{u}_s \cdot e^{-j\theta} \\ \underline{i}_s^\theta = \underline{i}_s \cdot e^{-j\theta} \\ \underline{\Psi}_s^\theta = \underline{\Psi}_s \cdot e^{-j\theta} \\ \underline{u}_s^\theta \cdot e^{j\theta} = R_s \underline{i}_s^\theta \cdot e^{j\theta} + \dfrac{\mathrm{d}}{\mathrm{d}t}(\underline{\Psi}_s^\theta \cdot e^{j\theta}) \end{cases} \tag{3.11}$$

$$\underline{u}_s^\theta \cdot e^{j\theta} = R_s \underline{i}_s^\theta \cdot e^{j\theta} + \frac{\mathrm{d}\underline{\Psi}_s^\theta}{\mathrm{d}t} \cdot e^{j\theta} + \underline{\Psi}_s^\theta \cdot e^{j\theta} \cdot j\frac{\mathrm{d}\theta}{\mathrm{d}t} \tag{3.12}$$

$$\underline{u}_s^\theta = R_s \underline{i}_s^\theta + \frac{\mathrm{d}\underline{\Psi}_s^\theta}{\mathrm{d}t} + j\omega_e \cdot \underline{\Psi}_s^\theta \tag{3.13}$$

A similar complex equation describes the rotor circuit with the difference that the reference frame rotation speed relative to the rotor is $\omega_e - \omega_{er}$ instead of $\omega_e(\omega_{er}$ is the electrical rotor angular speed). Moreover, the rotor voltage is always zero for squirrel cage induction motors.

$$\underline{u}_r^\theta = R_s \underline{i}_r^\theta + \frac{\mathrm{d}\underline{\Psi}_r^\theta}{\mathrm{d}t} + j(\omega_e - \omega_{er}) \cdot \underline{\Psi}_r^\theta = 0 \tag{3.14}$$

Equations (3.15) describe the relation between the electrical stator angular frequency $\omega_{es}$ and the stator current frequency $f_s$ on the one hand, and the relationship between the rotor angular speed $\omega_{er}$ and the rotor mechanical speed $\omega_r$ on the other hand. The variable '$p$' is the number of pairs of stator poles.

$$\begin{cases} \omega_{es} = p \cdot \omega_s = p \cdot 2\pi f_s \\ \omega_{er} = p \cdot \omega_r \end{cases} \tag{3.15}$$

The individual phase fluxes that are used to calculate the magnetic flux vectors are each composed of six components. The flux components are generated by the electromagnetic interaction between the three rotor windings and the three stator windings.

$$\begin{cases} \Psi_{sa} = \Psi_{sasa} + \Psi_{sasb} + \Psi_{sasc} + \Psi_{sara} + \Psi_{sarb} + \Psi_{sarc} \\ \Psi_{sb} = \Psi_{sbsa} + \Psi_{sbsb} + \Psi_{sbsc} + \Psi_{sbra} + \Psi_{sbrb} + \Psi_{sbrc} \\ \Psi_{sc} = \Psi_{scsa} + \Psi_{scsb} + \Psi_{scsc} + \Psi_{scra} + \Psi_{scrb} + \Psi_{scrc} \\ \Psi_{ra} = \Psi_{rasa} + \Psi_{rasb} + \Psi_{rasc} + \Psi_{rara} + \Psi_{rarb} + \Psi_{rarc} \\ \Psi_{rb} = \Psi_{rbsa} + \Psi_{rbsb} + \Psi_{rbsc} + \Psi_{rbra} + \Psi_{rbrb} + \Psi_{rbrc} \\ \Psi_{rc} = \Psi_{rcsa} + \Psi_{rcsb} + \Psi_{rcsc} + \Psi_{rcra} + \Psi_{rcrb} + \Psi_{rcrc} \end{cases} \tag{3.16}$$

In equation (3.16) each flux component is identified by four indices: the first two indicate the winding where the magnetic flux is measured while the last two indicate the winding that generates it. For instance, $\Psi_{sarb}$ is the flux generated into stator winding '*a*' by rotor winding '*b*'. The flux components related to stator phase '*a*' are described by (3.17). The names and the significance of the symbols are as follows:

- $l_{msr}$ – the mutual inductance between stator and rotor. It is proportional to the flux created by one rotor phase into one stator phase.
- $m_{\sigma s}$ – the stator mutual leakage inductance between two stator phases. It is proportional to the flux produced by one stator phase into another stator phase without influencing the rotor. It therefore models the magnetic field lines that intersect two stator windings without intersecting the rotor.
- $l_{ms}$ – the mutual inductance between stator phases. It is proportional to the flux created by one stator phase into another stator phase through the rotor. It models the magnetic field lines that are created by one stator phase but intersects both the rotor and the other stator winding.
- $l_{\sigma s}$ – the stator phase leakage inductance. It is proportional to the stator phase leakage magnetic flux. The corresponding magnetic field lines do not intersect any winding other than the stator winding which produces them.
- $\alpha$ – the angle between the stator *d*-axis and the rotor *d*-axis.

$$\begin{cases} \Psi_{sasa} = (l_{\sigma s} + l_{ms}) \cdot i_{sa} \\ \Psi_{sasb} = \left[ m_{\sigma s} + l_{ms} \cdot \cos\left(\dfrac{2\pi}{3}\right)\right] \cdot i_{sb} \\ \Psi_{sasc} = \left[ m_{\sigma s} + l_{ms} \cdot \cos\left(\dfrac{4\pi}{3}\right)\right] \cdot i_{sc} \\ \Psi_{sara} = (l_{msr} \cdot \cos\alpha) \cdot i_{ra} \\ \Psi_{sarb} = l_{msr} \cdot \cos\left(\alpha + \dfrac{2\pi}{3}\right) \cdot i_{rb} \\ \Psi_{sarc} = l_{msr} \cdot \cos\left(\alpha + \dfrac{4\pi}{3}\right) \cdot i_{rc} \end{cases} \tag{3.17}$$

The magnetic coupling between different windings is influenced by their relative position. The coupling is maximal when the angle between the two windings is zero and it is null at 90°. This geometric factor can be expressed by simple cosine functions due to the assumption that the magnetic field has a sinusoidal distribution. Adding the six components from (3.17) yields:

$$\Psi_{sa} = \left(l_{\sigma s} - m_{\sigma s} + \frac{3}{2} l_{ms}\right) \cdot i_{sa} + \frac{3}{2} l_{msr} \cdot \cos\alpha \cdot i_{ra} + \frac{\sqrt{3}}{2} l_{msr} \cdot \sin\alpha \cdot (i_{rc} - i_{rb}) \tag{3.18}$$

Equation (3.18) is obtained based on the property that the sum of the three phase currents is zero. Similar results are obtained for stator phases '*b*' and '*c*'.

$$\begin{cases} \Psi_{sb} = \left( l_{\sigma s} - m_{\sigma s} + \frac{3}{2} l_{ms} \right) \cdot i_{sb} + \frac{3}{2} l_{msr} \cdot \cos\alpha \cdot i_{rb} + \frac{\sqrt{3}}{2} l_{msr} \cdot \sin\alpha \cdot (i_{ra} - i_{rc}) \\ \Psi_{sc} = \left( l_{\sigma s} - m_{\sigma s} + \frac{3}{2} l_{ms} \right) \cdot i_{sc} + \frac{3}{2} l_{msr} \cdot \cos\alpha \cdot i_{rc} + \frac{\sqrt{3}}{2} l_{msr} \cdot \sin\alpha \cdot (i_{rb} - i_{ra}) \end{cases} \tag{3.19}$$

Eventually the stator flux space vector is calculated multiplying equations (3.18) and (3.19) with $l$, $\varepsilon$ and $\varepsilon^2$ and adding them together. The flux has two components: one depends on the stator currents and the other depends on the rotor currents.

$$\begin{cases} \underline{\Psi}_s^s = \underline{\Psi}_{ss}^s + \underline{\Psi}_{sr}^s \\ \underline{\Psi}_{ss} = \left( l_{\sigma s} - m_{\sigma s} + \frac{3}{2} l_{ms} \right) \cdot \underline{i}_s^s \\ \underline{\Psi}_{sr} = \frac{3}{2} l_{msr} \cos\alpha \cdot \underline{i}_r + \frac{\sqrt{3}}{2} l_{msr} \sin\alpha \cdot (-i_{rb} + i_{rc} + \varepsilon i_{ra} - \varepsilon i_{rc} - \varepsilon^2 i_{ra} + \varepsilon^2 i_{rb}) \end{cases} \tag{3.20}$$

The expression describing the flux component $\Psi_{sr}$ can be further transformed using the mathematical properties (3.21). The results are presented in (3.22), (3.23) and (3.24). Equation (3.24) becomes (3.25) in a general reference frame given by angle $\theta$.

$$\begin{cases} \varepsilon - \varepsilon^2 = j \cdot \sqrt{3} \\ -l + \varepsilon^2 = j \cdot \sqrt{3} \cdot \varepsilon \\ l - \varepsilon = j \cdot \sqrt{3} \cdot \varepsilon^2 \end{cases} \tag{3.21}$$

$$\underline{\Psi}_{sr}^s = \frac{3}{2} l_{msr} \cos\alpha \cdot \underline{i}_r + \frac{\sqrt{3}}{2} l_{msr} \sin\alpha \cdot j\sqrt{3} \cdot (i_{ra} + \varepsilon i_{rb} + \varepsilon^2 i_{rc}) \tag{3.22}$$

$$\underline{\Psi}_{sr}^s = \frac{3}{2} l_{msr} \cos\alpha \cdot \underline{i}_r^r + j \cdot \frac{3}{2} l_{msr} \sin\alpha \cdot \underline{i}_r^r = \frac{3}{2} l_{msr} \cdot \underline{i}_r^r \cdot e^{j\alpha} = \frac{3}{2} l_{msr} \cdot \underline{i}_r^s \tag{3.23}$$

$$\underline{\Psi}_s^s = \left( l_{\sigma s} - m_{\sigma s} + \frac{3}{2} \cdot l_{ms} \right) \cdot \underline{i}_s^s + \frac{3}{2} \cdot l_{msr} \cdot \underline{i}_r^s \tag{3.24}$$

$$\underline{\Psi}_s^\theta = \left( l_{\sigma s} - m_{\sigma s} + \frac{3}{2} \cdot l_{ms} \right) \cdot \underline{i}_s^\theta + \frac{3}{2} \cdot l_{msr} \cdot \underline{i}_r^\theta \tag{3.25}$$

The rotor flux expression is similar to the stator flux expression but each stator inductance is replaced by the corresponding rotor inductance. Thus, inducation motor equations, formulated for a reference frame defined by the angle $\theta(t)$ and the rotation speed $\omega(t)$, are:

$$\begin{cases} \underline{u}_s^\theta = R_s\,\underline{i}_s^\theta + \dfrac{\mathrm{d}\underline{\Psi}_s^\theta}{\mathrm{d}t} + j\omega_e\,\underline{\Psi}_s^\theta \\ \underline{u}_r^\theta = R_r\,\underline{i}_r^\theta + \dfrac{\mathrm{d}\underline{\Psi}_r^\theta}{\mathrm{d}t} + j(\omega_e - \omega_{er})\underline{\Psi}_r^\theta \\ \underline{\Psi}_s^\theta = \left(l_{\sigma s} - m_{\sigma s} + \dfrac{3}{2}\cdot l_{ms}\right)\cdot \underline{i}_s^\theta + \dfrac{3}{2}\cdot l_{msr}\cdot \underline{i}_r^\theta \\ \underline{\Psi}_r^\theta = \left(l_{\sigma r} - m_{\sigma r} + \dfrac{3}{2}\cdot l_{mr}\right)\cdot \underline{i}_r^\theta + \dfrac{3}{2}\cdot l_{msr}\cdot \underline{i}_s^\theta \end{cases} \tag{3.26}$$

The magnetic flux expressions in (3.26) are complicated because seven different inductances are involved. The mathematical technique of referring the rotor quantities to the stator is usually applied to the equations given in (3.26) in order to simplify the flux equations. The basic principle of referring the rotor quantities to the stator is to multiply rotor quantities with constant values in such a manner that the power transfer between stator and rotor is not altered. Thus, if the rotor current is multiplied by constant $k$ then the rotor voltage and the rotor flux are multiplied by $1/k$. On the other hand, the rotor resistance and the rotor inductance are multiplied by $1/k^2$. The constant $k$ that generates the simplest transformation of system (3.26) is given by (3.27) while the corresponding referred rotor quantities are (3.28). The equation linking all the referred quantities is (3.29).

$$k = \frac{l_{msr}}{l_{ms}} \tag{3.27}$$

$$\begin{cases} \underline{i}'^\theta_r = k\cdot \underline{i}_r^\theta = \dfrac{l_{msr}}{l_{ms}}\cdot \underline{i}_r^\theta \\ \underline{u}'^\theta_r = \dfrac{1}{k}\cdot \underline{u}_r^\theta = \dfrac{l_{ms}}{l_{msr}}\cdot \underline{u}_r^\theta \\ \underline{\Psi}'^\theta_r = \dfrac{1}{k}\cdot \underline{\Psi}_r^\theta = \dfrac{l_{ms}}{l_{msr}}\cdot \underline{\Psi}_r^\theta \\ R'_r = \dfrac{1}{k^2}\cdot R_r = \dfrac{l_{ms}^2}{l_{msr}^2}\cdot R_r \end{cases} \tag{3.28}$$

$$\underline{u}'^\theta_r = R_r\,\underline{i}'^\theta_r + \frac{\mathrm{d}\underline{\Psi}'^\theta_r}{\mathrm{d}t} + j(\omega_e - \omega_{er})\,\underline{\Psi}'^\theta_r \tag{3.29}$$

The referred rotor flux can be expressed as a function of the stator current and the referred rotor current vector (3.30).

$$\underline{\Psi}'^\theta_r = \frac{l_{ms}}{l_{msr}}\,\underline{\Psi}_r^\theta = \frac{l_{ms}^2}{l_{msr}^2}\left(l_{\sigma r} - m_{\sigma r} + \frac{3}{2}\,l_{mr}\right)\cdot \underline{i}'^\theta_r + \frac{3}{2}l_{ms}\cdot \underline{i}_s^\theta \tag{3.30}$$

The inductances $l_{ms}$, $l_{mr}$ and $l_{msr}$ are always related by equation (3.31). This relationship allows the rewriting of equation (3.30) as (3.32) and (3.33).

$$l_{ms}\cdot l_{mr} = l_{msr}^2 \tag{3.31}$$

$$\underline{\Psi}'^{\theta}_{r} = \frac{l_{ms}^2}{l_{msr}^2}(l_{\sigma r} - m_{\sigma r}) \cdot \underline{i'^{\theta}_{r}} + \frac{3}{2}l_{ms} \cdot \underline{i'^{\theta}_{r}} + \frac{3}{2}l_{ms} \cdot \underline{i}^{\theta}_{s} \tag{3.32}$$

$$\underline{\Psi}'^{\theta}_{r} = (L'_{\sigma r} + L_m) \cdot \underline{i'^{\theta}_{r}} + L_m \cdot \underline{i}^{\theta}_{s} = L'_r \cdot \underline{i'^{\theta}_{r}} + L_m \cdot \underline{i}^{\theta}_{s} \tag{3.33}$$

The significance of the symbols in the previous equations is:

- $L'_{\sigma r} = \frac{l_{ms}^2}{l_{msr}^2}(l_{\sigma r} - m_{\sigma r})$ – the total referred rotor leakage inductance.
- $L_m = \frac{3}{2}l_{ms}$ – the resulting stator–rotor mutual inductance.
- $L'_r = L'_{\sigma r} + L_m$ – the total referred rotor inductance.

Substituting the first equation (3.28) in (3.25), the stator flux can be written as:

$$\underline{\Psi}^{\theta}_{s} = \left(l_{\sigma s} - m_{\sigma s} + \frac{3}{2} \cdot l_{ms}\right) \cdot \underline{i}^{\theta}_{s} + \frac{3}{2} \cdot l_{ms} \cdot \underline{i}'^{\theta}_{r} \tag{3.34}$$

$$\underline{\Psi}^{\theta}_{s} = (L_{\sigma r} + L_m) \cdot \underline{i^{\theta}_{r}} + L_m \cdot \underline{i}^{\theta}_{s} = L_s \cdot \underline{i}^{\theta}_{s} + L_m \cdot \underline{i}'^{\theta}_{s} \tag{3.35}$$

The significance of the symbols is:

- $L_{\sigma s} = l_{\sigma s} - m_{\sigma s}$ – the total stator leakage inductance.
- $L_s = L_{\sigma s} + L_m$ – the total stator inductance.

Thus, (3.36) is the compact format of the induction motor equations initially presented in (3.26). This system of equations expresses the space vector model of the induction motor [159]. This model will be used in the neural induction motor control example presented in the second section of the book.

$$\begin{cases} \underline{u}^{\theta}_{s} = R_s \underline{i}^{\theta}_{r} + \frac{d\underline{\Psi}^{\theta}_{s}}{dt} + j\omega\underline{\Psi}^{\theta}_{s} \\ \underline{u}'^{\theta}_{r} = R'_r \underline{i}'^{\theta}_{r} + \frac{d\underline{\Psi}'^{\theta}_{r}}{dt} + j(\omega_e - \omega_{er})\underline{\Psi}'^{\theta}_{r} = 0 \\ \underline{\Psi}^{\theta}_{s} = L_s \underline{i}^{\theta}_{s} + L_m \underline{i}'^{\theta}_{r} \\ \underline{\Psi}'^{\theta}_{r} = L'_r \underline{i}'^{\theta}_{r} + L_m \underline{i}^{\theta}_{s} \end{cases} \tag{3.36}$$

Note: Usually, to simplify the notation, the apostrophe symbols are not included in the equations. Yet, the rotor quantities are implicitly referred to the stator. No apostrophe symbol is used in the rest of this book but they are implied for all rotor equations or parameters.

### 3.5.2 Induction motor control strategies

During the first one hundred years after its invention, the induction motor was known as a constant speed electrical machine. The advent of electrical power converters in the

1960s made possible the use of the induction motor as a variable speed machine. The recent development of the digital technology created the possibility of implementing complex control algorithms yielding high dynamic performance [229].

Correct control over the motor torque is a prerequisite of all the speed control strategies. The torque equation can be derived from power-based considerations and can be expressed as a function of the current and voltage space vectors. The total power consumed by the motor has three components: the power dissipated by the winding resistances $P_R$, the power stored in the internal magnetic fields $P_\mu$ and the mechanical power $P_M$. The motor torque is proportional to the mechanical power and inversely proportional to the rotor speed (3.37). The total motor power is the power consumed by all six stator and rotor windings so it can be calculated as in equation (3.38). Elementary algebraic calculations show that the rotor power and the stator power can be calculated as indicated by (3.39). The calculations can be performed in any reference frame defined by the time function $\theta(t)$. Now as:

$$\begin{cases} P = P_R + P_\mu + P_M \\ T = \dfrac{P_M}{\omega_r} = p \cdot \dfrac{P_M}{\omega_{er}} \end{cases} \tag{3.37}$$

therefore:

$$P = P_s + P_r = u_{sa}i_{sa} + u_{sb}i_{sb} + u_{sc}i_{sc} + u_{ra}i_{ra} + u_{rb}i_{sb} + u_{rc}i_{sc} \tag{3.38}$$

where:

$$\begin{cases} P_s = u_{sa}i_{sa} + u_{sb}i_{sb} + u_{sc}i_{sc} = \dfrac{2}{3}\,\mathrm{Re}\,\{\underline{u}_s^s \cdot \underline{i}_s^{s*}\} = \dfrac{2}{3}\,\mathrm{Re}\,\{\underline{u}_s^\theta \cdot \underline{i}_s^{\theta*} \\ P_r = u_{ra}i_{ra} + u_{rb}i_{rb} + u_{rc}i_{rc} = \dfrac{2}{3}\,\mathrm{Re}\,\{\underline{u}_r^r \cdot \underline{i}_r^{r*}\} = \dfrac{2}{3}\,\mathrm{Re}\,\{\underline{u}_r^\theta \cdot \underline{i}_r^{\theta*}\} \end{cases} \tag{3.39}$$

The equations (3.40) are obtained by substituting general equations (3.36) into (3.39). Thus, the three power components are calculated according to (3.41).

$$\begin{cases} P_s = \dfrac{2}{3}\,\mathrm{Re}\left\{\left[R_s\underline{i}_s^\theta + \dfrac{d\underline{\Psi}_s^\theta}{dt} + j\omega_e \cdot \underline{\Psi}_s^\theta\right] \cdot \underline{i}_s^{\theta*}\right\} \\ P_r = \dfrac{2}{3}\,\mathrm{Re}\left\{\left[R_r\underline{i}_r^\theta + \dfrac{d\underline{\Psi}_r^\theta}{dt} + j(\omega_e - \omega_{er}) \cdot \underline{\Psi}_r^\theta\right] \cdot \underline{i}_r^{\theta*}\right\} \end{cases} \tag{3.40}$$

$$\begin{cases} P_R = \dfrac{2}{3}\,\mathrm{Re}\,\{R_s\underline{i}_s^\theta \cdot \underline{i}_s^{\theta*} + R_r\underline{i}_r^\theta \cdot \underline{i}_r^{\theta*}\} \\ P_\mu = \dfrac{2}{3}\,\mathrm{Re}\left\{\dfrac{d\underline{\Psi}_s^\theta}{dt} \cdot \underline{i}_s^{\theta*} + \dfrac{d\underline{\Psi}_r^\theta}{dt} \cdot \underline{i}_r^{\theta*}\right\} \\ P_M = \dfrac{2}{3}\,\mathrm{Re}\,\{j\omega_e \cdot \underline{\Psi}_s^\theta \cdot \underline{i}_s^{\theta*} + j(\omega_e - \omega_{er}) \cdot \underline{\Psi}_r^\theta \cdot \underline{i}_r^{\theta*} \end{cases} \tag{3.41}$$

The imaginary number '$j$' in the expression of the mechanical power component $P_M$ can be eliminated using the general algebraic property (3.42).

$$\mathrm{Re}\{j \cdot z\} = -\,\mathrm{Im}\{z\} \tag{3.42}$$

The two components of the imaginary part in (3.42) can be rewritten as in (3.43), so that the mechanical power equation becomes (3.44).

$$\begin{cases} \mathrm{Im}\,\{\omega_e \cdot \underline{\Psi}_s^\theta \cdot \underline{i}_s^{\theta *}\} = \mathrm{Im}\,\{\omega_e \cdot (L_s \underline{i}_s^\theta + L_m \underline{i}_r^\theta) \cdot \underline{i}_s^{\theta *}\} = \omega_e L_m \cdot \mathrm{Im}\,\{\underline{i}_r^\theta \cdot \underline{i}_s^{\theta *} \\ \mathrm{Im}\,\{(\omega_e - \omega_{er}) \cdot \underline{\Psi}_r^\theta \cdot \underline{i}_r^{\theta *}\} = (\omega_e - \omega_{er}) L_m \cdot \mathrm{Im}\,\{\underline{i}_s^\theta \cdot \underline{i}_r^{\theta *}\} \end{cases} \tag{3.43}$$

$$P_M = -\frac{2}{3} L_m\, [\omega_e \cdot \mathrm{Im}\{\underline{i}_r^\theta \cdot \underline{i}_s^{\theta *}\} + (\omega_e - \omega_{er}) \cdot \mathrm{Im}\,\{\underline{i}_s^\theta \cdot \underline{i}_r^{\theta *}\}] \tag{3.44}$$

Based on the mathematical property (3.45) equation (3.44) is further transformed into (3.46).

$$\mathrm{Im}\,\{\underline{x} \cdot \underline{y}^*\} + \mathrm{Im}\,\{\underline{y} \cdot \underline{x}^*\} = 0 \tag{3.45}$$

$$P_M = \frac{2}{3}\, \omega_{er} L_m \cdot \mathrm{Im}\,\{\underline{i}_s^\theta \cdot \underline{i}_r^{\theta *}\} = \frac{2}{3}\, p \omega_r L_m \cdot \mathrm{Im}\,\{\underline{i}_s^\theta \cdot \underline{i}_r^{\theta *} \tag{3.46}$$

Therefore, the motor torque may be expressed by (3.47). It is seen that the motor torque depends only on the rotor current vector and on the stator current vector.

$$T = \frac{2}{3}\, p L_m \cdot \mathrm{Im}\,\{\underline{i}_s^\theta \cdot \underline{i}_r^{\theta *} \tag{3.47}$$

Alternatively, the torque can be expressed as equivalent functions of the stator magnetic flux and/or the rotor magnetic flux as shown in (3.48), where $\delta$ is the angle between the stator flux and the rotor flux.

$$\begin{cases} T = \frac{2}{3}\, p \cdot \mathrm{Im}\,\{\underline{\Psi}_r^\theta \cdot \underline{i}_r^{\theta *}\} & \text{(a)} \\ T = \frac{2}{3}\, p \cdot \mathrm{Im}\,\{\underline{i}_s^\theta \cdot \underline{\Psi}_s^{\theta *}\} & \text{(b)} \\ T = \frac{2}{3}\, p \cdot \dfrac{L_m}{L_s L_r - L_m^2} \cdot \mathrm{Im}\,\{\underline{\Psi}_s^\theta \cdot \underline{\Psi}_r^{\theta *}\} = \frac{2}{3}\, p \cdot \dfrac{L_m}{L_s L_r - L_m^2} \cdot |\underline{\Psi}_s^\theta| \cdot |\underline{\Psi}_r^\theta| \cdot \sin \delta & \text{(c)} \end{cases} \tag{3.48}$$

Relations (3.47) and (3.48) directly or indirectly underlie all induction motor control strategies. They can be classified as scalar control and vector control strategies (Fig. 3.7). The scalar control operates utilising simplified equations derived from the general space vector model (3.36). This approach involves only the space vector amplitudes and their corresponding frequencies and the simplified equations are valid only in steady-state operation. Consequently, scalar control is simple but generates poor response during transient operation [227]. In contrast, vector control operates directly with the space vector model of the motor and implements the equations given in (3.48). Therefore, it offers good results in both steady-state operation and transient operation. The group of vector control algorithms includes the direct torque control (DTC) method and the class of field-oriented control strategies. The theory of field-oriented control was developed by researchers at Siemens in 1968–1969. Since this time, researchers all over the world have implemented increasingly efficient practical systems based on this theory [229]. The actual motor speed is the most important information for any speed control algorithm. As illustrated in Fig. 3.7, there are two possible approaches to obtaining this measure: either to use a speed sensor or to calculate the speed based on the electrical motor quantities.

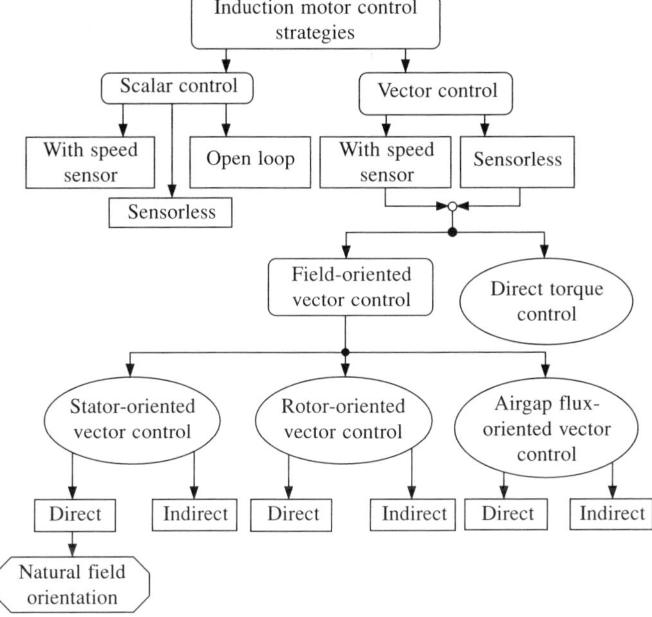


**Fig. 3.7** Classification of the induction control strategies

These two approaches are applicable to scalar control methods as well as to vector control methods but the use of vector control ensures better dynamic response. The interest in speed sensorless control emerged from practical applications where high control quality is required but the speed sensor is either difficult to use due to technical reasons, or too expensive. The speed sensorless control of the induction motor is currently one of the most intensively researched fields in electrical drives [228].

### 3.5.3 Scalar control

Scalar control uses the stator voltage amplitude $U_s = 2/3 \cdot |\underline{u}_s|$ and the stator frequency $f_s$ as input quantities and works well in steady-state and slow transient operation. This strategy varies the stator voltage and the stator frequency according to a function $U_s(f_s)$ so that the maximum torque available is large (and almost constant) at any stator angular frequency $\omega_{es}$.

In steady-state operation, the rotor flux has constant amplitude. Therefore, the rotor equation in rotor coordinates is:

$$R_r \underline{i}_r^r + \frac{d\underline{\Psi}_r^r}{dt} = R_r \underline{i}_r^r + j(\omega_{es} - \omega_{er})\, \underline{\Psi}_r^r = R_r \underline{i}_r^r + j(\omega_{es} - \omega_{er}) \cdot (L_r \underline{i}_r^r + L_m \underline{i}_s^r) = 0 \tag{3.49}$$

Under these conditions, the rotor current depends on the stator current space vector and on the slip angular frequency (the difference $\omega_{es} - \omega_{er}$) as indicated in (3.50).

$$\underline{i}_r^r = \frac{-j(\omega_{es} - \omega_{er})\, L_m \underline{i}_s^r}{R_r + j(\omega_{es} - \omega_{er}) L_r} \tag{3.50}$$

The initial motor torque expression (3.47) can be modified by substituting (3.50) in (3.47) which yields equation (3.51). Therefore, the motor torque is proportional to the stator current module squared.

$$T = \frac{2}{3} pL_m^2 \cdot \mathrm{Im}\left\{\frac{j(\omega_{es} - \omega_{er})\, \underline{i}_s^r \cdot \underline{i}_s^{r*}}{R_r + j(\omega_{es} - \omega_{er}) L_r}\right\} = \frac{2}{3} pL_m^2\, |\underline{i}_s^r|^2 \cdot \mathrm{Im}\left\{\frac{j(\omega_{es} - \omega_{er})}{R_r + j(\omega_{es} - \omega_{er}) L_r}\right\} \tag{3.51}$$

It is seen that the stator current depends on the stator voltage as indicated by (3.52), (3.53), (3.54), while the dependency between the torque and the stator voltage is obtained by combining equations (3.51) and (3.54) to give relationship (3.55).

$$\underline{u}_s^r = R_s \underline{i}_s^r + j\omega_{er} \underline{\Psi}_s^r + j(\omega_{es} - \omega_{er}) \underline{\Psi}_s^r = R_s \underline{i}_s^r + j\omega_{es} \underline{\Psi}_s^r \tag{3.52}$$

$$\underline{u}_s^r = R_s \underline{i}_s^r + j\omega_{es}\left[L_s \underline{i}_s^r - \frac{j(\omega_{es} - \omega_{er}) L_m^2 \cdot \underline{i}_s^r}{R_r + j(\omega_{es} - \omega_{er}) L_r}\right] \tag{3.53}$$

$$|\underline{i}_s^r| = \frac{|\underline{u}_s^r|}{\left|R_s + j\omega_{es} L_s + \dfrac{\omega_{es}(\omega_{es} - \omega_{er}) L_m^2}{R_r + j(\omega_{es} - \omega_{er}) L_r}\right|} \tag{3.54}$$

$$T = \frac{3}{2} pL_m^2 \cdot \frac{U_s^2}{\left|R_s + j\omega_{es} L_s + \dfrac{\omega_{es}(\omega_{es} - \omega_{er}) L_m^2}{R_r + j(\omega_{es} - \omega_{er}) L_r}\right|^2} \cdot \mathrm{Im}\left\{\frac{j(\omega_{es} - \omega_{er}) L_m^2}{R_r + j(\omega_{es} - \omega_{er}) L_r}\right\} \tag{3.55}$$

Figure 3.8 presents the torque–speed characteristic calculated according to (3.55) for a three-phase induction motor with the parameters $R_s = 0.371\ \Omega$; $R_r = 0.415\ \Omega$; $L_{s\sigma} = 2.72$ mH; $L_{r\sigma} = 3.3$ mH; $L_m = 84.33$ mH; $p = 1$; $P = 11.1$ kW. The motor is supplied by a three-phase 240 V/50 Hz supply. As the figure shows, the motor torque is zero at synchronous speed and has its maximum at a relatively high angular speed $\omega_M$ as compared to the rated stator angular frequency (314 rad/s). The motor normally operates at speeds between the synchronous angular speed and $\omega_M$. At high stator angular frequency, around the rated value, the stator resistance is negligible, thus, $|\underline{i}_s^r|$ in (3.54) depends only on the slip angular frequency ($\omega_{slp} = \omega_{es} - \omega_{er}$) and on the voltage angular frequency

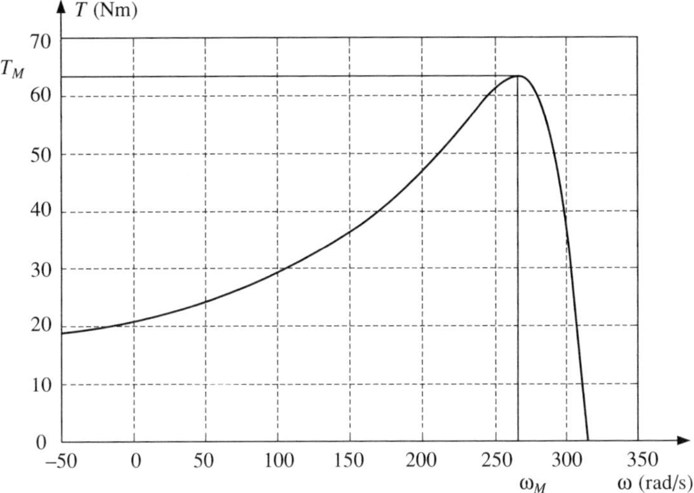


**Fig. 3.8** Induction motor mechanical characteristic ($P$ = 11.1 kW)

ratio ($U_s/\omega_s$). If this ratio is kept constant then the stator current amplitude and the motor torque depend solely on the slip angular frequency. Therefore, the maximum motor torque $T_M$ is independent of the stator angular frequency $\omega_{es}$. At low frequencies, however, the stator resistance has an important influence on the stator current and leads to a diminished maximum torque, with negative effects on the motor operation. The effect of the stator resistance on the motor torque can be counteracted by raising the stator voltage to compensate for the stator resistance. The function $U_s(\omega_s)$ that maintains $T_M$ constant at all frequencies can be derived from (3.55). The solution is a non-linear expression, difficult to implement into hardware. A linear approximation of this function is usually adopted in practical situations. The linear approximation $U_s(\omega_s)$ is defined by two points corresponding to the zero stator frequency and to the rated stator frequency:

- At zero stator frequency, the stator voltage has to generate a current equal to the stator current at rated stator angular frequency (314 rad/s) and maximum torque.
- At the rated stator frequency, the voltage attains its rated value.

The stator voltage amplitude is therefore defined by (3.56) where '$p$' is the number of stator pole pairs. This approximate solution does not provide a perfectly constant $T_{\max}$ but restricts its variation within a narrow interval.

$$U_s = R_s I_{s(\max T)} + \omega_{es} \frac{U_{s\mathrm{MAX}} - R_s I_{s(\max T)}}{p \cdot 2\pi \cdot 50} = U_{s0} + \Phi \cdot f_s \tag{3.56}$$

Speeds over the rated value can be obtained by increasing the stator frequency over the 50 Hz limit but in this case the voltage is maintained constant at its maximum value $U_{s\mathrm{MAX}}$. As a result, the maximum available torque decreases (it is inversely proportional to the frequency squared) and very high speeds cannot be obtained using this method. For instance, the maximum torque decreases by as much as 25 per cent from the rated value if the stator frequency is 100 Hz. The open-loop scalar control implements the strategy illustrated by (3.56). This offers an approximate control over the motor speed but the effects of the load torque variations cannot be compensated for due to the lack of any feedback information. A compensation of the average slip angular frequency can be performed instead so that the rotor speed equals the reference speed for the most frequent load torque value.

The control scheme can be implemented with a controlled rectifier as presented in Fig. 3.9, or with an uncontrolled rectifier. In the first case, the PWM inverter controls only the frequency of the output voltage, while the rectifier determines the output voltage amplitude. In the second case, the switching pattern inside the inverter is more complex and determines both the frequency and the amplitude of the output voltage.

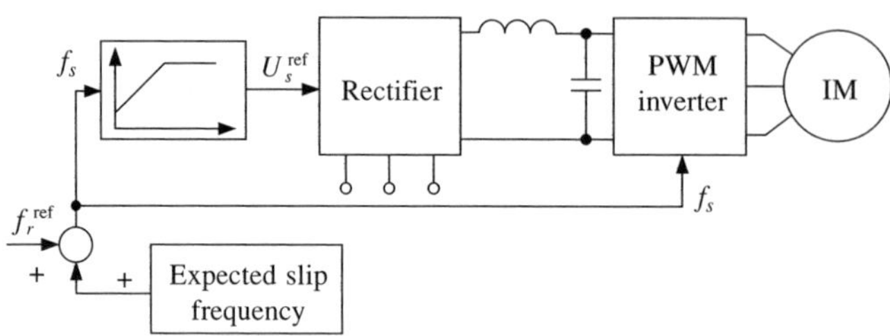


**Fig. 3.9** Open-loop scalar control scheme

The scalar control strategy with speed sensor can be implemented as in Fig. 3.10 using a controlled rectifier and a PWM inverter. As in the previous section, the controlled rectifier can be replaced by an uncontrolled rectifier if the inverter controls both the frequency and the amplitude of the output voltage. The voltage control loop modifies the d.c. voltage according to the required speed profile while the optimal slip frequency is calculated as a function of the current absorbed by the motor: the slip increases with the absorbed current. This type of slip-current correlation limits the current variations in the d.c. link during the transient motor operation.

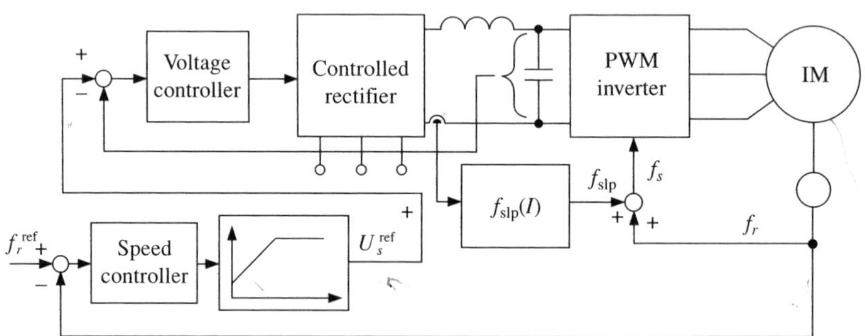


**Fig. 3.10** Scalar control scheme with speed sensor

An increase of the resistive load increases the current absorbed by the motor and decreases its speed. This lowers the d.c. link voltage. The speed controller responds by increasing the reference voltage while the slip calculator increases the motor slip. As demonstrated by equation (3.55) the motor torque increases with the increase of the stator voltage and with the increase of the slip angular frequency. On the other hand, the stator current depends on the stator voltage in the manner indicated in (3.54). Therefore, a torque increase can be obtained with a diminished current change if the slip angular frequency is changed accordingly. Conversely, when the load torque decreases the current drop in the d.c. link is limited and the temporary transformation of the motor into a generator is avoided, thereby reducing the strain on the power transistors in the PWM inverter.

The sensorless scalar control strategy is based on the possibility of calculating the slip frequency as a function of the stator frequency and the current in the d.c. link between the rectifier and the PWM inverter [183]. The equation underlying the slip angular frequency calculation can be derived from (3.54). The stator angular frequency is determined as the sum of the slip angular frequency and the calculated rotor angular speed corresponding to the actual voltage across the d.c. link. In general, the large d.c. link capacitor prevents the amplitude of the a.c. voltage from being increased as rapidly as the frequency, which is developed with practically no delay by simply feeding the right triggering pulses to the inverter transistors. Hence, it is customary to calculate the frequency control to the voltage control loop in the manner shown in Fig. 3.11 to prevent the motor from ever receiving the inappropriate voltage–frequency ratio.

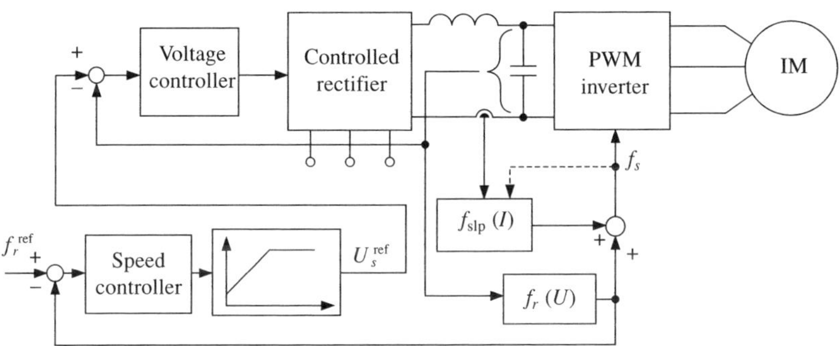


**Fig. 3.11** Sensorless scalar control scheme

Of the two control parameters, frequency control is by far the most sensitive as small changes in frequency produce large changes of slip frequency and hence large changes in current and torque. By slaving the frequency command to the d.c. bus voltage, the rate of frequency change is generally limited to a value to which the motor can respond without drawing excessive current or without regenerating.

### 3.5.4 Vector control

Vector control strategies use the space vector model of the induction motor to accurately control the speed and torque both in steady-state operation and in fast transient operation. The dynamic performance achieved by vector control strategies equals the dynamic performance offered by d.c. motor drives. In fact, with vector control, induction motor drives outperform d.c. drives because of higher transient current capability, increased speed range, and lower rotor inertia [46]. The class of vector control strategies encompasses field-oriented control methods and direct torque control methods. Field-oriented control methods use the rotor-oriented reference frame, the airgap-oriented reference frame or the stator-oriented reference frame (see Fig. 3.7). In each case, the reference frame real axis (axis '*d*') is oriented along the direction indicated by the corresponding magnetic flux. The rotor-oriented vector control simplifies the control system structure and generates very fast transient response. However, systems working with the stator flux vector or with the airgap flux vector have been successfully implemented as well [79], [43].

### 3.5.5 Rotor flux orientation

In the rotor flux-oriented reference frame, the rotor flux vector has no imaginary part so that the torque expression (3.48(a)) can be written as (3.57). The rotor flux and the rotor current depend on one another in the manner indicated by the equations in (3.58).

$$T = \frac{2}{3} p \cdot \mathrm{Re}\,\{\underline{\Psi}_r^{\theta}\} \cdot \mathrm{Im}\,\{\underline{i}_r^{\theta *}\} = -\frac{2}{3} p \cdot \Psi_{rd} \cdot i_{rq} \tag{3.57}$$

$$\begin{cases} \Psi_{rd} = L_r i_{rd} + L_m i_{sd} \\ \Psi_{rq} = L_r i_{rq} + L_m i_{sq} = 0 \\ R_r i_{rd} + \dfrac{\mathrm{d}\Psi_{rd}}{\mathrm{d}t} - (\omega_{e\Psi_r} - \omega_{er})\Psi_{rq} = R_r i_{rd} + \dfrac{\mathrm{d}\Psi_{rd}}{\mathrm{d}t} = 0 \\ R_r i_{rq} + \dfrac{\mathrm{d}\Psi_{rq}}{\mathrm{d}t} + (\omega_{e\Psi_r} - \omega_{er})\Psi_{rd} = R_r i_{rq} + (\omega_{e\Psi_r} - \omega_{er})\Psi_{rd} = 0 \end{cases} \tag{3.58}$$

Equations (3.59) and (3.60) can be derived from the previous system. They illustrate the influence of the stator current components over the rotor flux and on the rotor current component on axis '*q*' ($i_{rq}$). Thus, the modification speed of the rotor flux is limited by the rotor time constant $T_r = L_r/R_r$, while the rotor current component $i_{rq}$ can be changed rapidly as no time constant is involved in (3.60).

$$\frac{L_r}{R_r}\frac{\mathrm{d}\Psi_{rd}}{\mathrm{d}t} + \Psi_{rd} = L_m i_{sd} \tag{3.59}$$

$$i_{rq} = -\frac{L_m}{L_r} \cdot i_{sq} \tag{3.60}$$

As demonstrated by (3.59) and (3.60), the two quantities influencing the torque can be independently controlled by two uncoupled control loops. For high dynamic performance, the torque is controlled by keeping the rotor flux $\Psi_{rd}$ constant while varying the rotor current component $i_{rq}$. Keeping the rotor flux constant implies maintaining $i_{sd}$ at a constant value while the rotor current component $i_{rq}$ is controlled by the stator current component $i_{sq}$.

The control strategy requires the rotor flux orientation to be determined in order to calculate $i_{sd}$ and $i_{sq}$. The direct vector control method estimates the magnetic flux vector as a function of the stator voltage, the stator current and the rotor speed. There are three types of rotor flux estimators differing by the input data they use: the current–speed estimator ($I_s$, $\omega_{er}$), the current–voltage estimator ($I_s$, $U_s$) and the current–voltage–speed estimator ($I_s$, $U_s$, $\omega_{er}$). The indirect vector control method is simpler as it calculates only the argument $\theta$ of the rotor flux as a function of $i_{sd}$ and $i_{sq}$. The direct vector control is more robust than the indirect vector control but its performance depends on the type of flux estimator used.

The current–speed estimator is derived from the basic rotor equation and from the rotor flux expression as shown in (3.61), (3.62) and (3.63). The rotor flux is the solution of the integral equation (3.64). This estimator works well at low speeds but it is not precise at high speeds because in this case the speed measuring errors have a big influence on the calculation results.

$$\begin{cases} 0 = R_r \underline{i}_r^s + \dfrac{\mathrm{d}\underline{\Psi}_r^s}{\mathrm{d}t} - j\omega_{er}\underline{\Psi}_r^s \\ \underline{\Psi}_r^s = L_m \underline{i}_s^s + L_r \underline{i}_r^s \end{cases} \tag{3.61}$$

$$0 = R_r \cdot \frac{\underline{\Psi}_r^s - L_s \underline{i}_s^s}{L_r} + \frac{\mathrm{d}\underline{\Psi}_r^s}{\mathrm{d}t} - j\omega_{er}\underline{\Psi}_r^s \tag{3.62}$$

$$\frac{\mathrm{d}\underline{\Psi}_r^s}{\mathrm{d}t} = \left(-\frac{1}{T_r} + j\omega_{er}\right) \cdot \underline{\Psi}_r^s + \frac{L_s}{T_r}\underline{i}_s^s \tag{3.63}$$

$$\underline{\Psi}_{r(I,\omega)}^s = \int_0^t \left[\left(-\frac{1}{T_r} + j\omega_{er}\right) \cdot \underline{\Psi}_r^s + \frac{L_s}{T_r}\underline{i}_s^s\right] \mathrm{d}t \tag{3.64}$$

The current–voltage flux estimator is derived from the stator equation and the stator flux expression (see (3.65), (3.66), (3.67)). Therefore, the equation defining the current–voltage flux estimator is (3.68). This method offers accurate results at high speeds but the precision at low speeds is low.

$$\begin{cases} \underline{u}_s^s = R_s \underline{i}_s^s + \dfrac{\mathrm{d}\underline{\Psi}_s^s}{\mathrm{d}t} \\ \underline{\Psi}_s^s = L_s \underline{i}_s^s + L_m \underline{i}_r^s = L_s \underline{i}_s^s + \dfrac{L_m}{L_r}(\underline{\Psi}_r^s - L_m \underline{i}_s^s) \end{cases} \tag{3.65}$$

$$\underline{u}_s^s = R_s \underline{i}_s^s + L_s \frac{\mathrm{d}\underline{i}_s^s}{\mathrm{d}t} + \frac{L_m}{L_r}\left(\frac{\mathrm{d}\underline{\Psi}_r^s}{\mathrm{d}t} - L_m \frac{\mathrm{d}\underline{i}_s^s}{\mathrm{d}t}\right) \tag{3.66}$$

$$\frac{\mathrm{d}\underline{\Psi}_r^s}{\mathrm{d}t} = \frac{L_r}{L_m}(\underline{u}_s^s - R_s \underline{i}_s^s) + \frac{L_s L_r - L_m^2}{L_m} \cdot \frac{\mathrm{d}\underline{i}_s^s}{\mathrm{d}t} \tag{3.67}$$

$$\underline{\Psi}_{r(I,U)}^s = \frac{L_r}{L_m} \cdot \int_0^t (\underline{u}_s^s - R_s \underline{i}_s^s)\,\mathrm{d}t + \frac{L_s L_r - L_m^2}{L_m} \cdot \underline{i}_s^s \tag{3.68}$$

The current–voltage–speed estimator ($I_s$, $U_s$, $\omega_{er}$) combines the previous two solutions: equation (3.64) and equation (3.68). It generates good rotor flux estimates both at low speeds and high speeds.

$$\underline{\Psi}_{r(I,U,\omega)}^s = \frac{\underline{\Psi}_{r(I,\omega)}^s + \underline{\Psi}_{r(I,U)}^s}{2} \tag{3.69}$$

The rotor flux is the original choice for field orientation because in this reference frame the equations corresponding to the two axes ((3.59) and (3.60)) are completely independent.

As a result, this control method generates the best dynamic performance. On the other hand, the stator flux orientation has the advantage that the torque calculation uses the stator flux instead of the rotor flux as illustrated by (3.70) which is a consequence of (3.48(b)). The stator magnetic flux is much easier to calculate than the rotor magnetic flux because it depends on stator quantities (currents, voltages and resistance) that can be directly measured.

$$T = \frac{2}{3} p \cdot \mathrm{Im}\,\{\underline{i}_s^\theta\} \cdot \mathrm{Re}\,\{\underline{\Psi}_s^{\theta *}\} = \frac{2}{3} p \cdot \Psi_{sd} \cdot i_{sq} \tag{3.70}$$

$$\underline{\Psi}_s^\theta = \int_0^t (\underline{u}_s^\theta - R_s \underline{i}_s^\theta)\, \mathrm{d}t \tag{3.71}$$

A typical direct rotor field-oriented control scheme (see Fig. 3.12) contains two closed loops: one for $i_{sd}$ (controlling the motor magnetic flux) and the other for $i_{sq}$ (controlling the motor torque). The rotor flux orientation exploits the advantage that the two quantities can be controlled independently: the value of one stator current component does not have any influence over the value of the other current component. This property simplifies the control structure and generates good dynamic performance. One of the three flux observers previously described is used to determine the rotor magnetic flux. This information is used to calculate the reference frame transformations: from the stator reference frame to rotor reference frame, and from the rotor reference to stator reference frame.

The flux generating current component ($i_{sd}$) is maintained constant for speeds under

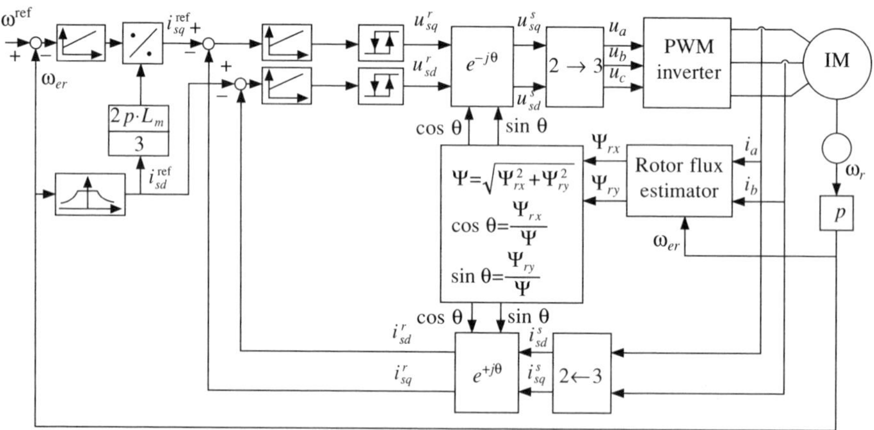


**Fig. 3.12** Direct rotor field-oriented control scheme

the rated value but is decreased for speeds above the rated value (in the so-called field weakening region). Regardless of the vector control strategy, it can be demonstrated that maintaining the magnetic flux constant at different stator frequencies implies that the stator voltage amplitude is approximately proportional to the stator frequency. As in the case of scalar control, the stator voltage amplitude is given by an equation similar to (3.56). Therefore, for speeds larger than the rated value the magnetic flux value cannot be kept constant because that would require high voltages that may damage the motor.

High speeds are obtained at the expense of the field weakening which decreases the efficiency of the motor.

The torque generating current component ($i_{sq}$) is calculated as a function of the required motor torque and the motor field. The reference current $i_{sq}^{\text{ref}}$ is proportional to the torque-to-field ratio. The torque is calculated in turn as a function of the difference between the reference speed and the actual speed of the motor.

In the case of indirect rotor field orientation, the flux orientation is calculated by integrating the stator angular frequency (3.72). The slip angular frequency is estimated as shown by equation (3.75) which is derived from the basic equations governing the rotor circuit (see (3.73) and (3.74)). In (3.75) it is implicit that the rotor flux amplitude is constant due to very good current controllers providing very fast (ideally instantaneous) dynamic response. Parameter detuning leads to a loss of rotor field orientation and to a deterioration of the system dynamic response. The rotor time constant $T_r$ especially should be updated through an estimator [169].

$$\theta(t) = \int_0^t (\omega_{\text{slp}} + \omega_r)\, \mathrm{d}t = \int_0^t \omega_s \cdot \mathrm{d}t \tag{3.72}$$

$$0 = R_r \underline{i}_r^r + \frac{\mathrm{d}\underline{\Psi}_r^r}{\mathrm{d}t} + j(\omega_{es} - \omega_{er}) \cdot \underline{\Psi}_r^r \tag{3.73}$$

$$\begin{cases} 0 = R_r i_{rq} + \omega_{\text{slp}} \Psi_{rd} + \dfrac{\mathrm{d}\Psi_{rq}}{\mathrm{d}t} = R_r i_{rq} + \omega_{\text{slp}} \Psi_{rd} = R_r i_{rq} + \omega_{\text{slp}} \Psi_r \\ 0 = R_r i_{rd} - \omega_{\text{slp}} \Psi_{rq} + \dfrac{\mathrm{d}\Psi_{rd}}{\mathrm{d}t} \end{cases} \tag{3.74}$$

$$\omega_{\text{slp}} = -\frac{R_r i_{rq}}{\Psi_r} = \frac{L_m R_r}{L_r \Psi_r} \cdot i_{sq} \approx \frac{1}{T_r} \cdot \frac{i_{sq}}{i_{sd}} \tag{3.75}$$

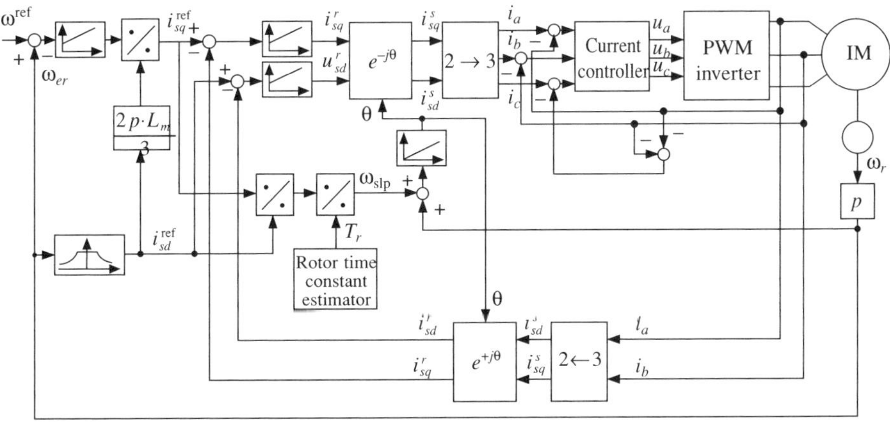


**Fig. 3.13** Indirect rotor field-oriented control scheme

### 3.5.6 Stator and airgap flux orientation

In the case of stator flux orientation, the flux equations take the form presented in (3.76). The magnetic flux vector and the stator current vector are the solutions of two coupled equations: (3.77) and (3.78) derived from (3.76). Therefore, the magnetic flux and the torque-generating current component cannot be controlled independently as in the case of rotor orientation. Here any modification of the magnetic flux has effects on the torque-generating current component. This slows the system transient response unless special compensation blocks are added to the control scheme.

$$\begin{cases} \Psi_{sd} = L_s i_{sd} + L_m i_{rd} \\ \Psi_{sq} = L_s i_{sq} + L_m i_{rq} = 0 \\ R_s i_{sd} + \dfrac{\mathrm{d}\Psi_{sd}}{\mathrm{d}t} - \omega_{e\Psi_s}\Psi_{sq} = R_s i_{sd} + \dfrac{\mathrm{d}\Psi_{sd}}{\mathrm{d}t} = u_{sd} \\ R_s i_{sq} + \dfrac{\mathrm{d}\Psi_{sq}}{\mathrm{d}t} + \omega_{e\Psi_s}\Psi_{sd} = R_s i_{sq} + \omega_{e\Psi_s}\Psi_{sd} = u_{sq} \end{cases} \tag{3.76}$$

$$\frac{\mathrm{d}\Psi_{sd}}{\mathrm{d}t} = u_{sd} - R_s i_{sd} \tag{3.77}$$

$$i_{sq} = \frac{u_{sq} - \omega_{e\Psi_s}\Psi_{sd}}{R_s} \tag{3.78}$$

The equations underlying the airgap flux orientation are identical to (3.77) and (3.78) but are expressed in a different reference frame. Both the stator and the airgap-oriented vector control strategy are similar to the rotor-oriented vector control in that the magnetic flux vector is kept constant for speeds below the rated value, while the torque is varied by modifying the corresponding current component ($i_{sq}$ in this case). Figure 3.14 presents an example of stator field orientation. The control method is similar to the rotor flux orientation but contains an additional flux controller. The flux controller is added to diminish the effects of the interaction between the magnetic flux vector and the torque-generating stator current component.

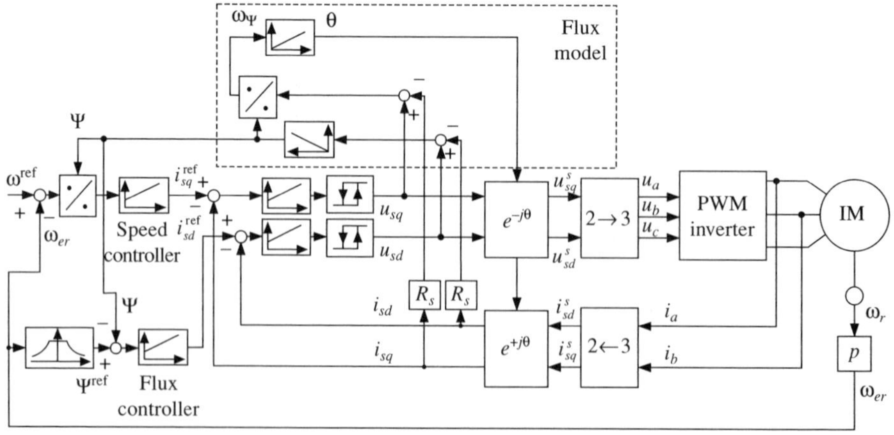


**Fig. 3.14** Direct stator field-oriented control scheme

### 3.5.7 Direct torque control

In a PWM inverter-fed machine, the vector $\underline{\Psi}_r$ is more filtered than $\underline{\Psi}_s$ and therefore $\underline{\Psi}_r$ rotates more smoothly. The motion of $\underline{\Psi}_s$, dictated by the stator voltage, is discontinuous, but the average velocity is the same with that of $\underline{\Psi}_r$ in steady state. The direct torque control (DTC) method is based on relation (3.48(c)). Therefore, the torque is controlled by varying the angle $\delta$ between the two flux vectors. Any DTC implementation contains a flux control loop and a torque control loop. The reference torque value is calculated by a speed controller, while the flux reference is determined as a function of the reference speed $\omega^{ref}$.

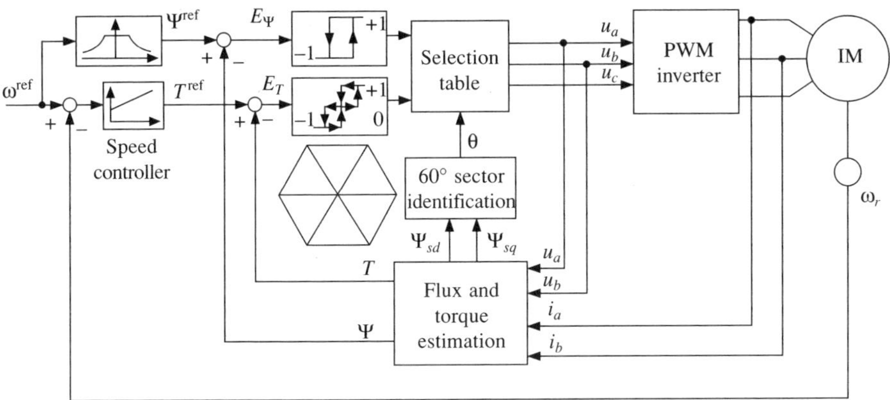


**Fig. 3.15** Direct torque control with speed sensor

The machine voltages and currents are sensed to estimate the torque and the stator flux vector. The flux vector estimation gives information about the 60° sector where $\underline{\Psi}_s$ is located. The errors $E_\Psi$ and $E_T$ generate digital signals through the respective hysteresis-band comparators. A three-dimensional look-up table then selects the most appropriate voltage vector ($u_a$, $u_b$, $u_c$) to satisfy the flux and torque demands.

DTC ensures fast transient response and generates simple implementations due to the absence of the closed-loop current control, traditional PWM algorithm and the vector transformations. It can be implemented with speed sensor as well as in sensorless configurations. However, the drawbacks of DTC are the pulsating torque, pulsating flux and the increased harmonic loss [46]. Recently a large number of papers have been published concerned with improving DTC control [162], [122], [123], [132], [56], [24], [55].

### 3.5.8 Sensorless vector control schemes

The speed estimation methods for induction motors are based on the possibility of being able to calculate the rotor speed as a function of stator currents and stator voltages. Therefore the physical speed sensor is replaced by a software or hardware implemented module that performs the necessary calculations. The relation between the voltage and current is influenced by both the motor speed and the winding parameters. These parameters are subject to alterations during the motor operation due to heating and magnetic saturation. Consequently, on-line parameter estimation procedures need to be implemented alongside speed estimation algorithms to ensure correct results under various operation conditions.

Complex mathematical methods have been developed to integrate the speed estimation with the electrical parameter estimation process and to achieve high accuracy and independence of the motor parameter variations. These methods combine the classical field orientation approach with extended Kalman filters [20], [164], Luenberger observers [153], [190], neural networks [38], [226] and fuzzy logic [231], [34]. A different approach makes use of the effects of the rotor saliencies on the stator currents and voltages [216] or the parasitic effects that originate from the discrete winding structure of a cage rotor. In both these two cases, the stator currents contain harmonics that depend on the rotor speed so that Fourier transforms are involved in the speed calculation. Most of these methods are more accurate at high speeds than at low speeds. As a result, the lowest speed at which the system works correctly is an important performance indicator.

The Kalman filter (KF) was developed by R. Kalman and R. Bucy in the early 1960s [138], [139]. The standard KF [224] is a recursive state estimator for multiple-input/multiple-output systems with noisy measurement data and with process noise (stochastic plant model). It uses the inputs and the outputs of the plant together with a state–space model of the system, to give optimal estimates of the system state. The space–state model is described by equation (3.79) where vector $x$ is the state of the system and vector $u$ contains the system inputs. The system output is given by (3.80). The matrices $v$ and $w$, known as the spectral density matrices, model the noise processes. The noise is supposed to be *white* and *Gaussian*.

$$\dot{x} = Ax + Bu + Fv \tag{3.79}$$

$$y = C \cdot x + w \tag{3.80}$$

The filter equation is given in (3.81), where $K$ is the gain matrix of the filter. $K$ is calculated as a function of the matrices $u$ and $w$ that describe the statistical properties of the noise processes. Equation (3.81) has the general form of a linear state–space observer. Thus, the KF is an optimal observer because it calculates the vector $x$ as a function of vector $u$ in such a manner that the adverse effect of the noise is minimised.

$$\dot{\hat{x}} = A\hat{x} + Bu + K(y - C\hat{x}) \tag{3.81}$$

In the standard linear form, the Kalman filter can only estimate the stator current *d-q* components, and the rotor current *d-q* components. To estimate the rotor speed and/or the rotor resistance (the critical electrical parameter for most of the control strategies), the time-varying variable is treated as a state variable. Consequently, a non-linear system model is generated. To use a non-linear model with the standard Kalman filter, the model must be linearised around the current operating point, giving a linear perturbation model. The result is the extended Kalman filter (EKF). A comparison of the performances of KF and EKF is presented in [164]. The applications using KFs and EKFs are very popular although they impose high computational demands on the digital equipment involved [200].

The sensorless vector control of induction motors continues to be investigated by many authors and several improvements have been proposed in recent years [160], [118], [140], [144], [192], [213], [230]. Many companies have launched their own sensorless vector control products [27]. The most representative products are shown in Table 3.2.

**Table 3.2** Representative a.c. sensorless vector control products

| Company | Product | Ratings kW | Vac input | Speed reg. (±%) | Torque reg. (±%) | Min. speed at 100% cont. torque |
|---|---|---|---|---|---|---|
| ABB | ACS 600 | 2.2–600 | 380–690 | 0.1–0.3 | 2 | 2 Hz |
| Allen–Bradley | 1336 Impact/ Force a.c. Drive | 0.75–485 | 230–600 | 0.5 | 5 | 0.5 Hz |
| Baldor Electric | 17H Encoderless Vector Control | 0.75–373 | 180–660 | 10% of slip | 3.5 | 100 rpm |
| Cutler–Hammer | AF93 | 1.5–15 | 340–528 | 0.5 | N/A | 50 rpm |
| Mitsubishi Electr. America | A200E<br>A024/A044 | 0.4–55<br>0.1–3.7 | 230–575<br>230–460 | 1.0<br>1–3 | N/A<br>N/A | <1 Hz<br>3 Hz |
| Siemens E & A | Master Drive 6SE70 | to 1500 | 208–690 | 0.1 | <2.5 | 0 |
| Square D | Altivar 66SV | 0.75–220 | 208–460 | 1.0 | N/A | 0.5 Hz |
| Yaskawa electr. America | VS–616G5 | 0.4–800 | 200–600 | 0.1 | 3 | 0.5 Hz |
| NFO Control AB | NFO Sinus Switch | 0.37–5.5 | 230–400 | 1 | 1 | 1 |

The natural field orientation (NFO) method, invented and patented by the Swedish company NFO Control AB, is one of the simplest and most efficient sensorless motor control strategies so far. NFO Control AB implemented this method into hardware alongside an improved PWM switching strategy and sell it under the name 'NFO Sinus Switch'. NFO is derived from the stator field-oriented vector control and it can be implemented for both speed sensor and sensorless systems but its advantages are fully exploited in the sensorless configuration. The corresponding control circuit is a simplification of the control scheme in Fig. 3.14. The essence of NFO is that the magnitude of the stator flux is not calculated by integration as in the case of stator flux orientation. The flux is set in open loop as a reference quantity that may be subject to change for field weakening [135], [136]. Thus, both the flux controller and the divider, that are present in Fig. 3.14 inside the speed control loop, are eliminated.

NFO can be implemented in several forms beginning with the basic configuration without current controllers, shown in Fig. 3.16, applicable to small drives. In this case, the voltage component $u_{sq}$ is determined by the speed controller while the voltage component $u_{sd}$ is calculated only as a function of the magnetising current $i_{ms}$ so that the correct stator magnetic flux is generated. The stator magnetising current is defined by (3.82).

$$\underline{\Psi}_s = L_s\underline{i}_s + L_m\underline{i}_r = L_m\underline{i}_{sm} \tag{3.82}$$

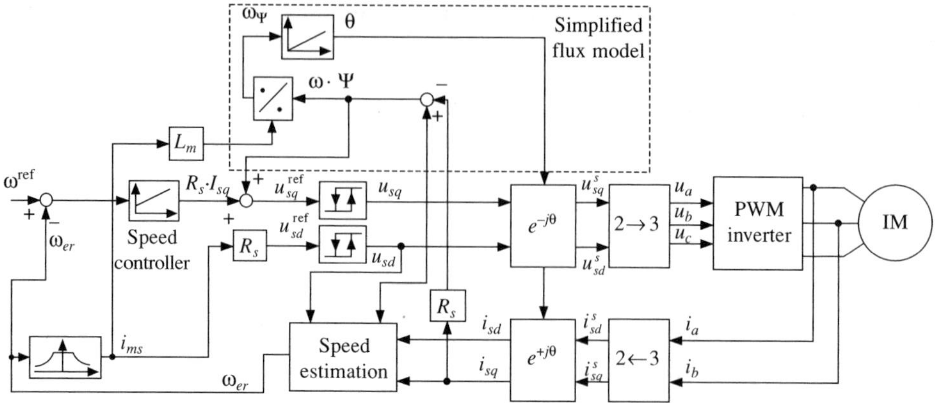


**Fig. 3.16** Natural field orientation (NFO)

The stator equations can be written using the quantity $i_{ms}$ as in system (3.83). One of the features of NFO is that the control scheme operates so that the modulus of $i_{sm}$ equals $i_{sd}$ (see Fig. 3.17). Therefore, the reference voltages are calculated according to (3.84).

$$\begin{cases} L_m \dfrac{\mathrm{d}i_{sm}}{\mathrm{d}t} = u_{sd} - R_s i_{sd} \\ \omega_{sm} L_m i_{sm} = u_{sq} - R_s i_{sq} \end{cases} \tag{3.83}$$

$$\begin{cases} u_{sd}^{\text{ref}} = R_s i_{sm} \\ u_{sq}^{\text{ref}} = R_s i_{sq} + \omega_{ms} L_m i_{sm} \end{cases} \tag{3.84}$$

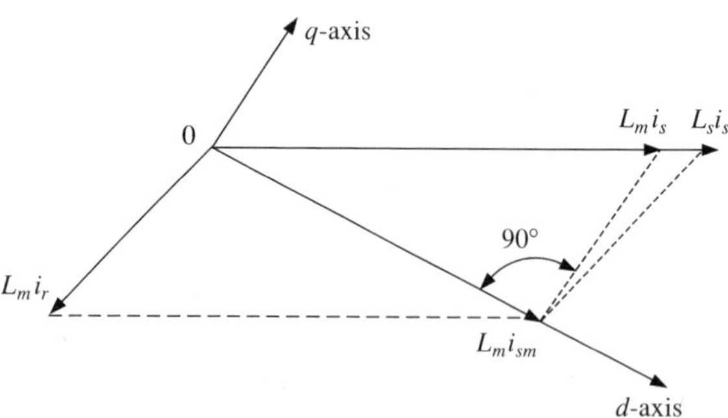


**Fig. 3.17** The stator and rotor current vectors in case of natural field orientation

The speed estimation is based on an inner voltage vector $\underline{e}_s$ defined according to (3.85). It is demonstrated [136] that the motor speed can be calculated using $\underline{e}_s$ as indicated in (3.86). This equation is valid whether or not $i_{sm}$ equals $i_{sd}$.

$$\begin{cases} \underline{e}_s = \underline{u}_s - \left( R_s + R_r \dfrac{1+\sigma_s}{1+\sigma_r} \right) \underline{i}_s - \sigma \dfrac{d\underline{i}_s}{dt} \\ \sigma_s = \dfrac{L_{\sigma s}}{L_m} \\ \sigma_r = \dfrac{L_{\sigma r}}{L_m} \\ \sigma = \dfrac{L_s L_r - L_m^2}{L_m^2} \end{cases} \tag{3.85}$$

$$\omega_r = \frac{e_{sq}}{pL_s \cdot \left( \dfrac{i_{sm}}{1+\sigma_s} - \sigma \cdot i_{sd} \right)} \tag{3.86}$$

At low speed, the magnitude of $\underline{e}$ is small. Therefore small errors in measuring the motor currents will lead to large relative errors in calculating the vector $\underline{e}$ (3.85) that will in turn reflect into large relative errors of the estimated rotor speed. Thus, the speed estimation precision is minimal at very small rotor speed. Most sensorless control strategies face the same problem, which is why the minimal speed that the system can efficiently control is one of the key parameters used in measuring the control system performance.

The space–vector concept has been described alongside the space–vector model of the three-phase induction motor. These concepts have been used to describe the main techniques for the control of induction motor drives. In the next chapter, neural network theory is briefly presented and neural control is considered with a view to assessing its applicability to produce efficient control systems for the envisaged induction motor applications.

## 3.6 Synchronous generators control

Synchronous generators are responsible for the bulk of the electrical power generated in the world today. They are mainly used in power stations and are predominantly driven either by steam or hydraulic turbines. These generators are usually connected to an *infinite bus* where the terminal voltages are held at a constant value by the 'momentum' of all the other generators also connected to it. Another common application of synchronous generators is their use in stand-alone or isolated power generation systems. The prime mover in such applications is usually a diesel engine.

The aim of the example presented in this section is to develop an improved control system for diesel engine driven stand-alone synchronous generator sets. Its primary objective is to design and build a working prototype that incorporates a new control strategy and the latest engineering innovations. The subsidiary objectives include ensuring that the prototype system:

- can be used effectively as a starting point for further studies into a new generation of controllers for stand-alone synchronous generator sets;
- incorporates a certain amount of artificial intelligence such that it is flexible and not specific to a particular type of engine-generator set;

- is designed using a systematic process which enables rapid prototyping of future improvements;
- takes advantage of modern digital electronic technology.

The following section discusses the control systems of synchronous generators, in an attempt to tie these developments to the design example presented in the second part of the book. Although a massive proportion of synchronous generators are electromagnetic, the use of permanent magnet synchronous machines as stand-alone generators has been studied for more than half a century. Permanent magnet synchronous generators (PMSGs) are more difficult to regulate and it is only with the recent developments in power electronics that they are seriously being considered for various applications [39], [191], [17]. One of the main advantages of the control system proposed in the examples section of this book is its ability to regulate stand-alone PMSGs as well as electromagnet generators. This functionality is duly demonstrated by the experiments presented in which a PMSG is used.

The study of synchronous generator control systems can roughly be divided into two parts: voltage regulation and speed governing. Both control elements contribute to the stability of the machine in the presence of perturbations. A reliable control system set is essential for the safe operation of generators. There are various methods of controlling a synchronous generator and suitability will depend on the type of machine, its application and the operating conditions. For instance, the voltage regulation of an electromagnet synchronous generator is usually achieved by controlling the field excitation current whereas permanent magnet generators do not have excitation systems and require a totally different strategy.

The voltage regulation system in an electromagnet synchronous generator is called an automatic voltage regulator (AVR). It is a device that automatically adjusts the output voltage of the generator in order to maintain it at a relatively constant value. This is achieved by comparing the output voltage with a reference voltage and, from the difference (or *error*), it makes the necessary adjustments in the field current to bring the output voltage closer to the required value. Older AVRs used in the early days belong to a class of electromechanical devices. They are generally slow acting and possess zones of insensitivity known as *dead bands*. There is a wide variety of electromechanical AVRs, ranging from vibrating contact regulators to carbon pile regulators [235]. However, they are now replaced with continuously acting electronic regulators that are much faster and do not possess dead bands, hence the term 'continuously acting'. Figure 3.18 shows a block diagram of an electronic AVR system [6].

The generator set comprises two sections: the *main* section and an *exciter*. Each section consists of an armature winding and a field winding. Electrical power is derived from the terminals of the main armature winding. The AVR maintains a closed-loop control of the terminal voltage by taking in a ratio of the main armature voltage as the input, comparing it to a preset reference value and producing a control signal to the exciter field based on the *error*. This induces the correct amount of current in the exciter armature and the current is transferred into the main field winding via a set of rotating diodes. The complete system is an efficient and well-established method of regulating the terminal voltage in a stand-alone electromagnet synchronous generator set. Power system generators also employ AVRs in their control systems, but they form only a part of a comprehensive control and regulation system. Most AVRs utilise analogue electronics to accomplish the control tasks. It is found to be cheap, simple and effective for the job.

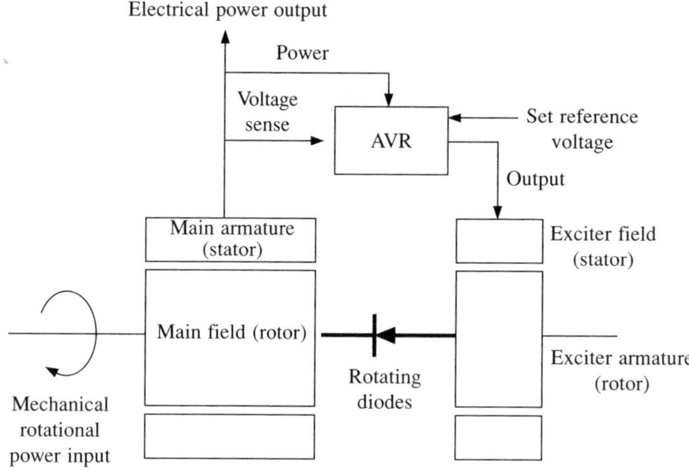


**Fig. 3.18** Block diagram of a synchronous generator and AVR

However, like all electronic applications, there is a growing interest in applying digital electronic technology to AVRs. The major advantages of digital AVRs over their analogue counterparts are the capability of realising sophisticated control functions and the ease of transmission and recording of information. The nature of digital systems allows complex algorithms to be executed either as a software-based system [164] or a hardware-based system [124]. Digital AVRs are capable of performing all the tasks of their analogue counterparts as well as a range of additional functions, including the following:

- They can have direct interaction with higher levels of controls. This allows them to be interfaced with the main computer of the building, power plant or supervisory system, enabling *interoperability*.
- The control parameters are more accessible to modifications with built-in logic switches and software tuning programs. Changes in the control parameters of analogue AVRs can be difficult and expensive to implement. Therefore the parameter values of analogue AVRs are usually fixed at the design and commission stage.
- Stabilisers can be configured to switch in and out the system without interrupting the service, depending on the need.
- They can operate at optimal or suboptimal conditions over a wider range of operation. This is especially true with the introduction of adaptive and intelligent controllers.

In 1976, Malik, Hope and Huber worked on a software-based digital AVR [124] for a power system generator, whereby the control functions of the AVR were stored in an on-line station mini computer. The flexibility of a software-based system extended the mathematical capabilities of the controller and allowed for one basic hardware design to be used with different control strategies, thus drastically reducing the cost of redesigning the controller to different specifications. The digital AVR used by Hirayama *et al.* (1993) [114] was capable of all the functions of an analogue AVR as well as possessing additional features such as advanced fault status indication, self-diagnostic features, verification of generator dynamics and the recording of transients. A more comprehensive monitor of the power systems was also possible by feeding the recorded transient data into a personal computer (PC) via an RS232C link. Dedicated software in the PC will then calculate the Bode diagram and indicate transient responses and rise times. Similarly, the digital excitation control systems (DECS) used in Godhwani and Basler's (1996) [98], [99] study into design methodology of controllers offer all the advantages of digital control. In their work, the PID settings for a generator controller were programmed into an electronically erasable programmable read only memory (EEPROM). Due to the fact that different systems require different PID settings for optimal performance, manufacturers of analogue controllers often have to provide multiple designs to accommodate variations in stability networks. Godwhani and Basler proposed a design methodology – direct design method of controller design – that allows the closed-loop poles to be placed at any desired location. The PID settings were then custom designed for each individual system using this method and stored into the EEPROM. This creates a single design which is flexible for the operation of a wide range of generator sizes (ranging from 10 kW to 50 MW).

As result of the highly non-linear nature of synchronous machines, it can be quite difficult to design a control system with high performance over different operating points. This is becoming an increasingly important issue especially in power systems. To address the problem, Marino [167] proposed *exact feedback linearisation*, a technique which received some recent attention [198]. There is also interest in adaptive AVRs [129], [95], [238] and power system controllers. There are currently several different approaches to adaptive control of synchronous generators in power system control such as:

- linear optimal control theory [175];
- self-tuning control theory [98], [185];
- fuzzy logic [114];
- adaptive neural network [145], [240], [150].

Although these works are mainly targeted for power systems applications, although some of the ideas presented can be incorporated into other applications. Stand-alone synchronous generators, systems with conventional AVRs and speed governors have always been operated at a fixed speed. The speed is determined by the desired power frequency and the number of pole-pairs in the machine. Running the generator at any other speed is not usually considered as a design option. One of the main problems in such a scheme is the task of maintaining the desired power frequency, especially under heavy load. Since the frequency is directly proportional to the speed, any change in speed would certainly disrupt the shape of the power waveform in that the frequency would vary in relation to the speed. It is only with recent developments in power devices and converter technology that variable-speed constant-frequency (VSCF) systems have been given serious attention. The matrix converter proposed in [232] initiated several studies into the a.c.–a.c. converter [241], [236], [215], but disparate issues such as commutation problems and the complexity of switching schemes generate reservations for its use in the present work. A more common solution for a.c.–a.c. power frequency conversion is to buffer the transition with a d.c. link, effectively producing an a.c.–d.c.–a.c. conversion scheme. In this book, a power electronic system implementing this scheme is called a *d.c. link converter*. The main components of a d.c. link converter include some form of rectifier, energy storage component(s) such as a smoothing capacitor and an inverter. Some schemes also incorporate a boost converter in the d.c. link [41].

In the last decade, d.c. link converters have received considerable attention in the area of VSCF wind energy systems. VSCF schemes are widely used in stand-alone wind energy conversion systems to solve the problem of frequency fluctuations which result from changes in wind velocity and load. Figure 3.19 shows a simplified block diagram of a typical VSCF wind energy conversion system that incorporates a d.c. link converter.

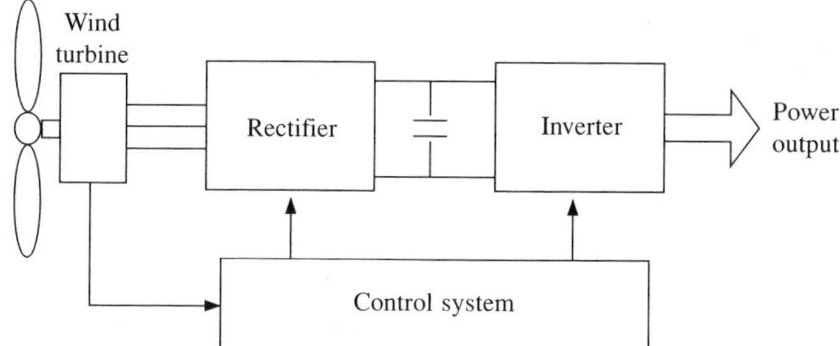


**Fig. 3.19** Simplified block diagram of a VSCF wind energy conversion system

Although most of these schemes are designed for induction generators [188], [202], [172] there are also numerous projects which involve synchronous generators [42]. Research in this area usually focuses around control systems for capturing maximum energy from varying wind velocity. Others include work such as the study of interface systems to improve the quality of power from VSCF wind turbines connected to a utility grid [134], [206], [57]. Unfortunately, the application of VSCF schemes in other systems such as stand-alone generators is somewhat limited. In an engine driven system, for example, the choice of speed can have great influence on the operational characteristics and efficiency of the engine. With the correct setting, it is possible for the system to be optimised for specific performances such as fuel consumption, exhaust emissions, vibrations and generator ratings. In a domestic or recreational application, where the generator system has to be placed close to the living space, reducing noise pollution may be an important consideration when selecting the operational speed. An example of a stand-alone generator system controller based on fuzzy logic is presented in the second section of this book.

# 4

# Elements of neural control

Neural control is a branch of the general field of intelligent control, which is based on the concept of artificial intelligence (AI). AI can be defined as computer emulation of the human thinking process. The AI techniques are generally classified as expert systems (ES), fuzzy logic (FL) and artificial neural networks (ANN). The classical expert systems are based on Boolean algebra and use precise calculations while fuzzy logic systems involve calculations based on an approximate reasoning. Fuzzy logic is a superset of conventional (Boolean) logic that has been extended to handle the concept of partial truth–truth values between 'completely true' and 'completely false' [88]. It was introduced by Dr Lotfi Zadeh of UC/Berkeley in the 1960s as a means to model the uncertainty of natural language. The truth of a logical expression in fuzzy logic is a number in the interval [0,1]. Fuzzy logic has emerged as a profitable tool for the control of complex industrial processes and systems. It is used for processes that have no simple mathematical model, for highly non-linear processes, or where the processing of linguistically formulated knowledge is to be performed. Although it was invented in the United States, the rapid growth of this technology originated from Japan and has now again reached the USA and Europe. The controllers based on this mathematical approach are known as fuzzy controllers.

The use of artificial neural networks (ANNs) is the most powerful approach in AI. ANNs are information processing structures which emulate the architecture and operational mode of the biological nervous tissue. Any ANN is a system made up of several basic entities (named neurones) which are interconnected and operate in parallel transmitting signals to one another in order to achieve a certain processing task [237]. One of the most outstanding features of ANNs is their capability to simulate the learning process. They are supplied with pairs of input and output signals from which general rules are automatically derived so that the ANN will be (in certain conditions) capable of generating the correct output for a signal that was not previously used. The neural approach can be combined with the fuzzy logic generating neuro-fuzzy systems that combine the advantages of the two control paradigms.

## 4.1 Neurone types

The operation of the artificial neurones is inspired by their natural counterparts. Each artificial neurone has several inputs (corresponding to the synapses of the biological neurones) and one single output, the axon. Each input is characterised by a certain weight indicating the influence of the corresponding signal over the neurone output. The neurone calculates an equivalent total input signal as the weighted sum of the individual input signals (4.1).

$$\text{net} = \sum_{i=1}^{n} w_i \cdot x_i \tag{4.1}$$

The resulting quantity is then compared with a constant value named the threshold level and the output signal is calculated as a function of their difference (net – $t$). This function is named the transfer function or the activation function. The input weights, the threshold level and the activation function are the parameters which completely describe an artificial neurone. Depending on the type of the artificial neurone the activation function may have several forms. There are analogue neurones using continuous real activation functions and discrete neurones whose activation functions are discontinuous. Bipolar neurones generate both positive and negative outputs while unipolar ones generate only positive values. In the case of bipolar analogue neurones, the most popular activation function is given by (4.2). The output varies continuously between –1 and +1, depending on the input signals that can have any real value (Fig. 4.1).

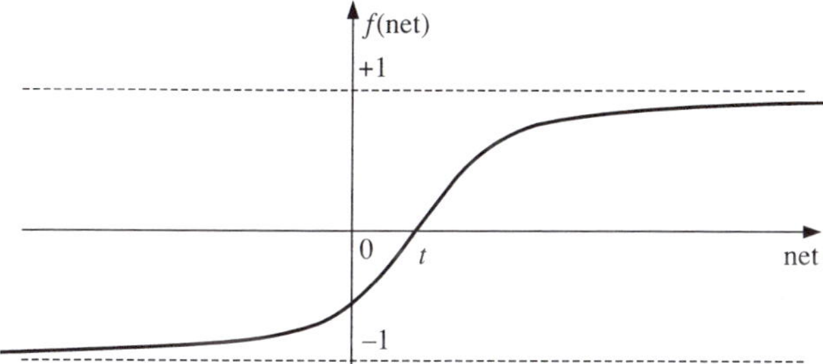


**Fig. 4.1** Sigmoidal activation function of bipolar analogue neurones ($\lambda = 1$)

Parameter $\lambda$ is a constant controlling the slope of the activation function's graph. Some authors consider $\lambda = 1$ to simplify the calculations while others operate with the more general format presented in (4.2) but the fundamental results and properties of the corresponding ANNs remain valid in both situations. The function in (4.2) is part of a larger transfer function class called 'sigmoidal functions'. What they have in common is the graph shape and the property to be derivable, which is essential in some applications.

$$f_1(\text{net}) = \frac{2}{1 + \lambda \cdot e^{-(\text{net} - t)}} - 1 \tag{4.2}$$

An alternative activation function is presented in (4.3). It is part of the sigmoidal functions group and, as shown in (4.4), it has the same limit values as function $f_1$. Unipolar analogue neurones are similar to bipolar ones with the difference that the output signals can only take values between 0 and +1 (Fig. 4.2). Their activation function is described by (4.5).

$$f_2(\text{net}) = \frac{1 - \lambda \cdot e^{-(\text{net} - t)}}{1 + \lambda \cdot e^{-(\text{net} - t)}} \tag{4.3}$$

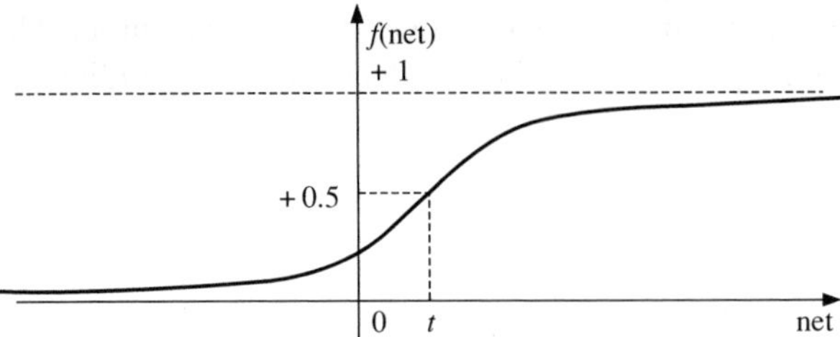


**Fig. 4.2** Sigmoidal activation function for unipolar analogue neurones ($\lambda = 1$)

$$\begin{cases} \lim\limits_{net \to +\infty} f_2(\text{net}) = +1 \\ \lim\limits_{net \to -\infty} f_2(\text{net}) = -1 \end{cases} \tag{4.4}$$

$$f_3(\text{net}) = \frac{1}{1 + \lambda \cdot e^{-(\text{net} - t)}} \tag{4.5}$$

Not all continuous activation functions are sigmoidal. The stepwise-linear activation function presented in (4.6) is not derivable in two points: net = $t - 1.0$ and net = $t + 1.0$ (Fig. 4.3).

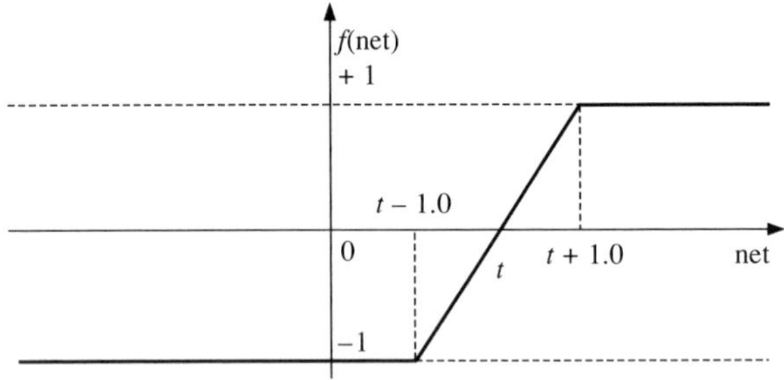


**Fig. 4.3** Non-sigmoidal activation function

$$f_4(\text{net}) = \begin{cases} -1 & \text{if net} < t - 1.0 \\ \text{net} - t & \text{if net} \in [t - 1.0; t + 1.0] \\ +1 & \text{if net} > t + 1.0 \end{cases} \tag{4.6}$$

Discrete neurones use threshold type activation functions. The bipolar discrete sort is associated with the activation function described in (4.7) while the unipolar type uses the activation function illustrated by (4.8). These two functions can be considered limiting cases ($\lambda \to \infty$) of the sigmoidal transfer functions presented in (4.2) and (4.5).

$$f_5(\text{net}) = \begin{cases} 1 \ \text{net} \geq t \\ -1 \ \text{net} < t \end{cases} \tag{4.7}$$

$$f_6(\text{net}) = \begin{cases} 1 \ \text{net} \geq t \\ 0 \ \text{net} < t \end{cases} \tag{4.8}$$

Over the last few years, more sophisticated types of neurones and activation functions have been introduced in order to solve different sorts of practical problems. In particular, radial basis neurones have proved very useful for many control system and system identification applications. These neurones use so-called radial basis activation functions. Equation (4.9) presents the most often used form for such a function, where '$x$' is the $n$-dimensional vector of input signals and '$t$' a constant vector of the same dimension while $\|\cdot\|$ is the Euclidean norm in the $n$-dimensional space.

$$f_7(x) = \exp(- \| x - t \|^2) \tag{4.9}$$

Practically $f_7$ shows how close vector '$x$' is to vector '$t$' in this $n$-dimensional space. The closer $x$ is to $t$, the larger is $f_7(x)$; if $x = t$ then $f_7(x) = 1$. The classical Gaussian bell is obtained for the unidimensional case while the two-dimensional case is illustrated by Fig. 4.4. Obviously, such a neurone type is very far from the biological model, but this is irrelevant since it proves useful for certain technical applications.

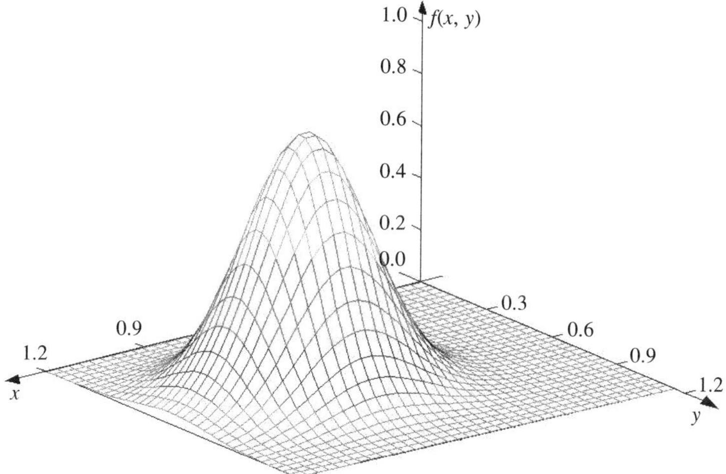


**Fig. 4.4** Radial basis activation function: two-dimensional case

## 4.2 Artificial neural networks architectures

Artificial neural networks differ by the type of neurones they are made of and by the manner of their interconnection. There are two major classes of neural networks: feed-forward ANNs and recurrent ANNs. Feed-forward artificial neural networks (FFANNs) are organised into cascaded layers of neurones. Each layer contains neurones receiving input signals from the neurones in the previous layer and transmitting outputs to the neurones in the subsequent layer. The neurones within a layer do not communicate to one another. The first network layer is named the input layer, while the last one is named the output layer. All the other neurone layers are known as the hidden layers of the neural network.

FFANNs do not have any memory of the past inputs so that they are used for applications where the output is only a function of the present inputs. Therefore, each input vector is simply associated with an output vector. If step activation functions are used, several analogue or discrete input vectors can be associated with a single discrete output vector. Such neural networks are used to solve classification problems. In a classification problem, the set of all possible input vectors is divided into several arbitrary subsets. Each subset is a class. The problem consists of finding out to which class a given input vector belongs. The neural network associates each class with a binary vector and generates the corresponding code for any input vector. Recurrent artificial neural networks include architectures where neurones in the same layer communicate (cellular neural networks) or architectures where some of the outputs of a FFANN are used as inputs (real-time recurrent networks, Hopfield networks). These neural architectures can be described either by continuous time models or by discrete time models.

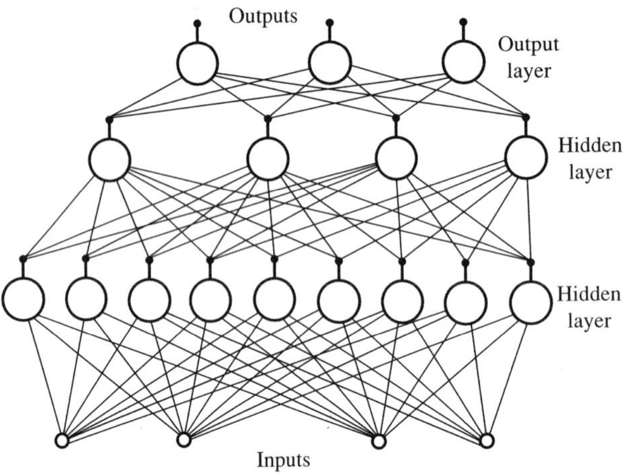


**Fig. 4.5** Feed-forward neural network architecture

The concept of cellular neural network (CNN) was first introduced by Chua and Yang (1988). It is a special class of recurrent neural networks, which consist of cells connected only to the cells in their neighbourhood (Fig. 4.6). Thus, the main feature of CNNs is the fact that information is directly exchanged just between neighbouring cells.

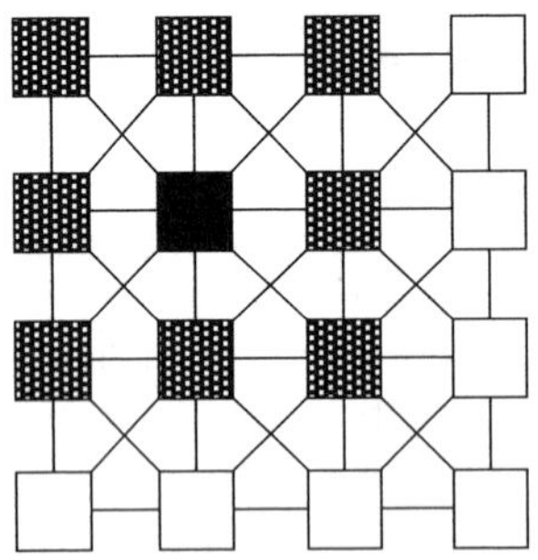

**Fig. 4.6** Cellular neural network

Due to this local interconnection property, CNNs have been considered particularly suitable for VLSI implementations for high speed parallel signal processing. CNNs are used in several application areas: image processing, artificial vision, associative memories, biological systems modelling, etc.

The real-time recurrent neural networks are the most adequate neural structures for modelling finite and infinite state machines. They explicitly implement the concept of 'internal state' as a set of neurone outputs which are used as future inputs of the FFANN contained inside the feedback architecture (Fig. 4.7).

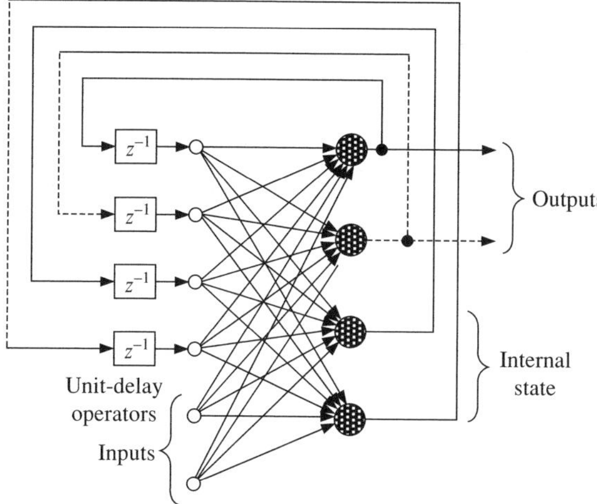


**Fig. 4.7** One-layer real-time recurrent neural network (discrete time model)

The discrete-time models contain delay units on the feedback connections, while the continuous-time models contain low-pass filters (usually first-order elements). Finite state machines are modelled by discrete-time models involving neurones with step activation functions, while infinite state machines are modelled by continuous-time models containing neurones with sigmoidal activation functions.

Hopfield networks are a particular case of recurrent neural networks that contain only one layer of neurones and there is no feedback loop between any neurone and itself. The connections between any two different neurones are symmetrical in Hopfield networks, that is the corresponding weights are equal. Furthermore, each neurone is connected to an external input signal. Each state of a Hopfield network can be characterised by a so-called 'energy function'. The evolution of the network's state determines a decrease of the energy function towards a local minimum. Each local minimum is associated to a stable state. In this respect, Hopfield networks can be configured in manners that allow solutions to be found for particular optimisation problems. This feature has been used for associative memory applications, and the optimisation of the power dispatch in the power systems [92].

## 4.3 Training algorithms

One of the most important features of neural networks is their ability to learn (to be trained) and improve their operation using a set of examples named training data set.

The training process is controlled by mathematical algorithms that fall in two main classes: constructive and non-constructive. The non-constructive training algorithms adapt only the connection weights and the threshold levels. The constructive algorithms modify all the network features including its architecture (neurones and even layers are added or eliminated as necessary). All the algorithms modify the neurone weights and thresholds based on calculations that analyse the network response to particular inputs. The modifications are performed in a manner that brings the network outputs closer to the expected ones.

Depending on the nature of the training data set, there are two categories of algorithms: supervised and unsupervised [110]. The supervised algorithms use a training data set composed of input–output pairs. The unsupervised algorithms use only the input vectors. In the case of supervised algorithms, the training process is controlled by an external entity (the 'teacher') that is able to establish whether the network outputs are adequate to the inputs and what is the size of the error. Then the network parameters are modified according to the particular correction method defining each training algorithm. In case of unsupervised methods (the Hebbian rule, the 'winner takes all' algorithm, etc.), there are no means to know what the expected outputs are. The network evolves as a result of the 'experience' gained from the previous input vectors. The weight values converge to a set of final values dictated by the input values used as training data set in conjunction with the particular training algorithm.

The unsupervised family of training algorithms is mainly used for signal and image processing, where pattern classification, data clustering or compression algorithms are involved. The control engineering problems are better tackled by supervised training methods, as the relationship between inputs and desired outputs is better defined and easier to control.

### 4.3.1 The error back-propagation algorithm

The most popular supervised training algorithm is the one named 'error back-propagation', or simply 'back-propagation'. It involves training a FFANN structure made up of sigmoidal activation function neurones. The back-propagation algorithm is a gradient method aiming to minimise the total operation error of the neural network. The total error is a function defined by equation (4.10) where $O_i^{\mathrm{ref}}$ is the column vector of the reference outputs and $O_i$ is the column vector of the actual network outputs corresponding to the input pattern number '$i$'. The total error Err is the sum of the errors corresponding to all $n_p$ input patterns.

$$\mathrm{Err} = \sum_{i=1}^{n_p} (O_i^{\mathrm{ref}} - O_i)^T \cdot (O_i^{\mathrm{ref}} - O_i) = \sum_{i=1}^{n_p} \| O_i^{\mathrm{ref}} - O_i \|^2 \tag{4.10}$$

For each training step, the vector of all neurone weights and threshold weights ($W$) is updated in such a way that the total error Err is decreased. The vector $W$ can be associated to a point in a $N_W$-dimensional space (the parameter space), where $N_W$ is the total number of weights and thresholds in the neural network. The most efficient way to perform the update is to shift the point $W$ along the curve indicated by the gradient of the total error ($\nabla$Err). This principle is illustrated by equation (4.11), where $W(t)$ is the parameter vector during the current training cycle, $W(t + 1)$ is the parameter vector for the next training cycle and $\eta$ is the learning-rate constant. Ideally, the algorithm stops when the total error is zero. In practice, it is stopped when the error is considered negligible.

$$W(t+1) = W(t) - \eta \cdot \nabla \mathrm{Err} = W(t) - \eta \cdot \left( \frac{\partial \mathrm{Err}}{\partial w_1} \frac{\partial \mathrm{Err}}{\partial w_2} \frac{\partial \mathrm{Err}}{\partial w_3} \cdots \frac{\partial \mathrm{Err}}{\partial w_{N_w}} \right)^T \tag{4.11}$$

For the practical calculation of the error gradient $\nabla$Err, the components in the vector $W$ are usually rearranged as a three-dimensional matrix. The matrix has a number of rectangular layers equal to the number of neurone layers in the neural network. Each rectangular layer is a two-dimensional matrix containing one line for each neurone in the corresponding layer of the neural network. Each line includes the input weights and its threshold level of a neurone. Therefore, the element $w_{jkm}$ in the three-dimensional matrix is the weight '$m$' of the neurone '$k$' situated in the layer '$j$' inside the neural network. The threshold level corresponds to the last element in each line and is not treated any differently to the input weights because it can be considered as an extra weight supplied with a constant input signal –1. In the case of a one-layer network, the components of the error gradient are calculated according to (4.12) where the input signal $x_m$ is the corresponding input signal.

$$\frac{\partial \mathrm{Err}}{\partial w_{1km}} = \frac{\partial}{\partial w_{1km}} \sum_{i=1}^{n_p} \| O_i^{\mathrm{ref}} - O_i \|^2 = -\sum_{i=1}^{n_p} 2(o_{ik}^{\mathrm{ref}} - o_{ik}) \cdot \frac{\mathrm{d}f}{\mathrm{d}(\mathrm{net}_{ik} - t_k)} \cdot x_m \tag{4.12}$$

As demonstrated in [243], bipolar sigmoidal activation functions given by (4.2) have the property (4.13), so that the equation (4.12) can be transformed into (4.14).

$$\frac{\mathrm{d}f}{\mathrm{d}(\mathrm{net} - t)} = \frac{1}{2}(1 - f^2) \tag{4.13}$$

$$\frac{\partial \mathrm{Err}}{\partial w_{1km}} = -\sum_{i=1}^{n_p} (o_{ik}^{\mathrm{ref}} - f_{i1k}) \cdot (1 - f_{i1k}^2) \cdot x_m \tag{4.14}$$

If the network has more than one layer then the input signal $x_m$ is generated by the neurone '$m$' in the layer two so that $x_m$ has to be replaced with $f_{2m}$. Equations (4.16) and (4.17) illustrate the calculations for the neurones in layers two and three.

$$\frac{\partial \mathrm{Err}}{\partial w_{1km}} = -\sum_{i=1}^{n_p} (o_{ik}^{\mathrm{ref}} - f_{i1k}) \cdot (1 - f_{i1k}^2) \cdot f_{2m} \tag{4.15}$$

$$\frac{\partial \mathrm{Err}}{\partial w_{2km}} = -\sum_{i=1}^{n_p} \sum_{x} (o_{ix}^{\mathrm{ref}} - f_{i1x}) \cdot (1 - f_{i1x}^2) \cdot w_{1xk} \cdot (1 - f_{i2k}^2) \cdot f_{i3m} \tag{4.16}$$

$$\frac{\partial \mathrm{Err}}{\partial w_{3km}} = -\sum_{i=1}^{n_p} \sum_{x} (o_{ix}^{\mathrm{ref}} - f_{i1x}) \cdot (1 - f_{i1x}^2) \cdot \sum_{y} w_{1xy} \cdot (1 - f_{i2y}^2) \cdot w_{2yk} \cdot (1 - f_{i3k}^2) \cdot f_{4m} \tag{4.17}$$

The previous calculations can be generalised for any number of layers [155], [243].

Each component of the gradient is determined following the iterative process (4.18). Similar results are obtained for all types of sigmoidal activation functions These equations justify the name of the training algorithm: the output errors of the FFANN affect the calculations referring to any weight because their influence propagates back to the inputs from one layer to the next in accordance with (4.18).

$$\begin{cases} \dfrac{\partial \mathrm{Err}}{\partial w_{jkm}} = \sum_{i=1}^{n_p} \sum_{x} (o_{ix}^{\mathrm{ref}} - f_{i1x}) \cdot \delta_{i1x} \\ \delta_{i1x} = (1 - f_{i1x}^2) \cdot \sum_{y} w_{1xy} \cdot \delta_{i2y} \\ \delta_{i2x} = (1 - f_{i2x}^2) \cdot \sum_{y} w_{2xy} \cdot \delta_{i3y} \\ \quad \cdots \\ \delta_{i(k-1)x} = (1 - f_{i(k-1)x}^2) \cdot \sum_{y} w_{(k-1)xy} \cdot \delta_{iky} \\ \delta_{ikx} = \begin{cases} 0 & x \neq m \\ (1 - f_{i(k-1)x}^2) \cdot f_{km} & x = m \end{cases} \end{cases} \tag{4.18}$$

The back-propagation algorithm faces the well-known problem of any non-linear optimising algorithm using the gradient method: it can become stuck in a local minimum of the objective function (the function 'Err' in this case). Therefore, the back-propagation algorithm is not guaranteed to generate a satisfactory solution for all input–output association problems and FFANN architectures. The training result depends on several factors [243]:

- Network architecture (number of layers, number of neurones in each layer).
- Initial parameter values $W(0)$.
- The details of the input–output mapping.
- Selected training data set (pairs of inputs and corresponding desired outputs).
- The learning-rate constant $\eta$.

Back-propagation is not a constructive algorithm; the network architecture has to be chosen in advance. Unfortunately, there is no clearly defined set of rules to be followed in order to decide which is the most appropriate architecture for a problem. Choosing the architecture is a result of a trial and error process supported by previous experience. However, it has been mathematically demonstrated that any input–output mapping can be learned by a FFANN with only one hidden layer, provided that the number of neurones in the hidden layer is large enough for the problem to be solved [173]. This means that if a neural network proves incapable of learning how to perform a certain task, the one possible solution is to increase the number of neurones in the hidden layer or layers.

A different solution is to restart the algorithm with another set of initial parameters $W(0)$. This solution is based on the assumption that the previous failure was generated by stopping at a local minimum. The trajectory of vector $W$ in the parameter space is dependent on its starting point $W(0)$, therefore, the situation may be avoided by changing the initial weights and thresholds.

Another important aspect is choosing an adequate training data set, so that if the number of different input values is finite, the training data set covers all the possibilities. Nevertheless, if this number is infinite (e.g. when the inputs are analogue signals), or if the number is too large, then only a selection of input combinations will be used to train the neural network. The quality of the training process is influenced by the way the training data set is generated. If the training data set adequately covers all the aspects of the input–output mapping, then the network will be able to generate correct answers for inputs that were not used during the training period. This property is called 'generalisation' and is made possible by the fact that any FFANN actually performs an interpolation in an $n$-dimensional space (where '$n$' is the length of the input vectors) [121]. The interpolation is carried out based on the information provided by the input vectors used during the training period. If the input–output mapping is continuous and smooth, then the network will easily generalise and yield correct answers as a result of a training performed with only few input vectors. If the input–output mapping is rugged and complicated, then a large number of input vectors is required for an adequate training process. The training process may require hundreds, thousands or even millions of steps of the type described by (4.11). The actual number depends on the nature of the input–output mapping and on the learning-rate constant $\eta$. A large value for $\eta$ accelerates the training process but also increases the chance that the vector $W$ oscillates around the final solution without ever reaching it. A small $\eta$ increases the chances to obtain the desired solution but also increases the necessary number of training cycles.

### 4.3.2 Algorithms derived from the back-propagation method

A series of new algorithms have been derived in the last two decades from the classical back-propagation method. They bring improvements to the training process by accelerating the convergence and improving the chances of finding a good solution for particular application types. The improvements proposed can be summarised as:

- The learning-rate constant $\eta$ is varied after each training cycle. It starts with a large value that is progressively diminished during the training process. Therefore, the training process is fast at the beginning but the final oscillations are avoided because $\eta$ decreases during the training process.
- Every adjustable network parameter has its own learning-rate constant $\eta_i$. The back-propagation algorithm may be slow, because the use of a unique learning-rate parameter may not suit all the complicated error variations in the $N_W$-dimensional parameter space. Thus, a learning-rate value that is appropriate for the adjustment of one weight is not necessarily appropriate for the adjustment of another. Thus, the learning algorithm is described by equation (4.19).

$$W(t+1) = W(t) - \left( \eta_1 \frac{\partial \mathrm{Err}}{\partial w_1} \eta_2 \frac{\partial \mathrm{Err}}{\partial w_2} \eta_3 \frac{\partial \mathrm{Err}}{\partial w_3} \cdots \eta_{N_W} \frac{\partial \mathrm{Err}}{\partial w_{N_W}} \right)^T \tag{4.19}$$

- If one learning rate is associated with each network parameter, then all the learning-rates are allowed to vary from one training cycle to the next. The variance may be calculated according to the first ballet point. More sophisticated methods may calculate the learning-rate constants based on the error function partial derivatives. Therefore, $\eta_i$ is large if the influence of $w_i$ over the error is small and $\eta_i$ is small otherwise (4.20).

$$\eta_i = \frac{1}{\left|\frac{\partial \text{Err}}{\partial w_i}\right| + K} \quad \text{where } K > 0 \tag{4.20}$$

- If the sign of the error derivative $\partial\text{Err}/\partial w_i$ oscillates for several consecutive iterations, the corresponding learning-rate parameter $\eta_i$ is decreased.
- The convergence of the training is accelerated by supplementing the current weight adjustment with a fraction of the previous weight adjustment, as shown in equation (4.21). This algorithm is named the momentum method [243] and the second term indicating the fraction of the most recent weight adjustment is called the momentum term. The momentum term $\alpha$ is a user-selected constant with values between 0.1 and 0.8.

$$W(t+1) = W(t) - \eta \cdot \nabla\text{Err} + \alpha[W(t) - W(t-1)] \tag{4.21}$$

Real-time recurrent neural networks need to be trained in such a manner that they learn a certain temporal correlation between inputs and outputs. A promising training method applicable to such situations and named the dynamic back-propagation training [112] has been derived from the classical one. The main feature of the new method is that input vectors are not applied randomly, but in rigorously defined series. The expected outputs depend both on the current input and on past inputs, while the error calculation is performed globally for the entire temporal series of input vectors.

### 4.3.3 Training algorithms for neurones with step activation functions

If the activation functions of the neurones in FFANN are not sigmoidal, the back-propagation algorithm cannot be used because the error function cannot be derived. However, two other recursive methods presented in (4.22) and (4.23) are applicable to the FFANNs with only one layer. These are recursive methods like the back-propagation algorithm, but in this case, the training process always has a finite number of cycles, provided that the desired input–output relation can be learned by a one layer network.

$$w_{jk}^{t+1} = w_{jk}^{t} + \eta \cdot \sum_{i=1}^{n_p} x_k \cdot (o_{ij}^{\text{ref}} - f_{ij}) \tag{4.22}$$

$$w_{jk}^{t+1} = w_{jk}^{t} + \eta \cdot \sum_{i=1}^{n_p} x_k \cdot (o_{ij}^{\text{ref}} - \text{net}_{ij}) \tag{4.23}$$

Finding the correct weights for a multilayer FFANN with step activation functions is a complicated problem. The two previous methods cannot be generalised for such networks and either constructive methods or genetic algorithms need to be used instead.

There are many other training algorithms than the ones presented here. However, the algorithms presented are used in the vast majority of control applications. The other algorithms known in the literature are used for mainly other types of applications like signal processing, associative memories, neurologic studies, etc.

### 4.3.4 The Voronoi diagram algorithm

The Voronoi diagram is a constructive algorithm applicable to FFANNs composed of neurones with a step activation function [47]. As previously mentioned, any FFANN containing step activation function neurones solves a classification problem. The Voronoi diagram is a graphical representation of the classification problem to be implemented by the FFANN. Let us consider the $m$-dimensional space of the input data and a set of points in this space, corresponding to a given set of input vectors. The Voronoi diagram (also known as Thiessen polygons or Dirichelet tessallation) is a partitioning of the $m$-dimensional space into convex regions called Voronoi cells, each of which defines the region of influence of one given point in its interior. Any Voronoi cell can be defined as the intersection of a finite number of half-spaces and is therefore delimited by a finite number of hyperplanes.

Each hyperplane can be modelled by one neurone with a step activation function such as (4.7) or (4.8). In the unipolar situation, the neurone generates the output signal '1' for the inputs corresponding to points on a given side of the hyperplane, while '0' is generated for all the other inputs. As illustrated by (4.24), there is a one-to-one correspondence between the algebraic parameters defining the hyperplane and the neurone parameters. The same applies to bipolar neurones, but the output '0' is replaced by '–1'.

$$\begin{aligned}
&\left(\sum_{j=1}^{m} w_j \cdot x_j\right) - t \geq 0 \Leftrightarrow \text{net} - t \geq 0 \Leftrightarrow f(\text{net}) = 1 \\
&\left(\sum_{j=1}^{m} w_j \cdot x_j\right) - t < 0 \Leftrightarrow \text{net} - t < 0 \Leftrightarrow f(\text{net}) = 0
\end{aligned} \tag{4.24}$$

A Voronoi cell is defined by its borders. Consequently, a point in the input data space belongs to a certain Voronoi cell only if all the corresponding neurones simultaneously generate the required outputs. Thus, the set of convex cells in a Voronoi diagram can be modelled by a FFANN with two layers. The input layer contains the neurones modelling the hyperplanes and the second layer contains one neurone for each convex cell. All the neurones defining the borders of a particular cell feed the corresponding neurone in the second layer. The classes defined by a classification operation are not necessarily convex. Therefore, one class may be the union of several Voronoi cells. As a result, a third layer is necessary in the corresponding neural network. The third layer contains one neurone for each class of input vectors. Each neurone is connected only to those neurones in the second layer implementing Voronoi cells that are part of the given input vector class.

Figure 4.8 illustrates a Voronoi diagram example built for a neural network with two inputs and one output. Thus, the diagram is two-dimensional and the hyperplanes are straight lines. The shaded areas cover the Voronoi cells that belong to the class '1', the other cells are part of class '0'. There are four Voronoi cells in Fig. 4.8: $r_a$, $r_b$, $r_c$, $r_d$ that belong to class '1', and they are bounded by nine lines modelled by neurones n1 through n9. Therefore, the first neurone layer contains nine neurones, the second contains four neurones (one neurone for each of the Voronoi cells) whereas the third layer contains a single 4-input neurone (Fig. 4.9). The outputs of the neurone in the third layer are '1' when $X_1$ and $X_2$ correspond to a point in one of the shaded areas and '0' for all the other cases.

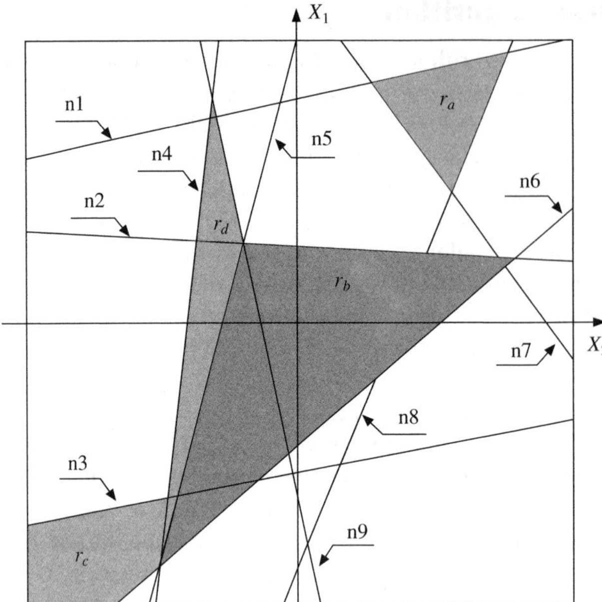


**Fig. 4.8** The Voronoi diagram for a 2D example

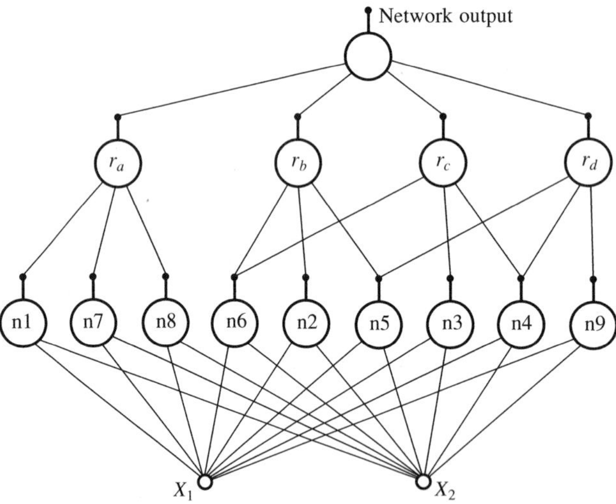


**Fig. 4.9** The neural architecture based on the Voronoi diagram in Fig. 4.8

Very efficient computer algorithms have been developed for the construction of Voronoi diagrams in high dimensional space [89]. They are able to solve this class of problem in linear time and this performance provides tremendous impetus for further research on this topic.

## 4.4 Control applications of ANNs

In recent years, neural solutions have been suggested for many industrial systems using either feed-forward or recurrent neural networks. Most of the published papers describe control system applications built around a feed-forward neural network included inside a traditional feedback control system. The ANN is usually made up of sigmoidal activation function neurones and back propagation is normally used to train the network either on-line or off-line. Some applications use neurones with a radial base activation function. The ANN may play different roles: plant identification [105], [212], non-linear controller [131], [225], and fault signalling [127], [126]. The neural plant identification technique can be applied to induction motor sensorless speed estimation, for example in [38] where the plant parameter to be identified is the rotor speed.

Typical neural networks used for identification purposes are multilayer feed-forward structures containing neurones with sigmoidal activation function. There are two configurations for plant identification: the forward configuration and the inverse configuration [243]. In case of forward configuration, the neural network receives the same input vector $x$ as the plant, and the plant output provides the reference output $O^{\text{ref}}$ during the training (Fig. 4.10(a)). During the identification, the norm of the error vector $\|O^{\text{ref}} - O\|$ is minimised using the back-propagation algorithm. As illustrated in Fig. 4.10(b), the inverse plant identification employs the plant output $y$ as the network input, while the neural network generates an approximation of the input vector of the plant. The norm of the error vector to be minimised through learning is therefore $\|x - O\|$.

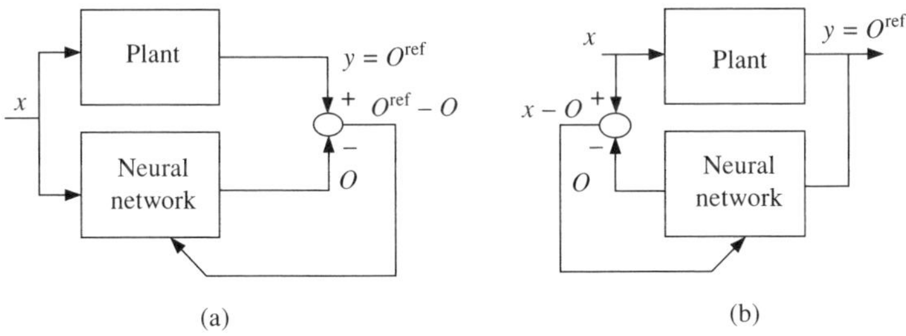


**Fig. 4.10** Neural network configuration for plant identification: (a) forward plant identification; (b) inverse plant identification

Feed-forward neural networks generate instantaneous response, thus they can model the steady-state operation of the plant but are not directly capable of modelling its dynamic behaviour. To account for the plant dynamics, the FFANN has to be supplied with a series of past inputs of the plant. Such an approach requires that the neural network is interfaced with a shift register that stores the time series of input vectors (see Fig. 4.11). The shift register is updated at each operation step. An update consists of storing the most recent input vector and discarding the oldest input vector.

An alternative solution is to use recurrent neural networks. This solution is purely neuronal in that it does not require a shift register. However, most of the control systems have used the first solution so far, because the dynamic back-propagation algorithm requires more computation resources than its static counterpart.

Both identification configurations have advantages and disadvantages. Forward plant

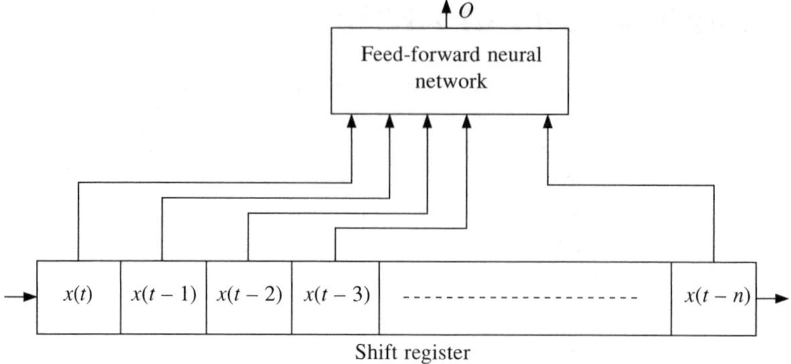


**Fig. 4.11** Neural network interfacing for modelling the plant dynamics

identification is always feasible, but it does not immediately allow for the construction of the plant control. In contrast, plant inverse identification facilitates simple plant control. However, the identification itself is not always feasible because in some cases more than one vector $x$ corresponds to a certain vector $y$ (or series of such vectors).

Figure 4.12 presents a basic control system using a neurocontroller. There are two alternatives: either the neural network is trained only off-line in an inverse identification configuration, as presented in Fig. 4.10(b), or it is initially trained off-line but the training continues on-line in the control system. For training purposes, the back-propagation algorithm is the most appropriate. Shift registers are used, both during the off-line identification process and inside the control system, to enable the modelling of the dynamic plant behaviour. The neurocontroller input consists of the most recent plant outputs plus the output reference for the current time. Therefore, at each operation step, it generates a control vector $O$ that causes the plant to produce the expected output $y^{\text{ref}}$.

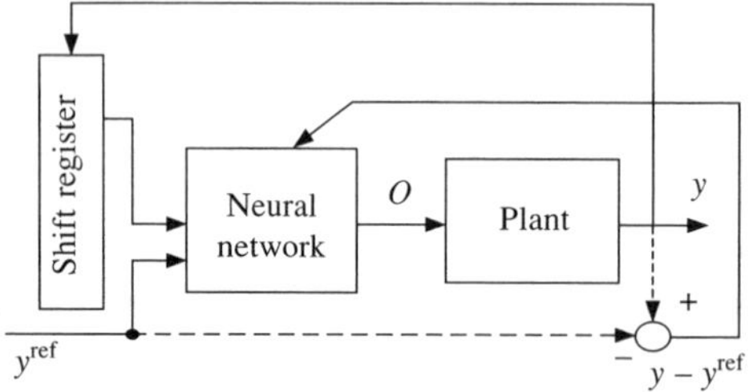


**Fig. 4.12** Basic control system configuration using a feed-forward neurocontroller

The fault signalling applications are part of the larger class of classification applications. The task of the neural network is to analyse the input data and to generate information about the operation of the plant: normal operation, or abnormal operation. In the second case, it may give further details about the abnormality: short circuit, surpassing voltage or speed limits, etc. The neural network is of the feed-forward type and is trained off-line using experimental data that reflects all possible operation modes of the plant.

## 4.5 Neural network implementation

Hardware implemented neural networks are essentially arrays of interconnected processing units that operate concurrently. Each unit has a simple internal structure that, in some cases, includes a small amount of local memory. The most important design issues concerning any neural network hardware implementation are the degree of parallelism, the information processing performance, the flexibility and the silicon area. There are several categories of neural network hardware implementation [161]:

- Analogue implementation.
- Digital implementation.
- Hybrid implementation.
- Optical implementation.

### 4.5.1 Analogue hardware implementation

Analogue neural networks can exploit physical properties of silicon devices to perform network operations obtaining very high processing speed. However, analogue design can be very difficult because of the need to compensate for parameter variations with temperature, manufacturing conditions, etc. One approach is to implement neurones using common operational amplifiers and resistors [243]. The operational amplifier implements the activation function, while the resistors determine the weight values (Fig. 4.13). The amplifier output voltage $V_{out}$ depends on the input voltages $V_+$ and $V_-$ that, in turn, depend both on the input voltages and on the resistors' values. Ohm's laws are used to perform all the necessary calculations.

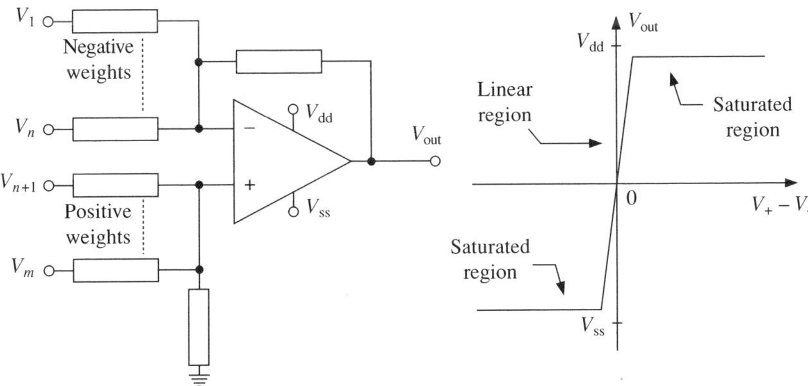


**Fig. 4.13** Neurone implementation using operational amplifiers and resistors

The implementation style using resistors ensures very good linearity but it is not flexible because the weight values are set during the manufacturing process and they cannot be altered afterwards. Creating a changeable analogue synapse involves the complication of analogue weight storage. The simplest approach is to replace the fixed value resistors by MOS transistors that can operate as voltage adjustable switches. Each transistor is controlled by a voltage $V_{gs}$ produced by the charge stored on a capacitor, which periodically has to be refreshed. The influence of $V_{gs}$ upon the resistance between the source and the drain of each transistor is illustrated in Fig. 4.14. Thus, the dependence between $I_{ds}$ and $V_{ds}$ is not linear but it can be used as an acceptable approximation of a linear function within certain ranges of currents and voltages. More sophisticated multiplication mechanisms (such as Gilbert multipliers) need to be used if very good linearity is required over a large range of voltages.

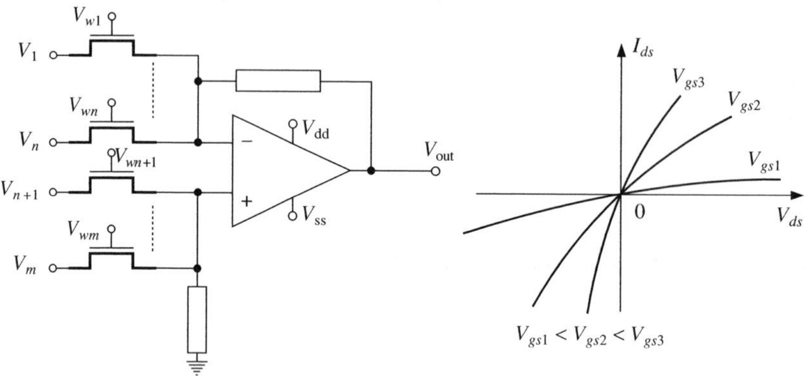


**Fig. 4.14** Neurone implementation with electrically tuneable weights

The number of operational amplifiers that can be integrated on a chip is limited. Therefore, the implementation methods that use operational amplifiers are applicable only to small-scale neural networks. To obtain high integration densities, the implementation of the activation function is performed with very simple circuits. A minimalist design style is adopted in the analogue approach described in [21]: each activation function is modelled by a circuit containing a single MOS transistor. The design methodology is based on current-mode subthreshold CMOS circuits, according to which the signals of interest are represented as currents. The current mode approach offers signal processing at the highest bandwidth for a given power consumption. In [171] a different approach is described: the basic building block is a transconductance amplifier (Fig. 4.15). In its basic form, the amplifier contains three MOS transistors and transforms a differential input voltage $V_{in} = V_1 - V_2$ into a differential output current $I_{out} = I_1 - I_2$. The relationship between input and output is non-linear and is a good approximation of a sigmoidal activation function.

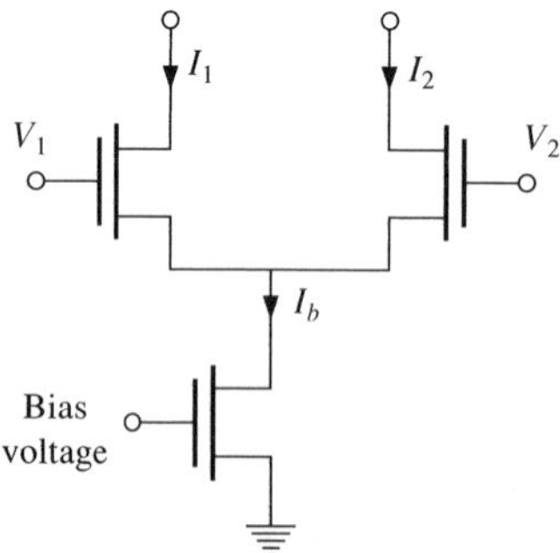


**Fig. 4.15** Circuit diagram of a differential transconductance amplifier

The first analogue commercial chip was the Intel 80170NW ETANN (electrically trainable analogue neural network) [12]. It contains 64 neurones and 10 280 weights. The non-volatile weights are stored as charge on floating transistor gates and a Gilbert multiplier provides 4-quadrant multiplication. A flexible design, including internal feedback and division of the weights into $64 \times 80$ banks, allows multiple configurations including three layers of 64 neurones/layer, and two layers with 128 inputs and 64 neurones/layer. No on-chip training was provided, so the connection with a PC is necessary. The PC performs the training process and then transmits the resulting weight values to the neural chip.

New implementation technologies and possible applications of analogue neural chips continue to be investigated and several successes have been reported in the literature [176], [156], [163], [109], [168].

### 4.5.2 Digital hardware implementation

The digital neural network category encompasses many subcategories including slice architectures, single instruction multiple data (SIMD) approach, systolic array devices, radial basis function (RBF) architectures, ASIC and FPGA implementations. For designers, digital technology has the advantage of mature fabrication techniques and digital chips are easily embedded into most applications. However, digital calculations are usually slower than in analogue systems, especially when performing the multiplications between weights and input signals. Moreover, analogue inputs must first be converted into digital format. The most common performance rating used to compare digital neural implementations is the connection-per-seconds (CPS), which is defined as the rate of multiplication and accumulate operations during normal operation.

Slice architectures for neural networks provide basic building blocks to construct networks of arbitrary size and precision. For example, the NeuroLogix NLX-420 Neural Processor Slice has 16 processing elements and a speed of 300 MCPS. A common 16-bit input bus is multiplied by different weights in each parallel processing element. The weights are initially read from outside the chip. The 16-bit weights and inputs can be selected by the user as 16 1-bit values, four 4-bit values, two 8-bit values or one 16-bit value. The 16 neuronal inputs are processed by a user-defined piecewise continuous activation function to produce a 16-bit output. Internal feedback allows the implementation of multilayer networks and multiple chips can be interconnected to build large networks.

A far more elaborate approach is to place many small processors on a chip. Two architectures dominate such designs: single instruction with multiple data (SIMD) and systolic arrays. For SIMD design, each processor executes the same instruction in parallel, but on different data. In systolic arrays, the basic processors are connected in a matrix architecture. Each processor does one calculation step before passing its result on to the next processor in a pipelined manner. A systolic array system can be built with Siemens MA-16. The MA-16 provides fast matrix operations using $4 \times 4$ processor matrices with a 16-bit interconnecting bus. The overall performance is 400 MCPS. The multiplier and accumulator outputs have 48-bit precision. Weights are stored on-chip and neurone activation functions are generated off-chip via look-up tables. Multiple chips can be cascaded.

The networks with RBF neurones provide fast learning and straightforward implementation. The comparison of input vectors to stored training vectors can be done quickly if non-Euclidean distances (such as the Manhattan norm shown in (4.25)) are calculated with no multiplication. One of the commercially available products is the Nestor NI1000 chip. The Nestor NI1000, developed jointly by Intel and Nestor, contains 1024 stored vectors of 256 5-bit elements. The chip has two on-chip learning algorithms, but it is relatively slow: 40 kCPS.

$$\| X - Y \|_{\text{Manhattan}} = \sum_i | x_i - y_i | \tag{4.25}$$

Digital ASIC and FPGA solutions require that the ANN is fully designed and trained for a particular application before its actual hardware implementation. The operation of the ANN is usually described in terms of Boolean functions or in terms of logic operations and threshold gates (TGs). The threshold gate is a more general concept than a logic gate. Any logic gate can be considered a particular case of a TG but TGs can perform more complex information processing tasks than logic gates. They have inputs with different integer weights that make them very suitable for neurone hardware implementation. Unfortunately, the technology limits the weights to small integer values: $0, \pm 1, \pm 2, \pm 3$. The direct use of TGs to implement neurones generates compact hardware structures, but this approach can only be used for a limited number of ASIC technologies. It cannot be used for FPGA implementation because they are not available inside the complex logic blocks (CLBs) of FPGA chips. However, the indirect use of TGs is possible because a TG can be emulated by a digital structure composed of no more than a few AND, OR and NOT interconnected logic gates.

Designing an ANN for a specific application involves the use of either training algorithms or constructive algorithms. Constructive algorithms are the preferable approach in many situations because they are able to determine both the network architecture and the neurone weights and are guaranteed to converge in finite time. A large number of constructive algorithms, reviewed in [36], have been developed in the last decade. They are divided into three categories: geometric ([47], [193]), network-based [205] and algebraic [120]. Several VLSI friendly algorithms have been created in order to bring closer the design stage and the implementation stage. These algorithms consider some specific aspects of VLSI implementation technology: the precision of the input weights and the neurone fan-in. These factors lead to important limitations that need to be taken into account when designing a neural network. One of the first VLSI friendly algorithms used the concept of an 'adaptive tree network' [22]. Further research in this direction has been extended by using a combination of AND gates and OR gates, alongside threshold gates (TGs) [25].

### 4.5.3 Hybrid implementation techniques

Hybrid design attempts to combine the advantages of analogue and digital techniques. The use of analogue implementation is attractive for reasons of compactness, speed and the absence of quantisation effects. The advantage of digital signals is their robustness. These signals are not affected by disturbances and the calculations performed in digital format always yield precise results.

The pulse modulation technique is one of the most promising principles that can be used to develop efficient hybrid architectures. Using pulse modulation, the internal signals of the neural network are modelled as pulse streams whose parameters are varied in accordance with the neurone states. Depending on the parameter that is varied, there are three theoretical alternatives: the pulse amplitude modulation, the pulse width modulation and the pulse frequency modulation [110].

In case of pulse amplitude modulation, the amplitude of the pulses is modulated in time in a manner that reflects the variation of the corresponding neurone signal. This technique is not satisfactory in neural networks because the information is transmitted as analogue voltage levels, which makes it susceptible to processing errors due to circuit parameter variations.

The pulse width modulation method alters the pulses' duration according to the amplitude of the neural signal. The pulse width modulated signal is robust since the information is coded as a set of time intervals and no analogue voltage is used. However, if several signals in the neural network have almost similar values, then a large number of pulse edges occur almost simultaneously. The existence of this synchronism represents a drawback in VLSI networks since many synapses tend to draw current from the internal supply lines simultaneously. It follows that the internal supply lines have to be oversized to accommodate the high instantaneous currents that may be produced by the use of pulse width modulation.

Pulse frequency modulation maintains both the amplitude and the width of the pulses constant but modifies the frequency of the pulses. This modulation scheme generates robust signals as well. Moreover, different signals modelling the equal analogue quantities are usually phase-shifted, which leads to avoiding the synchronism of the pulse edges. Thus, the power requirement is averaged over time as a result of using pulse frequency modulation. Hybrid neural networks combining pulse frequency modulation and neurones implemented in analogue technology have been successfully designed and implemented [178], [58].

Another reason for producing hybrid neural network implementations is the need to interface the neural architectures with existing digital equipment. In such a situation, the external inputs and outputs are digital, to facilitate the integration into the digital systems, while internally some or all of the processing is done in analogue technology. The AT&T ANNA chip, for example, is externally digital and all the internal signals are in digital format, but it uses capacitor charge to store the neurone weights [195]. The charge is periodically refreshed by a specialised internal mechanism. The chip structure includes multiplying digital-to-analogue converters (MDACs), electronic devices capable of multiplying a digital value with an analogue signal. The MDACs are used to perform the multiplications between the weights and the input signals of each neurone. Conversely, the Bellcore CLNN-32 chip uses digital 5-bit weights, but the neurone inputs and outputs are analogue signals [19]. As in the case of the ANNA chip, the multiplications between weights and signals involve the use of MDACs. The overall performance of the Bellcore chip is 100 MCPS. Thus, the MDACs allow the neural network designer freely to combine analogue and digital technologies in an optimal fashion for a given application problem.

### 4.5.4 Software versus hardware implementations

Software implementation uses a classical von Neuman machine (a general-purpose microprocessor or a DSP). This approach can be used to implement any kind of neural network structures and any training algorithm. However, neural networks simulated on von Neuman machines run in a series fashion which does not allow them to be used in real-time applications. The operation speed of the neural network is inversely proportional to the number of neurones. Consequently, very large neural networks can only be efficiently software implemented if special hardware resources are also available: either large general-purpose parallel machines or cheaper alternatives such as specialised coprocessors, or accelerator cards for personal computers.

The hardware approach overcomes the speed limitations of software implemented neural networks. True parallel operation mode is achieved in this case, making the calculation speed independent of the network complexity. The actual speed of hardware implementation solutions depends on the technology. The highest speed is achieved using optical implementations while the lowest speed is obtained with the electronic digital architectures. Several optical neural processors have been reported in the literature [94], [90]; however, optical technology has not yet attained maturity. The approach is still too expensive, too imprecise and too rigid, so electronic implementations are preferred in most cases.

The training process is faster in the case of specialised chips when compared to software implementations but only relatively simple training strategies are currently implemented into hardware. Thus, a limited number of training algorithms can be performed on-line. If the practical application does not require on-line training, the training process can be performed off-line in a software system, and then the resulting weights can be downloaded into the neural chip. Alternatively, the obtained weights can be used to produce an ASIC or FPGA implementation. FPGA implementations are preferable as they allow fast prototyping and furthermore, some FPGA chips are able to change their structure on-line. This feature supports the design of a large range of new on-line training algorithms for digital implemented neural networks.

# 5

# Neural FPGA implementation

This chapter describes a new strategy for implementing neural networks into digital hardware using logic gates and determines the resulting implementation complexity to prove its superiority when compared to results previously presented in the literature. The strategy is illustrated by a complete implementation example: the neural network controlling the current through the stator windings of the induction motor. Experimental results are presented to demonstrate the validity of the adopted design and implementation principles.

## 5.1 Neural networks design and implementation strategy

The FFANN design and implementation manner adopted in this book is adapted to applications that require high operation speed, accurate control over the network outputs, low cost digital hardware and fast prototyping. FPGA chips are ideal for fast prototyping but the low cost versions still have a limited number of available logic gates. Therefore, the amount of required hardware resources needs to be minimised by optimising the number of neurones and by a compact implementation of each neurone. The classical FFANN design method using neurones with sigmoidal activation function and the back-propagation training algorithm is not appropriate in this context because the resulting number of neurones is large and the sigmoidal activation function requires a considerable amount of hardware resources for implementation. Therefore, neural networks designed using the constructive Voronoi algorithm and consisting of neurones with step activation functions were used instead. The constructive algorithm ensures the minimisation of the neurone number, while the step activation function simplifies the implementation size of each neurone.

### 5.1.1 General implementation principles

The hardware resources offered by FPGA chips are limited to logic gates and flip-flops. The implementation strategy developed in this book uses exclusively logic gates to transform any FFANN into a digital hardware structure. The strategy exploits the equivalence between the operation of logic gates and the operation of particular types of neurones. $N$-input AND gates and $n$-input OR gates are assimilated to $n$-input unipolar binary neurones (the input and output values can only be '0' or '1') having positive input weights. The difference between the two logic gate types consists in the relationship between their input weights and the threshold level.

An OR gate output is activated whenever at least one of the inputs is active (is '1'). Thus, the threshold level of the corresponding neurone is positive, but lower than the smallest input weight, as illustrated by (5.1).

$$0 < t_{OR} \leq \min_i \{w_i\} \tag{5.1}$$

The output of an $n$-input AND logic gate is activated only when all the '$n$' inputs are active. Therefore, the threshold level in this case can be as large as the total sum of all the input weights. However, it cannot be higher than this sum because otherwise the output cannot be activated in any conditions at all.

$$t_{AND} \leq \sum_{i=1}^{n} w_i \tag{5.2}$$

On the other hand, the threshold level of the corresponding neurone must be higher than the total sum of any combination of '$n - 1$' input weights. This condition is expressed by (5.3):

$$t_{AND} > -\min \{w_i\} + \sum_{i=1}^{n} w_i \tag{5.3}$$

As a result, the threshold levels for the two sorts of neurones are confined within the interval limits shown in (5.4). Conversely, any neurone with binary input signals ('0' and '1') whose parameters comply with one of two conditions (5.4), behaves either as an AND gate or as an OR gate.

$$\begin{cases} t_{AND} \in \left( -\min_i \{w_i\} + \sum_{i=1}^{n} w_i ; \sum_{i=1}^{n} w_i \right] & \text{(a)} \\ t_{OR} \in \left( 0; \min_i \{w_i\} \right] & \text{(b)} \\ w_i > 0 \quad \forall i = 1, 2, \ldots, n \end{cases} \tag{5.4}$$

Neurones whose parameters do not comply with any of the two relations (5.4) can be implemented as a configuration containing several interconnected logic gates. The details of the hardware configuration depend on the relationship between the input weights and the threshold level. The number of necessary gates increases with the complexity of this relationship. To simplify the logical analysis, the adopted implementation strategy decomposes the complex neurones into a pyramidal structure of simpler subneurones. Each subneurone can be further decomposed into higher-order subneurones until each of them can be implemented with a small number of logic gates.

As explained in Chapter 4, the Voronoi algorithm produces a FFANN with up to three layers of neurones with step activation functions. The algorithm version that produces unipolar neurones is adopted because unipolar neurones are more adequate for hardware implementation than bipolar neurones. The network accepts analogue input signals but generates digital output signals. The neurones in the input layer have analogue inputs and binary outputs, while the rest of the neurones operate only with binary signals.

Therefore, the neurones in layers two and three are appropriate for direct digital hardware implementation. The neurones in the first layer need to be converted first into a digital form that uses bit patterns as inputs instead of analogue signals.

The most appropriate binary codification to be used for neurone input quantities is the complementary code (also named 'two's complement' and symbolised by $C_2$). It is used mainly in computer technology for integer number representations, but it can be readily adapted for real values in the interval [–1; +1). Considering an $n$-bit representation '$b_{n-1}\ b_{n-2}\ b_{n-3} \ldots b_1 b_0$', the corresponding integer value ($I_n$) is given by:

$$I_n = -2^{n-1} \cdot b_{n-1} + \sum_{i=0}^{n-2} 2^i \cdot b_i \tag{5.5}$$

The largest positive number, which can be represented on '$n$' bits, is $2^{n-1} - 1$ while the smallest number is $-2^{n-1}$. Real values between –1.0 and +1.0 can be represented dividing all the integer values $I_n$ by $2^{n-1}$. Thus, equation (5.6) illustrates the complementary code extended to real numbers:

$$R_n = \frac{I_n}{2^{n-1}} = -\,b_{n-1} + \sum_{i=0}^{n-2} 2^{-n+1+i} \cdot b_i \tag{5.6}$$

The large-scale utilisation of complementary code in digital technology is due to the advantages of simple hardware implementation of addition and subtraction. A hardware implemented neural control system contains not only neural networks but also traditional digital structures. Therefore, the use of the same codification manner for the two modules is an important advantage because it simplifies the interface between them. Thus, the new implementation strategy consists of two parts. In the first phase, the initial FFANN mathematical model is digitised, so that the neurones in the input layer operate only with binary signals. The input signals of the converted FFANN consist of bit strings coding the values of the initial analogue inputs. In the second phase, all the neurones are converted into a set of interconnected logic gates. The implementation into logic gate structures is performed neurone by neurone. Each neurone corresponds to a hardware configuration containing at least one logic gate.

### 5.1.2 Model digitisation

The equations underlying the conversion of the analogue neurones into equivalent digital neurones can be demonstrated by decomposing this process in two successive stages. The first stage is to replace the analogue input signals by binary patterns. The second stage brings additional corrections to the neurone mathematical model, so that the resulting neurones use the complementary code extended to real numbers described by (5.6).

The principles underlying the digitisation process involve two basic concepts: the codification style and the neurone behaviour. The codification style, illustrated in Fig. 5.1, is defined as the correspondence between the initial analogue input signals and the binary input codes used by the digital neurone. On the other hand, the neurone behaviour is described by the relationship between the analogue inputs and the neurone output signal. The initial neurone behaviour has to be maintained unchanged during the two stages of the digitisation process. To achieve this, the neurone parameters (input weights

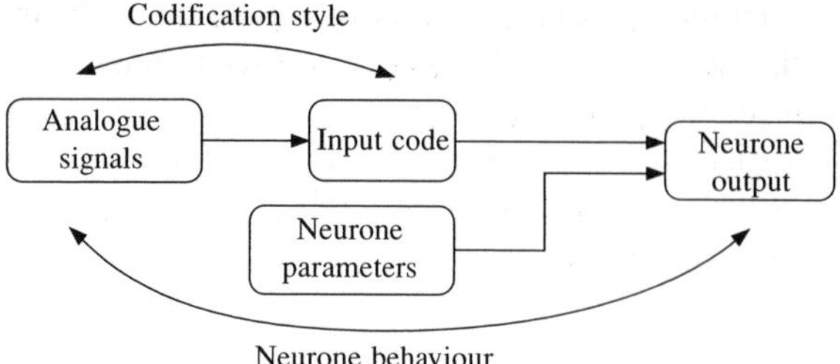


**Fig. 5.1** Basic concepts related to the neurone digitisation process

and the threshold levels) need to be modified at each conversion stage, in a manner that counteracts the effects of replacing the analogue input signals with binary patterns.

The minimal condition to attain this aim is to perform the changes such that the sign of the activation function argument is kept constant. This principle is expressed by equation

$$\text{sign}\left(\sum_{i=1}^{m} w_i \cdot x_i - t\right) = \text{sign}(\text{net} - t) = \text{constant} \tag{5.7}$$

However, for reasons of mathematical simplicity, a more restrictive condition is used instead, namely the argument 'net – $t$' of the activation function is itself kept constant rather than only the sign of it:

$$\sum_{i=1}^{m} w_i \cdot x_i - t = \text{net} - t = \text{constant} \tag{5.8}$$

### *5.1.2.1 Conversion stage one*

The first step, illustrated in Fig. 5.2, transforms the analogue neurones generated by means of the Voronoi algorithm into digital neurones. The newly obtained neurones receive binary patterns on their inputs instead of analogue signals. The task is achieved

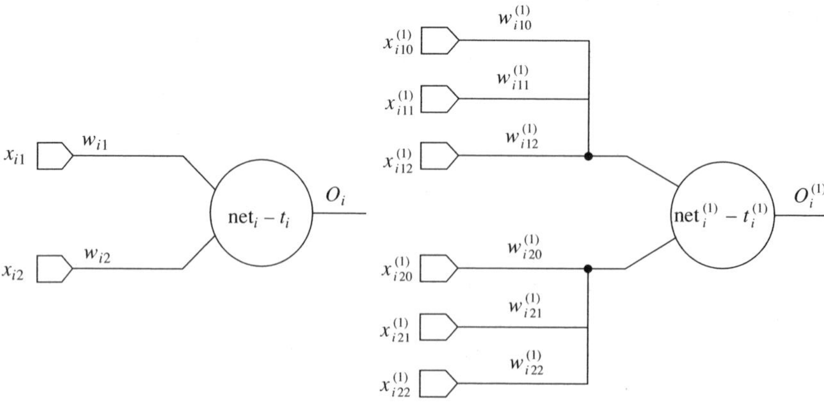


**Fig. 5.2** The neurone model before and after stage one of the conversion

by keeping the threshold level unchanged while splitting each input defined by its initial weight $w_{ij}$ into $n_b$ subinputs, whose weights $w_{ijp}$ ($p = 0, 1, \ldots n_b - 1$) are calculated as:

$$\begin{cases} w_{ijp}^{(1)} = \dfrac{2^{p+1}}{2^{n_b}} \cdot w_{ij} \quad \forall p < n_b - 1 \\ w_{ij(n_b-1)}^{(1)} = -\, w_{ij} \\ t_i^{(1)} = t_i \end{cases} \tag{5.9}$$

The superscript '(1)' in equations (5.9) shows that the corresponding quantities have been calculated during the first conversion stage. Likewise, the superscript '(2)' identifies the quantities calculated during the second conversion stage.

The result of the previous calculations is that the initial '$m$' inputs are turned into '$m$' input clusters, each cluster containing '$n_b$' subinputs. The symbol '$w_{ij}$' stands for the weight number '$j$' of the neurone '$i$' in the network, while '$w_{ijp}^{(1)}$' represents the weight of subinput '$p$' in cluster '$j$' pertaining to neurone '$i$'. The index $p = 0$ corresponds to the least significant binary figure, while $p = n_b - 1$ corresponds to the most significant one.

According to the previous considerations, only those neurone parameter changes that maintain argument '$\text{net}_i - t_i$' of the activation function constant are allowed. The argument corresponding to the neurone after the first conversion stage is calculated as

$$\text{net}_i^{(1)} - t_i^{(1)} = \sum_{j=1}^{m} \sum_{p=0}^{n_b-1} w_{ijp}^{(1)} \cdot x_{jp}^{(1)} - t_i^{(1)} = \sum_{j=1}^{m} \left( -w_{ij} \cdot x_{jp}^{(1)} + \sum_{p=0}^{n_b-2} w_{ij} \cdot \frac{2^{p+1}}{2^{n_b}} \cdot x_{jp}^{(1)} \right) - t_i^{(1)} \tag{5.10}$$

where $x_{jp}^{(1)}$ ($p = 0, 1, 2, \ldots n_b - 1$) are the bits of the binary code received by each new neurone input.

Equation (5.10) can be transformed into

$$\text{net}_i^{(1)} - t_i^{(1)} = \sum_{j=1}^{m} w_{ij} \cdot \left( -x_{j(n_b-1)}^{(1)} + \sum_{p=0}^{n_b-2} 2^{-n_b+p+1} \cdot x_{jp}^{(1)} \right) - t_i^{(1)} \tag{5.11}$$

The expression between parentheses corresponds to the extended complementary code definition given in equation (5.6). Therefore, (5.11) is further transformed into

$$\text{net}_i^{(1)} - t_i^{(1)} = \sum_{j=1}^{m} w_{ij} \cdot x_j - t_i^{(1)} = \sum_{j=1}^{m} w_{ij} \cdot x_j - t_i = \text{net}_i - t_i \tag{5.12}$$

where $x_j$ is an analogue input value of the initial neurone. This proves that the condition expressed by (5.7) is fulfilled. Thus, during conversion stage one the codification style based on the complementary code has been introduced and the required modifications of the neurone parameters have been performed so that the neurone behaviour has been maintained unchanged.

### 5.1.2.2 Conversion stage two

The conversion of the neural network into logic gate architecture is based on the relations

(5.4) and on the possibility to transform any neurone into an equivalent structure containing interconnected elements that comply with (5.4). Such transformations are possible only if all the neurone weights are positive. The stage one neurones may have both positive and negative weights. The second conversion stage aims to replace these neurones with equivalent ones having only positive weights. The simplest way to eliminate negative input weights is to use only the module of their values. Consequently, the relationship between stage one neurone weights and their stage two counterparts is expressed by

$$w_{ijp}^{(2)} = | w_{ijp}^{(1)} | \tag{5.13}$$

Adopting this method means that supplementary parameter alterations are required in order to counteract the neurone behaviour alteration which is caused by changing the sign of some input weights. As the weight values have already been changed according to (5.13), the neurone behaviour can be corrected by changing the threshold level and/or the codification style.

It can be demonstrated that no change of the threshold level can counteract the effect of the input weight alterations. Thus, the change of the threshold level needs to be carried out in such a manner that equation

$$\text{net}_i^{(2)} - t_i^{(2)} = \sum_{j=1}^{m} \sum_{p=0}^{n_b-1} | w_{ijp}^{(1)} | \cdot x_{ijp}^{(2)} - t_i^{(2)} = \sum_{j=1}^{m} \sum_{p=0}^{n_b-1} w_{ijp}^{(1)} \cdot x_{ijp}^{(1)} - t_i^{(1)} = \text{net}_i^{(1)} - t_i^{(1)} \tag{5.14}$$

is fulfilled for any input bits $x_{ijp}$. However, if the input signals to the stage two neurones are the same as the inputs to stage one neurones ($x_{ijp}^{(2)} = x_{ijp}^{(1)}$), then there is no constant value $t_i^{(2)}$ that allows (5.14) to be valid for any combination of input signals. To prove this, the value of $t_i^{(2)}$ can be calculated as

$$t_i^{(2)} = t_i^{(1)} + \sum_{j=1}^{m} \sum_{p=0}^{n_b-1} (| w_{ijp}^{(1)} | - w_{ijp}^{(1)}) \cdot x_{ijp}^{(1)} \tag{5.15}$$

which is derived from equation (5.14). The value calculated according to (5.15) is dependent on the input bits $x_{ijp}^{(1)}$ and therefore is not a constant as the threshold level should be.

Equation (5.15) demonstrates that no acceptable solution exists when the codification style of stage one neurone is identical to the codification style of stage two neurone. Therefore, the codification style needs to be altered as well. A simple solution to this problem can be found if the input bits corresponding to negative input weights at stage one neurones are reversed at stage two neurones. The modification can be readily implemented into hardware with NOT logic gates as shown in Fig. 5.3.

The relationship between the input bits of stage two neurones and stage one neurones is expressed by function

$$x_{ijp}^{(2)} = \begin{cases} x_{ijp}^{(1)} & \text{if } w_{ijp}^{(1)} > 0 \\ 1 - x_{ijp}^{(1)} & \text{if } w_{ijp}^{(1)} < 0 \end{cases} \tag{5.16}$$

The two situations in (5.16) can be compressed into equation

$$x_{ijp}^{(2)} = \frac{1 - \text{sign}\,(w_{ijp}^{(1)})}{2} + \text{sign}\,(w_{ijp}^{(1)}) \cdot x_{ijp}^{(1)} \tag{5.17}$$

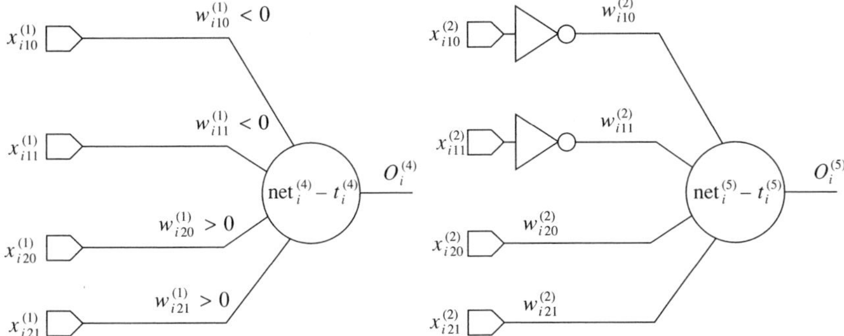


**Fig. 5.3** The neurone '*i*' before and after stage two of the conversion

where the 'sign' function is defined by

$$\text{sign}\,(x) = \begin{cases} +1 & x > 0 \\ 0 & x = 0 \\ -1 & x < 0 \end{cases} \tag{5.18}$$

Using (5.13) and (5.17), the argument of the transfer function for stage two neurones can be calculated as

$$\text{net}_i^{(2)} - t_i^{(2)} = \sum_{j=1}^{m} \sum_{p=0}^{n_b-1} \left[ \frac{|w_{ijp}^{(1)}| - \text{sign}(w_{ijp}^{(1)}) \cdot |w_{ijp}^{(1)}|}{2} + \text{sign}(w_{ijp}^{(1)}) \cdot |w_{ijp}^{(1)}| \cdot x_{ijp}^{(1)} \right] - t_i^{(2)} \tag{5.19}$$

and

$$\text{net}_i^{(?)} - t_i^{(?)} = \sum_{j=1}^{m} \sum_{p=0}^{n_b-1} w_{ijp}^{(1)} \cdot x_{ijp}^{(1)} + \sum_{j=1}^{m} \sum_{p=0}^{n_b-1} \left( \frac{|w_{ijp}^{(1)}| - w_{ijp}^{(1)}}{2} \right) - t_i^{(2)} \tag{5.20}$$

Given the requirement of equality between the two activation function arguments, the threshold level can be calculated based on equation

$$\sum_{j=1}^{m} \sum_{p=0}^{n_b-1} w_{ijp}^{(1)} \cdot x_{ijp}^{(1)} + \sum_{j=1}^{m} \sum_{p=0}^{n_b-1} \left( \frac{|w_{ijp}^{(1)}| - w_{ijp}^{(1)}}{2} \right) - t_i^{(2)} = \sum_{j=1}^{m} \sum_{p=0}^{n_b-1} w_{ijp}^{(1)} \cdot x_{ijp}^{(1)} - t_i^{(1)} \tag{5.21}$$

Therefore, the result is

$$t_i^{(2)} = t_i^{(1)} + \sum_{j=1}^{m} \sum_{p=0}^{n_b-1} \frac{|w_{ijp}^{(1)}| - w_{ijp}^{(1)}}{2} \tag{5.22}$$

The threshold level $t_i^{(2)}$ is constant in equation (5.22) because it depends exclusively on constant quantities. The parameters of stage one neurones depend on the initial parameters as described by (5.9). Consequently, (5.22) can be successively transformed as:

$$t_i^{(2)} = t_i + \sum_{j=1}^{m} \frac{|w_{ij}| + w_{ij}}{2} + \sum_{j=1}^{m} \sum_{p=0}^{n_b-2} \frac{\frac{2^{p+1}}{2^{n_b}} \cdot |w_{ij}| - \frac{2^{p+1}}{2^{n_b}} \cdot w_{ij}}{2} \tag{5.23}$$

$$t_i^{(2)} = t_i + \sum_{j=1}^{m} \frac{|w_{ij}| + w_{ij}}{2} + \sum_{j=1}^{m} \sum_{p=0}^{n_b-2} \frac{2^p}{2^{n_b}} \cdot (|w_{ij}| - w_{ij}) \tag{5.24}$$

$$t_i^{(2)} = t_i + \sum_{j=1}^{m} \frac{|w_{ij}| - w_{ij}}{2} + \sum_{j=1}^{m} \sum_{p=0}^{n_b-2} \frac{2^p}{2^{n_b}} \cdot (|w_{ij}| - w_{ij}) + \sum_{j=1}^{m} w_{ij} \tag{5.25}$$

$$t_i^{(2)} = t_i + \sum_{j=1}^{m} \sum_{p=0}^{n_b-1} \frac{2^p}{2^{n_b}} \cdot |w_{ij}| - \sum_{j=1}^{m} \sum_{p=0}^{n_b-1} \frac{2^p}{2^{n_b}} \cdot w_{ij} + \sum_{j=1}^{m} w_{ij} \tag{5.26}$$

$$t_i^{(2)} = t_i + \frac{2^{n_b} - 1}{2^{n_b}} \cdot \sum_{j=1}^{m} |w_{ij}| - \frac{2^{n_b} - 1}{2^{n_b}} \cdot \sum_{j=1}^{m} w_{ij} + \sum_{j=1}^{m} w_{ij} \tag{5.27}$$

$$t_i^{(2)} = t_i + (1 - 2^{-n_b}) \cdot \sum_{j=1}^{m} |w_{ij}| + 2^{-n_b} \cdot \sum_{j=1}^{m} w_{ij} \tag{5.28}$$

Thus, the parameters of the final digital neurones can be calculated as a function of the initial analogue neurone parameters by combining (5.28) with (5.13) and (5.9), the result being

$$\begin{cases} w_{ijp}^{(2)} = \dfrac{2^{p+1}}{2^{n_b}} \cdot |w_{ij}| \; p = 0, 1, 2, \ldots n_b - 1 \\ t_i^{(2)} = t_i + (1 - 2^{-n_b}) \cdot \sum_{j=1}^{m} |w_{ij}| + 2^{-n_b} \cdot \sum_{j=1}^{m} w_{ij} \end{cases} \tag{5.29}$$

As shown in Fig. 5.4, the final implementation solution uses a codification style that involves two binary codes. The first one is the complementary code. This code is transformed by a set of NOT gates into the second code, which is directly used by the neurone obtained after the second conversion stage. This neurone model has only positive input weights so that it can be transformed into a digital structure containing exclusively AND logic gates and OR logic gates.

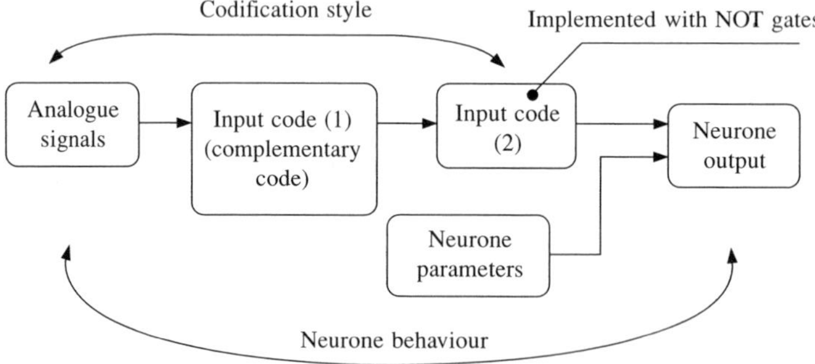


**Fig. 5.4** Neurone conversion solution

### 5.1.3 Digital model implementation using logic gates

The FFANN implementation into a hardware structure is performed separately for each neurone. The implementation method requires that at first the array of input weights $w_{ijp}^{(2)}$ is sorted in descending order. The sorted array contains a total number of $A = m \times$

$n_b$ elements ($w_1^s, w_2^s, w_3^s \ldots w_A^s$), where '$m$' is the number of analogue input signals and $n_b$ is the number of bits used for each input code. The sorted weights correspond to the input signals $x_1^s, x_2^s, \ldots x_A^s$. An iterative conversion procedure is used to analyse the input weights and to generate the corresponding netlist description of the logic gate implementation. As mentioned in section 5.1.1, the iterative procedure decomposes the initial neurone into a pyramidal structure of interconnected subneurones. The structure comprises a top subneurone, a layer of first-order subneurones, a layer of second-order subneurones, etc. The subneurones have all the properties of normal neurones but they have fewer inputs than the initial neurone. Some subneurones are implementable by very simple logic gate configurations. The rest are further decomposed into second-order and third-order subneurones until all of them are implemented.

#### *5.1.3.1 Preliminary considerations*

A series of interrelated basic concepts needs to be defined before describing the iterative hardware implementation process: **terminal weight group, group threshold level, dominant weight, cumulated weight, critical weight, non-critical weight, significant weight, insignificant weight**.

A **terminal weight group** (or simply a **terminal group**) is a set of weights comprising the last $N$ consecutive elements in the sorted array. Therefore any terminal weight group can be uniquely identified by the symbol $G_t(F)$ where '$F$' is the index of its first element. There are a number of $A$ overlapping terminal weight groups in the sorted array: $G_t(1), G_t(2), G_t(3), \ldots, G_t(A)$. Terminal weight group $G_t(1)$ encompasses all the weights in the array. The weights of each first-order subneurone generated by the iterative implementation algorithm are the weights of a terminal group. However, no terminal group generates a first-order subneurone in the final implementation. Thus, the number of first-order subneurones in the pyramidal logic gate structure is situated in the interval [0; $A$].

The **group threshold level** $T_t$ is a quantity calculated by the conversion algorithm for each **terminal group of weights** that is to be converted into a subneurone. The group threshold level equals the threshold level of the subneurone to be generated. The same terminal group can be analysed by the implementation algorithm more than once in different contexts. Each time it can be associated with a different threshold level.

If a weight value is larger than the group threshold level, then it is named a **dominant weight** of the corresponding terminal group. Any dominant weight is related to a **dominant input** that, if active, is able to activate the neurone output signal (force it to '1'), even if all the other input signals are inactive ('0'). The dominant weights in a subneurone are always the first in the corresponding terminal group, because the initial array of weights was sorted in descending order. Consequently, the number $D$ of dominant inputs can be determined using condition (5.30), and if the largest weight in a terminal group is not dominant, no weight is dominant in that terminal group.

$$\begin{cases} w_{F+i}^s \geq T_t & \forall\, 0 \leq i < D \\ w_{F+i}^s < T_t & \forall\, i \geq D \end{cases} \tag{5.30}$$

The **cumulated weight** of a terminal group $G_t(F)$ is defined as the sum of all its component weights. The cumulated weight equals the 'net' value of the neurone when all its inputs are active ('1') in the same time. This is the maximum 'net' value of the corresponding subneurone. If the cumulated weight is smaller than the group threshold level, then the subneurone output is always inactive, regardless of the input signals.

$$W_t(F) = \sum_{i=F}^{A} w_i^s = \max\left\{\sum_{i=F}^{A} w_i^s \cdot x_i^s\right\} \tag{5.31}$$

The output of a subneurone can be activated either by dominant inputs or, if no dominant input is active, by combinations of several non-dominant inputs. Some of these non-dominant inputs are included in all the combinations capable of activating the output. They are named **critical inputs** and they correspond to **critical weights**. Activating these inputs does not necessarily ensure that the subneurone output is active. They only bring the 'net' value of the subneurone close to the **group threshold**, so that the output can be activated in conjunction with less important input signals (the importance of an input signal is proportional to its corresponding weight). As the initial array is sorted in descending order, the critical weights always follow the dominant weights in any terminal group.

Thus, the critical weights can be determined by subtracting all the dominant weights from the cumulated weight. The result has to be larger than the group threshold. Each of the remaining weights is then subtracted from the previous result, obtaining a series of increasing values. Those values that are smaller than the group threshold level correspond to critical weights. This method is summarised in (5.32) where $D$ is the number of dominant weights and $C$ is the number of critical weights in the given terminal group.

$$\begin{cases} W_t(F) - \sum_{i=0}^{D-1} w_{F+i}^s > T_t \\ W_t(F) - w_{F+D+j}^s - \sum_{i=0}^{D-1} w_{F+i}^s < T_t \quad \forall\, 0 \le j < C \\ W_t(F) - w_{F+D+j}^s - \sum_{i=0}^{D-1} w_{F+i}^s \ge T_t \quad \forall\, j \ge C \end{cases} \tag{5.32}$$

Thus, if all dominant inputs are '0' and at least one of the critical weights is '0' at the same time, then the neurone output cannot be active. On the other hand, the subneurone output can be active when all the dominant inputs are inactive, but all the critical inputs are active.

In some cases, the critical inputs are sufficient to activate the neurone output. In other cases, the critical inputs can activate the output only in conjunction with certain combinations of less important inputs, because the sum of the critical weights is lower than the threshold level. These less important inputs, involved in activating the subneurone output, are named **non-critical inputs** and they correspond to **non-critical weights**. As opposed to critical inputs, none of the non-critical inputs is essential for the subneurone activation. If a non-critical input is inactive, its task can be performed by groups of other non-critical inputs, so that the 'net' value is maintained above the threshold level and the subneurone is kept active. However, if all non-critical inputs are deactivated at the same time, the subneurone output is deactivated as well. A subneurone with $D$ dominant weights and $C$ critical weights has non-critical weights as well, if and only if the conditions (5.33) are fulfilled. These conditions signify that the neurone output can be activated by non-dominant inputs but the task cannot be performed by critical inputs alone.

$$\begin{cases} \sum_{i=D}^{A} w_{F+i}^{s} \geq T_t \\ \sum_{i=0}^{C-1} w_{F+D+i}^{s} < T_t \end{cases} \tag{5.33}$$

The three previous input categories (dominant, critical and non-critical) are unequally important for the subneurone operation, but all influence the output signal. These types of inputs have **significant weights**. **Insignificant inputs** do not influence the subneurone output at all. The insignificant inputs have **insignificant weights**, which are very small and do not affect the relation between the subneurone 'net' value and the group threshold level, regardless of the corresponding input signals. Consequently, these inputs are not implemented into hardware.

The effect of sorting the initial array of input weights is that the weights of the same type are grouped together. Furthermore, the groups are arranged in a standard sequence: dominant, critical, non-critical and insignificant, as illustrated by Fig. 5.5 on the particular case of a neurone with 12 arbitrarily chosen input weights.

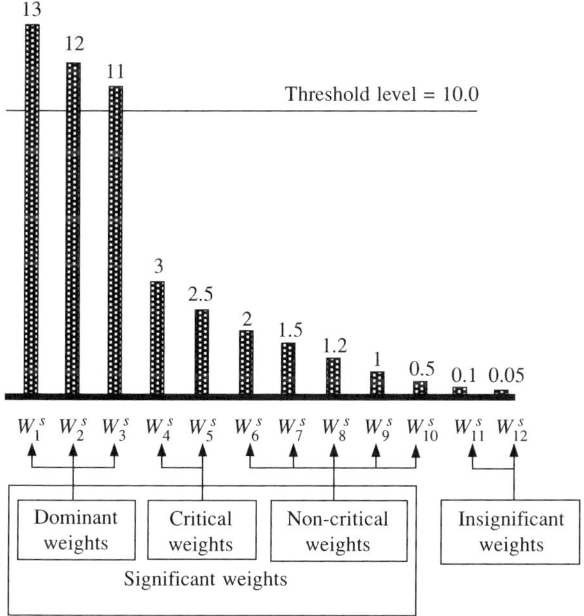


**Fig. 5.5** The neurone weight types and their relative position in the sorted array of weights

One or several weight types can be absent from the sequence. For instance, a neurone complying with condition (5.4(a)) is implementable with an AND logic gate and has only critical weights, because if one of the AND inputs is '0' (inactive) the logic output is '0' as well. Similarly, the neurones complying with condition (5.4(b)) are implementable with OR logic gates and have only dominant input weights.

#### *5.1.3.2 The implementation process – detailed description*

In this section, the hardware implementation of the digital neurones is described in detail, using the concepts and the formulas from the previous section. The implementation process is divided into three procedures (Fig. 5.6):

- The first one carries out a preliminary neurone check. It analyses the sign of its threshold level '$t$'. If the sign is negative or zero, the neurone output is always active regardless of the input signals and the neurone implementation is a simple connection between *V*cc (+5 V) and its output.
- If the threshold level is positive, the array of weights is sorted in descending order and then the second procedure is called. This is a recursive implementation procedure that repeatedly calls itself and builds the required pyramidal structure, gate by gate.
- Eventually the third procedure is called which, according to the principles discussed in section 5.1.2 (at conversion stage two), attaches inverter gates to those inputs in the sorted array that correspond to negative weight values at conversion stage one $(w_x^s \Leftrightarrow w_{ijp}^{(1)} < 0)$.

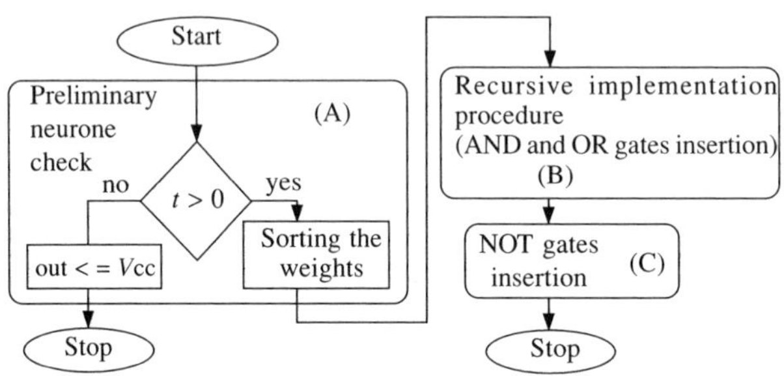


**Fig. 5.6** The hardware implementation process

The recursive implementation procedure (B) has two input parameters that are recalculated for each call of the procedure. The two parameters are the current terminal group defined by its starting index $F$, and the associated threshold level of the terminal group $T_t$. The parameters at the first call are $F = 1$ and $T_t = t$. Thus, the process starts by analysing the terminal group $G_t(F) = G_t(1)$, which comprises all the weights in the array in conjunction with the neurone threshold level '$t$'. The operation of the recursive implementation can be described in ten steps.

**Step 1** The number $D$ of dominant inputs and the number $C$ of critical inputs are calculated by means of (5.30) and (5.32). Condition (5.33) is used to determine whether the subneurone has non-critical inputs. Table 5.1 presents all the possible situations and the next algorithm step to be performed in each case.

**Step 2** The neurone has no significant input and therefore its output is always inactive. The hardware implementation reduces to a simple connection between the neurone output and the circuit ground. End of the procedure (B).

**Step 3** The subneurone has only dominant inputs and it is implemented as a $D$-input OR gate. End of the procedure (B).

**Table 5.1** Subneurone implementation cases

| Dominant inputs | Critical inputs | Non-critical inputs | Next algorithm step |
|---|---|---|---|
| $D = 0$ | $C = 0$ | $N = 0$ | step 2 |
| $D > 0$ | $C = 0$ | $N = 0$ | step 3 |
| $D = 0$ | $C > 0$ | $N = 0$ | step 4 |
| $D > 0$ | $C > 0$ | $N = 0$ | step 5 |
| $D = 0$ | $C = 0$ | $N > 0$ | step 7 |
| $D > 0$ | $C = 0$ | $N > 0$ | step 8 |
| $D = 0$ | $C > 0$ | $N > 0$ | step 6 |
| $D > 0$ | $C > 0$ | $N > 0$ | step 9 |

**Step 4** The subneurone has only critical inputs and it is implemented as a $D$-input AND gate. End of the procedure (B).

**Step 5** The subneurone can be activated either by one of the dominant inputs or by all the critical inputs together. Therefore, the current subneurone can be decomposed into a simpler subneurone plus one higher-order subneurone. The first has $D + 1$ dominant inputs and is connected to the $D$ dominant inputs of the initial subneurone, while input $D + 1$ is fed by the second subneurone. The output of the second subneurone is activated only when the initial subneurone is activated due to the critical input signals. Therefore, it is implemented as a $C$-input AND logic gate. End of the procedure (B).

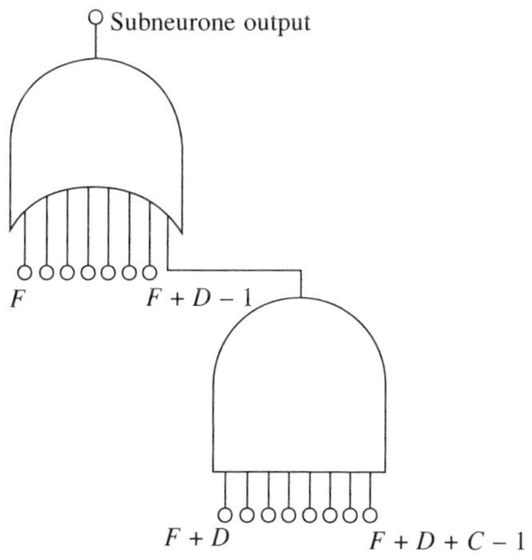


**Fig. 5.7** Subneurone implementation at step 5

**Step 6** The subneurone has critical and non-critical inputs. Therefore, the subneurone output is active if all the critical inputs are active simultaneously with certain combinations of non-critical inputs. The subneurone can be decomposed into a higher-order subneurone supplying a simple subneurone implementable as an AND gate with $C + 1$ inputs. The first $C$ gate inputs are connected to the current subneurone critical inputs, while the last input is connected to the output of the higher-order subneurone which analyses the remaining input combinations.

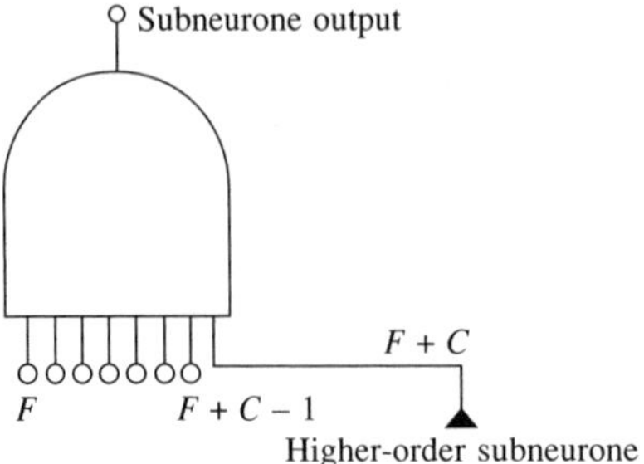


**Fig. 5.8** Subneurone implementation at step 6

The recursive implementation procedure needs to be recalled to generate the implementation of the higher-order subneurone. The new parameters are given in (5.34). The new threshold level is lower than the previous one because the remaining input signals need to cover only the difference between the previous threshold and the sum of the $C$ critical weights already implemented by the AND gate. Go to step 10.

$$\begin{cases} F = F + C \\ T_t = T_t - \sum_{i=0}^{C-1} w^s_{F+i} \end{cases} \tag{5.34}$$

**Step 7** The subneurone has only non-critical inputs. Thus, there are several combinations of input signals capable to activate the subneurone output. The combinations are classified into a number of categories. Each category is associated with a terminal group and comprises all the combinations that involve the first input in the given terminal group. In some terminal groups, the different combinations share only the first input but in others, they share more than one input. It is necessary to calculate the number $K$ of combination categories and the number $S$ of shared inputs, apart from the first one in each category. The first requirement is achieved, as shown in (5.35), by calculating the cumulated weight of the smaller terminal groups included in the current one and comparing the result with the current threshold level.

$$\begin{cases} W_t(F + i) \geq T_t & \forall\, 0 \leq i < K \\ W_t(F + i) < T_t & \forall\, i \geq K \end{cases} \tag{5.35}$$

For each terminal group $G_t(F + j)$ ($j = 0, 1, 2, \ldots K - 1$), the number $S(j)$ of shared inputs is determined according to (5.36). To calculate $S(j)$, individual weights are subtracted from the cumulated weight of the group and the result is compared with $T_t$. One input is shared by all the combinations in the current category, if and only if the subtraction result is smaller than the threshold level. Otherwise, there are input combinations capable of boosting the 'net' value of the neurone above the threshold level without using the tested input. The number $S(j)$ does not include the first input in the corresponding terminal group. According to the definition, this input is implicitly used by all the combinations in the same category, so that the input weight $w^s_{F+j+i}$ is not even tested in (5.36).

$$\begin{cases} W_t(F+j) - w^s_{F+j+i+1} < T_t & \forall 0 \le i < S(j) \quad [\text{if } S(j) > 0] \\ W_t(F+j) - w^s_{F+j+i+1} \ge T_t & \forall\, i \ge S(j) \end{cases} \tag{5.36}$$

Therefore, the current subneurone is implemented as an OR gate with $K$ inputs as illustrated in Fig. 5.9. The OR gate inputs are fed by AND gates with $S(j) + 2$ inputs ($j = 0, 1, 2, \ldots K - 1$) that model the $K$ different combination categories. The first $S(j) + 1$ inputs of each AND gate are connected to **all** the shared inputs of the combinations in the respective category (including this time the first input in the corresponding terminal group). Input $S(j) + 2$ is connected to the output of a higher-order subneurone that analyses the contribution of the remaining inputs to the total net value. Go to step 10. The recursive implementation procedure is recalled $K$ times for each high-order subneurone. The parameters for each call are calculated according to (5.37). The principles that underlie these calculations are similar to those applying to the parameters in (5.34). Go to step 10.

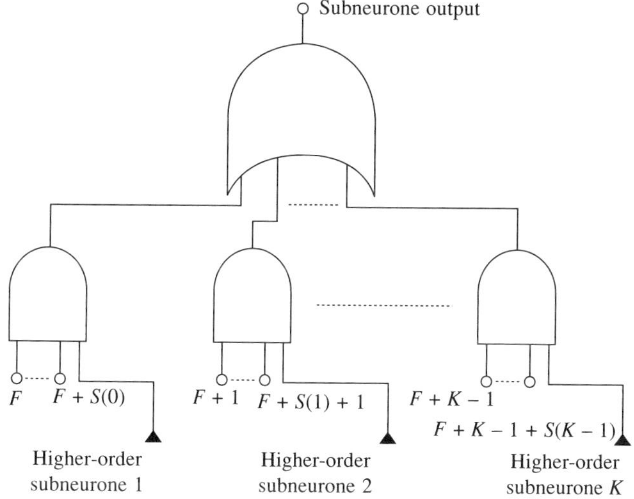


**Fig. 5.9** Subneurone implementation at step 7

$$\begin{cases} F_j = F + S(j) + 1 \\ T_{tj} = T_t - \sum\limits_{i=0}^{S(j)} w^s_{F+j+i} \end{cases} \tag{5.37}$$

**Step 8** The neurone has dominant inputs and non-critical inputs. The combinations of non-critical inputs able to activate the neurone output fall into a number of $K$ categories. Number $K$ is determined using method (5.38), which is similar to (5.35) but takes into account the existence of the $D$ dominant inputs. Thus, the index of the first non-critical input is, in this case, $F + D$ instead of $F$, so that the initial index $F + i$ in (5.35) has to be replaced with $F + D + i$.

$$\begin{cases} W_t(F+D+i) \ge T_t & \forall\, 0 \le i < K \\ W_t(F+D+i) < T_t & \forall\, i \ge K \end{cases} \tag{5.38}$$

Similarly, the number $S(j)$ of shared inputs in each category of input combinations is calculated according to method (5.39), which is derived from (5.36) by replacing each '$F$' with '$F + D$' to take into account the existence of the dominant inputs.

$$\begin{cases} W_t(F + D + j) - w^s_{F+D+j+i+1} < T_t & \forall 0 \le i < S(j) \quad [\text{if } S(j) > 0] \\ W_t(F + D + j) - w^s_{F+D+j+i+1} \ge T_t & \forall i \ge S(j) \end{cases} \tag{5.39}$$

As shown in Fig. 5.10, the neurone is implemented by an OR gate with $D + K$ inputs interconnected with $K$ AND gates. The first $D$ inputs of the OR gate are connected to the subneurone dominant inputs, while the rest of the inputs are supplied by the AND gates.

As in the previous cases, the recursive procedure is recalled $K$ times to implement the $K$ higher-order subneurones in Fig. 5.10. The parameters for each call are given in (5.40). Go to step 10.

$$\begin{cases} F_j = F + D + S(j) + 1 \\ T_{tj} = T_t - \sum_{i=0}^{S(j)} w^s_{F+D+j+i} \end{cases} \tag{5.40}$$

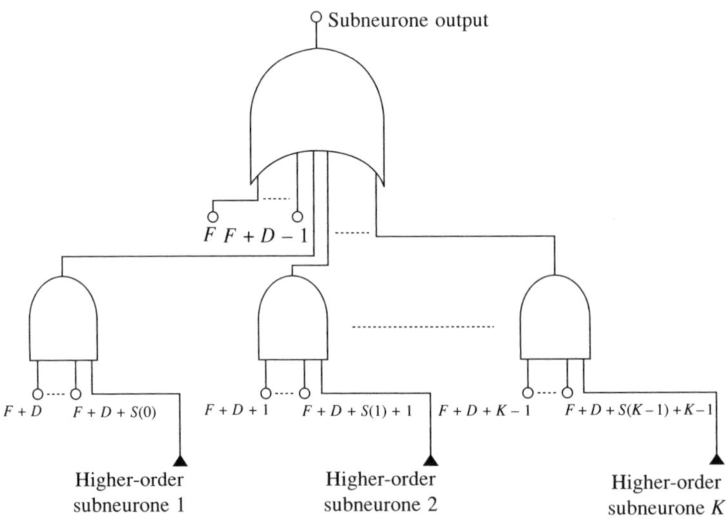


**Fig. 5.10** Subneurone implementation at step 8

**Step 9** The neurone contains all three types of significant inputs: dominant, critical and non-critical. It is implemented by a $D$ + 1-input OR logic gate cascaded with a $C$ + 1-input AND gate and a higher-order subneurone as shown in Fig. 5.11. The higher-order subneurone analyses the combinations of non-critical inputs and activates the current subneurone output when a valid combination is received on the inputs simultaneously with all the critical inputs being active. To generate the higher-order subneurone implementation, the recursive procedure is called with the parameters calculated in (5.41). Go to step 10.

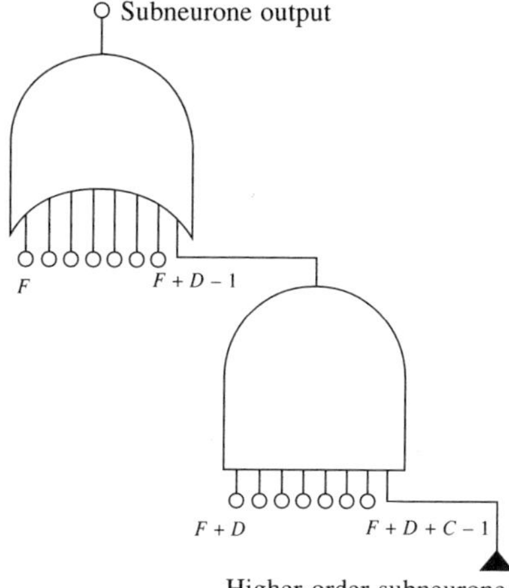


**Fig. 5.11** Subneurone structure at the step 9

$$\begin{cases} F = F + D + C \\ T_t = T_t - \sum_{i=0}^{C-1} w^s_{F+D+i} \end{cases} \tag{5.41}$$

**Step 10** The execution of the implementation process returns to the point where the present procedure call has been performed. This point can be inside this procedure at steps 6, 7, 8 or 9, or it can be at the stage where the recursive process itself was initiated. In the first case, according to computer programming principles, the old parameters $F$ and $T_t$ are restored and the execution resumes at the stage where this call was initiated. In the second case, the execution of the present procedure stops.

### 5.1.3.3 Neurone implementation example

For a better understanding of the implementation algorithm, a complete example is presented in Fig. 5.13. The neurone has $A = 12$ input weights and positive threshold level '$t$'. The weights are sorted in descending order and the recursive implementation procedure is initiated with parameters $F = 1$ and $T_t = t = 10$, as shown in Fig. 5.12. The number of dominant and critical inputs is calculated at step 1 of the recursive implementation procedure. The result is $D = 3$, $C = 0$. The three dominant inputs correspond to the dominant weights $w^s_1$, $w^s_2$, $w^s_3$ in Fig. 5.13. Condition (5.33) is used to demonstrate that the neurone has non-critical inputs as well. Thus, according to Table 5.1, the next step to be performed is step 8. The number $K$ of non-critical input combinations is calculated using relations (5.38). The result is $K = 3$. The first two groups contain weight combinations sharing only one input each, while in the third group, four inputs are shared. Therefore, the output of the neurone implementation is generated by the 6-input OR gate g1 connected to the three dominant inputs and to three AND gates (g2, g3, g4). Gates g2 and g3 have two inputs while g4 has five inputs.

As illustrated in Fig. 5.12, the iterative procedure recalls itself three times to generate the subneurones corresponding to the three previously mentioned groups of weights. First, the procedure is recalled with parameters $F = 5$ and $T_t = t - w^s_4 = 1.9$ to generate

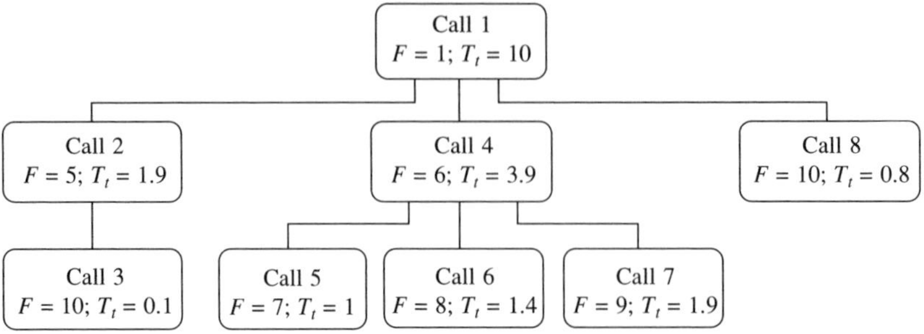


**Fig. 5.12** The recursive implementation process for the neurone in Fig. 5.13

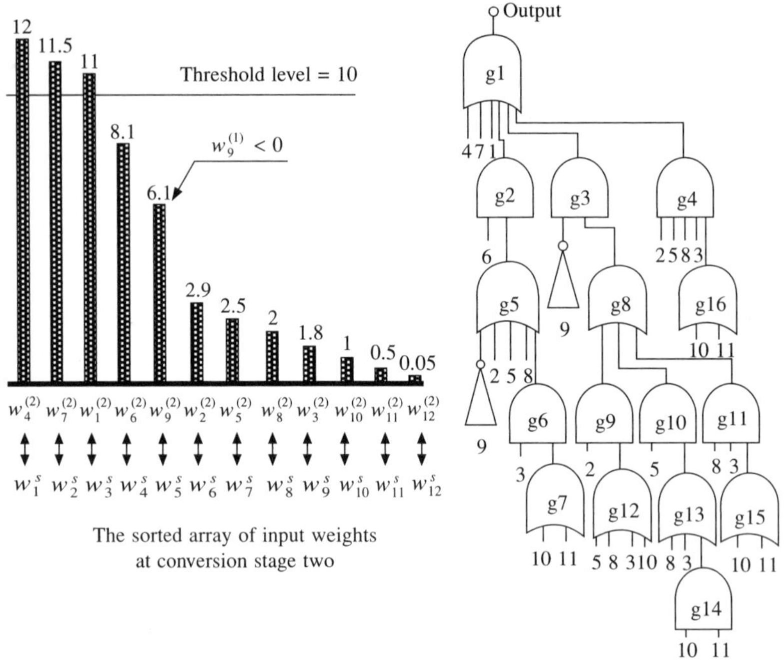


**Fig. 5.13** Digital mathematical model to gate structure conversion example

the implementation of the subneurone connected to gate g2. This subneurone has four dominant inputs (related to the weights $w_5^s$, $w_6^s$, $w_7^s$, $w_8^s$), one critical input (corresponding to $w_9^s = w_3^{(2)}$) and two non-critical inputs (corresponding to $w_{10}^s$ and $w_{11}^s$). Step 9 is carried out and gates g5 and g6 are inserted into the hardware structure. The remaining three inputs belong to a higher-order subneurone that requires the iterative procedure to be called for the third time.

The procedure parameters are redefined as $F = 10$ and $T_t = t - 1.9 - w_9^s = 0.1$ during call number three. As a result, the corresponding subneurone contains two dominant inputs plus one insignificant input (corresponding to $w_{12}^s$) and it is implemented by the 2-input OR gate g7. At this stage, calls number 2 and 3 of the iterative procedure are finished. The control is handed over to call number 1 which initiates the call number 4 with the parameters $F = 6$, $T_t = t - w_5^s = 3.9$ in order to generate the implementation of the subneurone connected to the AND gate g3. The new subneurone has only non-critical weights falling in $K = 3$ categories and it is implemented by logic gates g8, g9, g10 and g11. This subneurone is connected to three third-order subneurones, which are analysed during procedure calls 5, 6 and 7, and their implementations contain the gates g12 to g15.

The end of procedure call 7 brings procedure call 4 to an end as well. The control is passed back to procedure call 1, which initiates the call number 8 with parameters $F$ = 10 and $T_t = t - w_6^s - w_7^s - w_8^s - w_9^s = 0.8$, and implements the subneurone connected to the AND gate g4. This second-order subneurone has two dominant inputs (corresponding to $w_{10}^s$ and $w_{11}^s$) and one insignificant input (related to $w_{12}^s$), so that it is implemented by the 2-input OR gate g16. The end of procedure call number 8 is followed by the end of procedure call number 1, which stops the entire recursive process. At this point the third procedure is called (procedure $C$ in Fig. 5.6), and inverter gates are connected to the inputs related to the weight $w_9$. After this stage is finished, the neurone hardware implementation is complete. It is seen that the weight $w_{12}$ is insignificant due to its small value and therefore the corresponding input was not necessary in any combination of inputs. Thus, the neurone implementation consists of eight subneurones and requires a total of 18 logic gates arranged on six layers.

This example illustrates the complicated calculations necessary to transform even a simple neurone model into a system of interconnected logic gates. ANNs containing several neurones require an amount of calculations that can be efficiently performed only by specialised software instruments. Such instruments have been developed and they are presented in the next section.

## 5.2 Universal programs – FFANN hardware implementation

The solution adopted in this book is universally applicable. It implies a three stage automatic analysis of the FFANN mathematical model and the generation of a VHDL model describing the corresponding hardware structure. The task is carried out by a set of three interconnected C++ programs, given in Appendix A and illustrated in Fig. 5.14, which communicate by means of simple ASCII files.

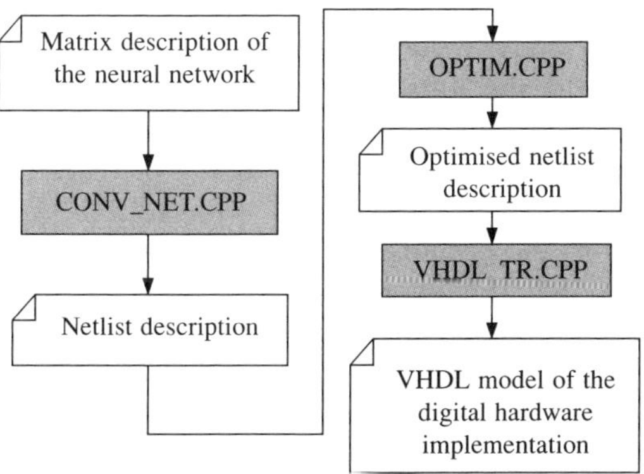


**Fig. 5.14** The data flow between mathematical description and the VHDL model of an FFANN

The first program, CONV_NET, transforms the input mathematical model into a preliminary netlist description of the hardware implementation. The mathematical model consists of a set of matrices containing the parameters of the neurones in the neural network. Each matrix refers to one neural layer and each row in a matrix contains the parameters of a single neurone. The first elements of a row are the neurone weights while the last one is the threshold level. The transformation starts with the model digitisation, performed according to equations (5.29), and then applies the algorithm illustrated by Fig. 5.6. The program allows the user to set the number of bits used by the analogue inputs, and the maximal number of inputs per logic gate. If a larger number of inputs are required at a certain stage of the conversion, a pyramidal interconnection of simpler gates will be used to replace the required gate (Fig. 5.15).

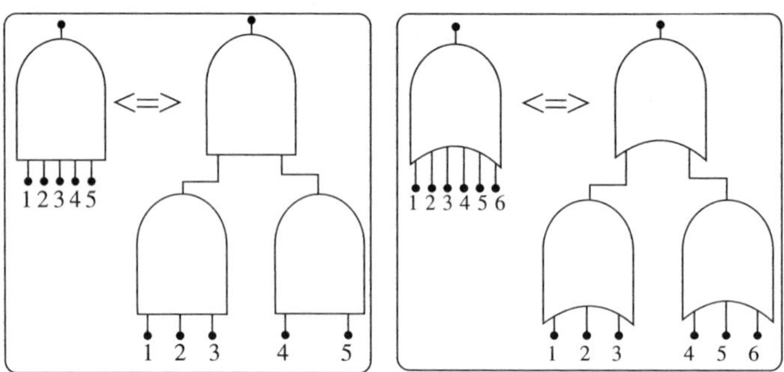


**Fig. 5.15** Examples of fan-in reduction using interconnections of simpler logic gates

The second program, OPTIM, minimises the netlist description by eliminating the redundant components. The netlist optimisation requires that three memory tables containing the circuit nodes and gates are built. Each table contains data about a specific type of logic gate (NOT, AND, OR). The tables are thoroughly explored to find groups of redundant gates (gates of the same type connected to the same input nodes). Each group is replaced with a single logic gate whose output signal is distributed to all the circuit nodes previously connected to the outputs of the eliminated gates. For instance, the gates g7, g15 and g16 in Fig. 5.13 are redundant and can be replaced by a single 2-input OR gate. The elimination of any gate changes the circuit configuration. Gates that were initially connected to different nodes can be connected to the same nodes after the elimination of a number of redundant gates. This creates the opportunity for further elimination of redundant gates. Therefore, after the equivalent gates are removed, the tables are updated and the process is restarted as shown in Fig. 5.16. The optimisation process stops only when no additional modification can be made in any of the three tables.

The third program, VHDL_TR, transforms the optimised netlist description into a VHDL model of the hardware implemented neural network. The obtained VHDL file can be synthesised using any commercially available software package specialised in FPGA design. The file contains a single VHDL entity (the network) whose corresponding architecture comprises a number of internal signals and a list of assignment statements. Each statement models one or several identical logic gates by associating a logical expression with an internal signal or with an output signal.

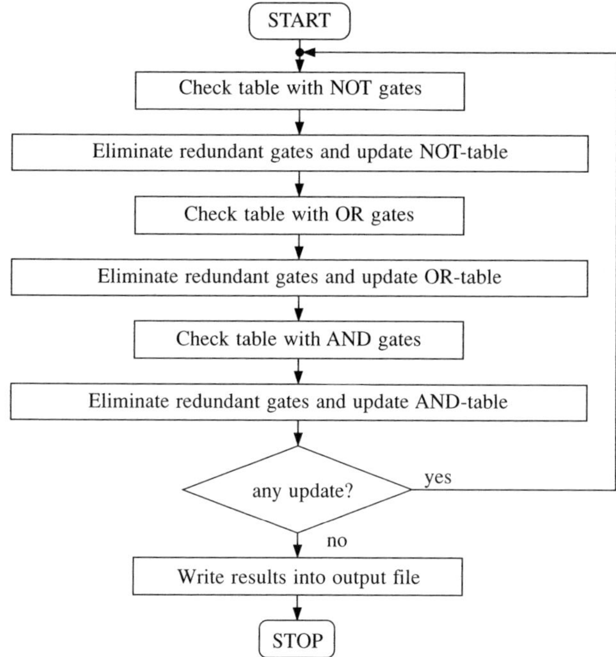


**Fig. 5.16** The general flowchart underlying the optimisation program

For exemplification, the VHDL model of the neurone in Fig. 5.13 is presented below. The model has been generated using the three universal C++ programs. It is important to note that the index of the components inside the input port 'd_in' vary between 0 and 11 instead of 1 to 12 as it was in Fig. 5.13.

```
-- Code Fragment 5.1
LIBRARY ieee;
USE ieee.std_logic_1164.all;

ENTITY network1 IS
  PORT(d_in : IN std_logic_vector(11 DOWNTO 0); --the 12 input signals
    d_out: OUT std_logic_vector(0 DOWNTO 0));--the single output
END network1;
ARCHITECTURE arch_network1 OF network1 IS
  SIGNAL n1,n2,n3,n4,n5,n7,n8,n9,n10,
         n11,n12,n13,n16: std_logic;
BEGIN
n16<= NOT d_in(8);                                  -- the NOT gate
n1<= d_in(5) AND n4;                                -- gate g2
n2<=n16 AND n7;                                     -- gate g3
n8<=d_in(1) AND n11;                                -- gate g9
n9 <=d_in(4) AND n12;                               -- gate g10
n5<=d_in(2) AND n15;                                -- gate g6
n13 <=d_in(9) AND d_in(10);                         -- gate g14
n7<=n8 AND n9 AND n10;                              -- gate g8
```

```
n10 <=d_in(7) AND d_in(2) AND n15;                  -- gate g11
n3<=d_in(1) AND d_in(4) AND d_in(7) AND d_in(2) AND n15;
n15<=d_in(9) OR d_in(10);                           -- gates g7, g15, g16
n12 <=d_in(7) OR d_in(2) OR n13;                    -- gate g13
n11 <=d_in(4) OR d_in(7) OR d_in(2) OR d_in(9);     -- gate g12
n4<=n16 OR d_in(1) OR d_in(4) OR d_in(7) OR n5;     -- gate g5
d_out(0)<=d_in(3) OR d_in(6) OR d_in(0) OR n1 OR n2 OR n3;
END arch_network1;

CONFIGURATION conf_network1 OF network1 IS
  FOR arch_network1
  END FOR;
END conf_network1;
```

The optimisation performed by OPTIM has three important effects on the previous VHDL model:

- A single expression models the group of the redundant gates g7, g15 and g16. The other logic gates are modelled by individual logic expressions.
- The internal signals n6, n14 and n15 are absent in the list at the beginning of the network architecture. They have been removed alongside with the gates g7, g15 and g16.
- The three types of logic operators (NOT, AND, OR) occur in three distinct sections of the network architecture description. Inside each section, the logic expressions are sorted in ascending order according to the number of logic operators involved. This feature is just a side effect of the optimisation algorithm but it simplifies the inspection of the obtained VHDL model (for instance, counting the total number of gates of a certain type or with a certain fan-in).

## 5.3 Hardware implementation complexity analysis

There are two important cost functions characterising the ANN hardware implementations: the input–output delay and the required chip area. For most applications, the delay time is satisfactory but the chip area is critical because the hardware resources are always limited. The input–output delay is approximately proportional to the implementation depth, which is defined as the number of layers of elementary circuit units: TGs or logic gates. Several approximate methods have been proposed to determine the required chip area of an ANN, depending on the envisaged implementation technology. They imply the calculation of the number of neurones, the number of implementation units (logic gates or threshold gates, depending on the technology) [36], the total input number of all implementation units [108], the sum of all input weights and thresholds [36], etc. In the case of FPGA implementations, the total number of gates is the most suitable means to determine the implementation complexity.

It is difficult to calculate in advance the number of logic gates required by a pyramidal logic structure with $n$ inputs, like the one in Fig. 5.13, because the result depends on the fan-in of each individual logic gate. The calculations are simple only if it is possible to achieve the implementation with logic gates having the same fan-in Δ. In this case, the implementation complexity is given by equation (5.42), where $\lceil x \rceil$ is the ceiling function (the smallest integer greater than or equal to $x$).

$$N_{\mathrm{LG}\Delta} = \left\lceil \frac{n-1}{\Delta - 1} \right\rceil \tag{5.42}$$

Usually the implementation algorithms require logic gates with different fan-ins, so that (5.42) is applicable only to a limited number of practical situations. However, any Δ-input AND gate or Δ-input OR gate can be replaced by a number of Δ gates of the same type, but having only two inputs. If the fan-in is restricted to $\Delta = 2$, then equation (5.42) is simplified as (5.43).

$$N_{\mathrm{LG2}} = n - 1 \tag{5.43}$$

Any logic circuit can be built using exclusively 2-input logic gates. Therefore, the number of equivalent 2-input gates in the neural network implementation is a universal measure of its hardware complexity and offers a means to compare different implementation algorithms. As opposed to the rest of the logic gates, the NOT gates always have $\Delta = 1$. However, they are not taken into account when estimating the implementation complexity because the NOT logic operator can be integrated into 2-input logic gates as shown in Fig. 5.17. Note that the total number of inputs '$n$' in (5.42) and (5.43) is larger than the number of the neurone binary inputs ($A = n_a \times n_b$) because some input signals drive more than one gate in the pyramidal structure.

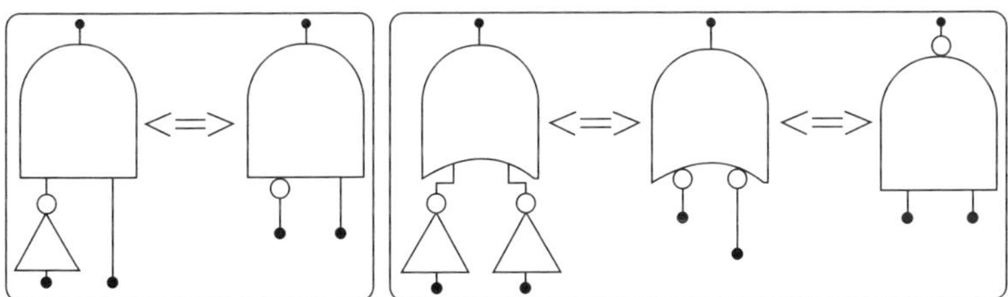


**Fig. 5.17** The integration of NOT operator in complex logic gates: NOT-AND and NOT-OR

### 5.3.1 Results previously reported in the literature

An efficient neural network implementation strategy is one that minimises the number of equivalent 2-input gates in the corresponding digital circuit. Only a few complexity minimising algorithms have been developed so far for digital hardware implementations. The most relevant two of them are proposed in [214] and [37] and lead to the same order of implementation complexity but generate different circuit depths. The results presented in [37] are converted here in numbers of 2-input gates and then a comparison is performed between these results and the hardware complexity generated by the new implementation strategy proposed in this book.

In [37] the neural network is treated as a set of $k$ Boolean functions $F_{n,m,i}$ ($i = 1, 2, \ldots. k$), with $n$ inputs and a cumulated total of $m$ groups of '1' in the truth table. In the one-dimensional case a group of '1' is a set of successive $n$-bit input strings, whose corresponding function outputs are all '1'. The approach can be extended from one dimension to several dimensions, as shown in Fig. 5.18. The number of truth table dimensions equals the number $n_a$ of inputs of the analogue neurone modelled.

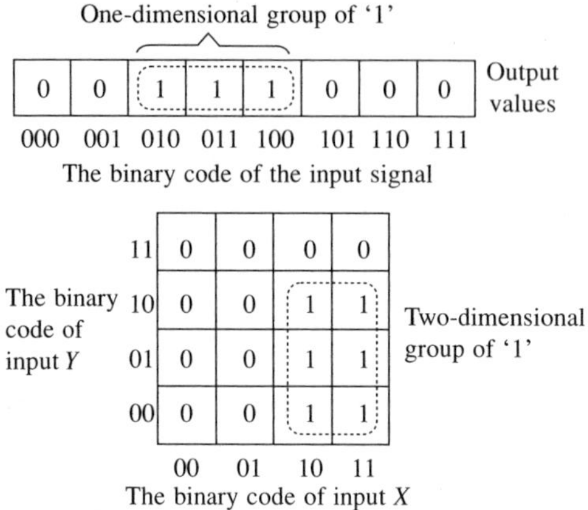


**Fig. 5.18** Groups of '1' in one-dimensional and two-dimensional truth tables

The groups of '1' are generated with a constructive method so that they optimally cover those points in the $d$-dimensional input data space that have to activate the outputs of the corresponding network. Therefore, the constructive method used in [37] is a particular case of the Voronoi algorithm, where all the Voronoi cells are hypercubes bounded by hyperplanes parallel to the axes of the input data space. Three implementation alternatives are compared:

(i) Direct function implementation in disjunctive normal form (DNF) (initially proposed [33]).

(ii) A more sophisticated strategy which involves the use of $n$-bit comparators alongside AND gates and OR gates. The comparators model the $n_a$-dimensional hyperplanes parallel to $n_a - 1$ axes of the input data space. Each of them performs comparisons between one of the $n_a$ analogue input signals and a constant. The second layer is made up of $2n_a$-input AND gates. Each AND gate implements a hypercube-shaped Voronoi cell corresponding to a group of '1'. The third layer is made up of OR gates combining the information provided by different AND gates.

(iii) A synthesis of the previous two methods that replaces some of the comparators with DNF terms of the Boolean function. This method analyses the size of the groups of '1'. The small cells are more efficiently implemented in DNF format, while large groups are better implemented by comparators. Thus, some of the comparators are replaced by a number of AND gates and NOT gates.

Any $n$-bit comparator between a variable quantity and a constant value can be implemented with up to '$n - 1$' 2-input gates [35]. A neural network with $n_a$ analogue inputs coded on $n_b$ bits each requires up to $n_a \cdot (2^{n_b} - 1)$ comparators, which is equivalent to $n_a \cdot (n_b - 1) \cdot (2^{n_b} - 1)$ 2-input logic gates. The redundancy across different comparators can be reduced by optimisation algorithms. The optimisation is limited by the number of comparators. The comparators' outputs are independent signals and therefore are generated by separated logic gates. Thus, if there are $n_a \cdot (2^{n_b} - 1)$ comparators then the complexity of the circuit cannot be decreased below $n_a \cdot (2^{n_b} - 1)$ 2-input logic gates.

The set of Boolean functions $F_{n,m,i}$ $(i = 1, 2, \ldots k)$ contains a total of $m$ groups of '1' in the truth tables. Consequently, the implementation complexity of the second neurone layer in the corresponding ANN is up to $(2 \cdot n_a - 1) \cdot m$ equivalent 2-input logic gates. On the other hand, the hardware complexity of the third layer is up to $m - k + C_k^2$. This result is a generalisation of the particular cases illustrated in Fig. 5.19. Thus, if the neural network has only one output ($k = 1$) then the third layer is implemented as a pyramid of OR logic gates with the complexity of $m - 1$ equivalent 2-input gates. If the neural network has two outputs then the situation is more complex. Thus, the outputs are generated by two different pyramids that can share some of the '$m$' inputs (Fig. 5.19(c)) or they can be completely separate (Fig. 5.19(b)). When the two pyramids share part of the '$m$' inputs the resulting implementation contains three subpyramids and two extra OR gates generating the actual output signals. As a result, the hardware complexity is larger than in Fig. 5.19(b).

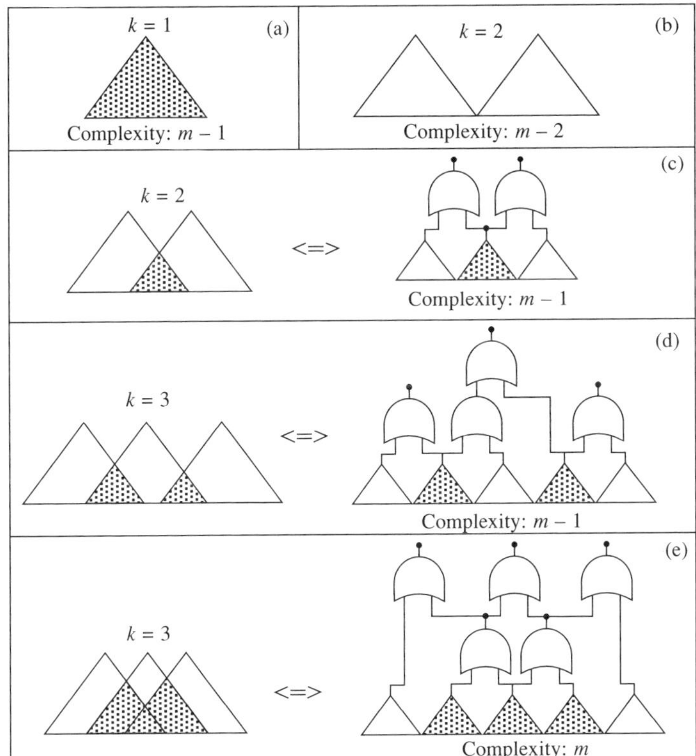


**Fig. 5.19** Analysis of the third layer complexity (typical situations)

When the number of neural network outputs is larger than two, several situations are possible depending on the number of shared clusters of input signals between different OR-gate pyramids. Two possibilities are illustrated in Fig. 5.19(d) and Fig. 5.19(e) for the situation when $k = 3$. Generally, the hardware complexity corresponding to the third neural network layer increases with the number of shared clusters of inputs. The maximal number of input clusters is $C_k^2$, therefore, a number of $k + C_k^2$ subpyramids are contained in the corresponding hardware implementation. Furthermore, each output is generated by a pyramidal OR-gate structure with $k$ inputs and a complexity of $k - 1$ 2-input logic gates. Therefore the total complexity of the third layer can be calculated as

$$m - k - C_k^2 + k \cdot (k - 1) = m - k + C_k^2 \tag{5.44}$$

Thus, the upper limit of the implementation complexity for method (ii) in [37] is:

$$N_{\text{LG2(ii)}} = n_a \cdot (2^{n_b} - 1) + (2 \cdot n_a - 1) \cdot m + m - k + \frac{k(k-1)}{2} \tag{5.45}$$

It is demonstrated in [37] that the hybrid method (iii) generates implementations with up to four times less complexity. The complexity level given in (5.45) is thereby reduced to:

$$N_{\text{LG2(iii)}} = \frac{n_a \cdot (2^{n_b} - 1)}{4} + \frac{n_a \cdot m}{2} + \frac{k(k-1)}{8} - \frac{k}{4} \tag{5.46}$$

The complexity of the second layer can be calculated as a function of the number $N_{\text{neur}}$ of neurones in the first layer of the ANN generating the Boolean functions $F_{n,m,i}$ $(i = 1, 2, \ldots k)$. Two different situations are illustrated in Fig. 5.20(a) and (b). In Fig. 5.20(a), the decomposition of the central region of the diagram into Voronoi cells is demonstrated and the implementation of one cell is illustrated. In Fig. 5.20(b), the neural network contains only one neurone and therefore each Voronoi cell is defined by two comparators

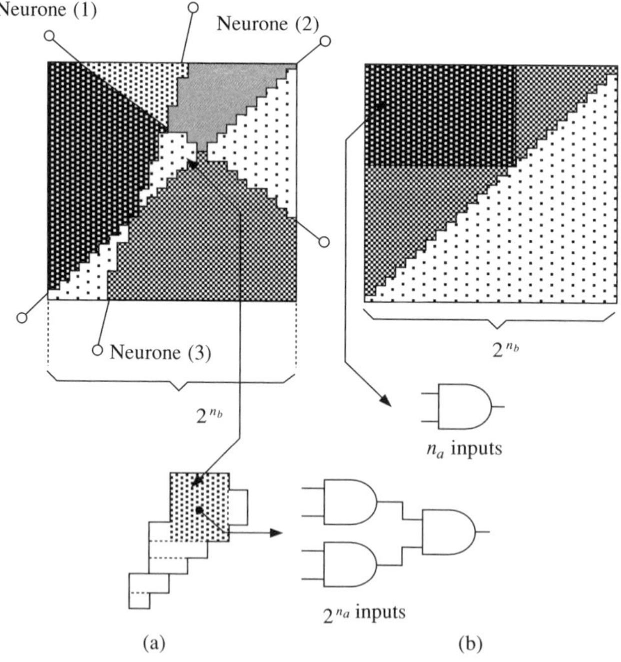


**Fig. 5.20** The estimation of the second layer complexity in a two-dimensional case

supplying a single 2-input AND gate. This result can be generalised for an $n_a$-dimensional input data space as follows: each neurone modelling an oblique hyperplane in the input data space is part of a number of up to $2^{n_b \cdot (n_a - 1)}$ groups of '1'. The corresponding Voronoi cell requires up to $2 \cdot n_a$ comparison results but only $n_a$ of them are linked to a particular neural hyperplane. If required, the other $n_a$ signals are related to a different neurone. Therefore the maximal number of inputs of the AND-gate structure implementing the second layer is $N_{\text{neur}} \cdot n_a \cdot 2^{n_b \cdot (n_a - 1)}$.

In these conditions, equation (5.46) becomes (5.47). The result in (5.47) is an upper limit because it is possible that more than one neurone includes a certain group of '1'.

$$N_{\text{LG2 oblique}} = \frac{n_a \cdot (2^{n_b} - 1)}{4} + \frac{N_{\text{neur}} \cdot n_a \cdot 2^{n_a n_b - n_b} - 1}{2} + \frac{k(k-1)}{8} - \frac{k}{4} \quad (5.47)$$

The most general parameter used to compare the general properties of the implementation algorithms is the **order of complexity** of the generated hardware structure. The order of complexity is a concept initially used in software engineering and computer sciences, but it has been extended for assessing the size of neural hardware implementations [35], [36], [37]. The order of complexity associated with an implementation algorithm is an expression that shows how the implementation complexity varies with the increase of the network parameters (number of neurones, number of interconnections, etc.). The increase can be linear, polynomial, exponential, factorial, etc. The order of complexity is obtained considering that all network parameters involved in the exact hardware complexity equation have very large values. In such a situation, one of the terms has a much larger value than the others, so that the overall complexity can be approximated by this term alone. The order of complexity is defined as the expression of the most significant term in the hardware complexity equation, after all the constant factors have been eliminated. The elimination of constant factors is justified by the fact that they do not affect the relation between two different orders of complexity.

For instance, an exponential order of complexity always implies larger implementations than a linear order of complexity. Provided that the size of the network ($N_{\text{neur}}$) is sufficiently large, (5.48) is fulfilled and the exponential expression generates larger values than any linear expression, regardless of the constants $K_1$ and $K_2$ involved. The same considerations apply to any pair of complexity orders.

$$K_1 \cdot 2^{N_{\text{neur}}} > K_2 \cdot N_{\text{neur}} \Leftrightarrow \frac{2^{N_{\text{neur}}}}{N_{\text{neur}}} > \frac{K_2}{K_1} \quad (\text{if } N_{\text{neur}} \text{ is sufficiently large}) \quad (5.48)$$

For implementations where both $n_a$ and $n_b$ are large numbers, the second term in (5.47) is the largest and the approximation in (5.49) is valid. Thus, the implementation size undergoes an exponential increase with $n_a$ and $n_b$ which makes the implementation of sizeable neural networks very difficult.

$$N_{\text{LG2 oblique}} \approx \frac{N_{\text{neur}} \cdot n_a \cdot 2^{n_a n_b - n_b}}{2} \Leftrightarrow O(N_{\text{neur}} \cdot n_a \cdot 2^{n_a n_b - n_b}) \quad (5.49)$$

## 5.3.2 The analysis of the new implementation method

The logic structure generated by the implementation algorithm adopted in this book, initially presented in section 5.1.3.2, is analysed now from a geometrical point of view, in order to determine the corresponding hardware complexity. The analysis is first performed for neurones with two analogue inputs (*X* and *Y*), and then the results are generalised for any number of analogue inputs. The hardware complexity is initially assessed without taking into account any hardware optimisation. Then the improvements brought about by the optimisation algorithm presented in section 5.2 are considered as well, and the hardware complexity after optimisation is discussed.

#### *5.3.2.1 Implementation without optimisation*

If the neurone has only two analogue inputs then the input data space is two-dimensional and the hyperplanes are reduced to simple lines. The input data space is divided in four quadrants depending on the values of the most significant bits in *X* and *Y* input binary codes. The half-space where the neurone output is active covers a number of one, two, three or four of these quadrants. The quadrants can be either partially covered or totally covered. Each quadrant is in its turn divided into four other subquadrants defined by the second significant bit in each input code. The division process can be carried out for $n_b$ times because each input code contains $n_b$ bits. Each division in four subquadrants corresponds to a subneurone inside the complete hardware implementation. The subneurone models the boundary between the active region and the inactive region inside the subquadrant. If the symmetrical situations are ignored, there are only eight types of relative positions between the hyperplane and the four quadrants. They are analysed in Fig. 5.21 alongside the corresponding subneurone implementations. The presented results apply to the subneurones of orders larger than one but smaller than $n_b$. The analysis of the first-order subneurones generates results similar to the findings shown in Fig. 5.21, but the bits '0' and '1' in the truth tables are reversed. This situation is caused by the use of two's complement codification where the most significant bit of positive numbers is '0', while for negative numbers it is '1'. Therefore, all subneurones of order $i < n_b$ in the pyramidal structure are fed with the signals of zero, one, two or three higher-order subneurones. The subneurones of order $n_b$ are not fed by other subneurones because there are no more bits available in the input codes to generate such subneurones. In this case, as shown in Fig. 5.22, the higher-order subquadrants are either completely included or completely excluded from the active region of the current subquadrant. Such subquadrants are named elementary subquadrants because they cannot be further divided into higher-order subquadrants.

If the area bounded by the hyperplane is more than half the surface of a higher-order subquadrant then it is completely included in the active region. Otherwise, it is completely excluded. There are five types of $n_b$-order subneurones defined by the number of subquadrants that are included in the active region (Fig. 5.22). Two of them have the hardware complexity $N_{LG2} = 1$ (H and J), while the other three have the complexity $N_{LG2} = 0$ (G, I, K), because they do not require any logic gate for their implementation.

In a data space with more than two dimensions, there are more subneurone types than in Fig. 5.21. As Fig. 5.21 demonstrates, the number of subneurones equals every time the number of subquadrants are crossed by the hyperplane. Therefore, the highest hardware complexity is obtained when the number of subquadrants crossed by the hyperplane is maximal. A hyperplane in a $n_a$-dimensional space can cross up to $2^{n_a} - 1$ quadrants. Therefore, this is the maximum number of higher-order subneurones that can feed the current subneurone. In such a situation, the pyramidal logic gate structure can be considered as a binary tree with $n_a + 1$ layers of nodes and $2^{n_a}$ leaves: one leaf for each subquadrant

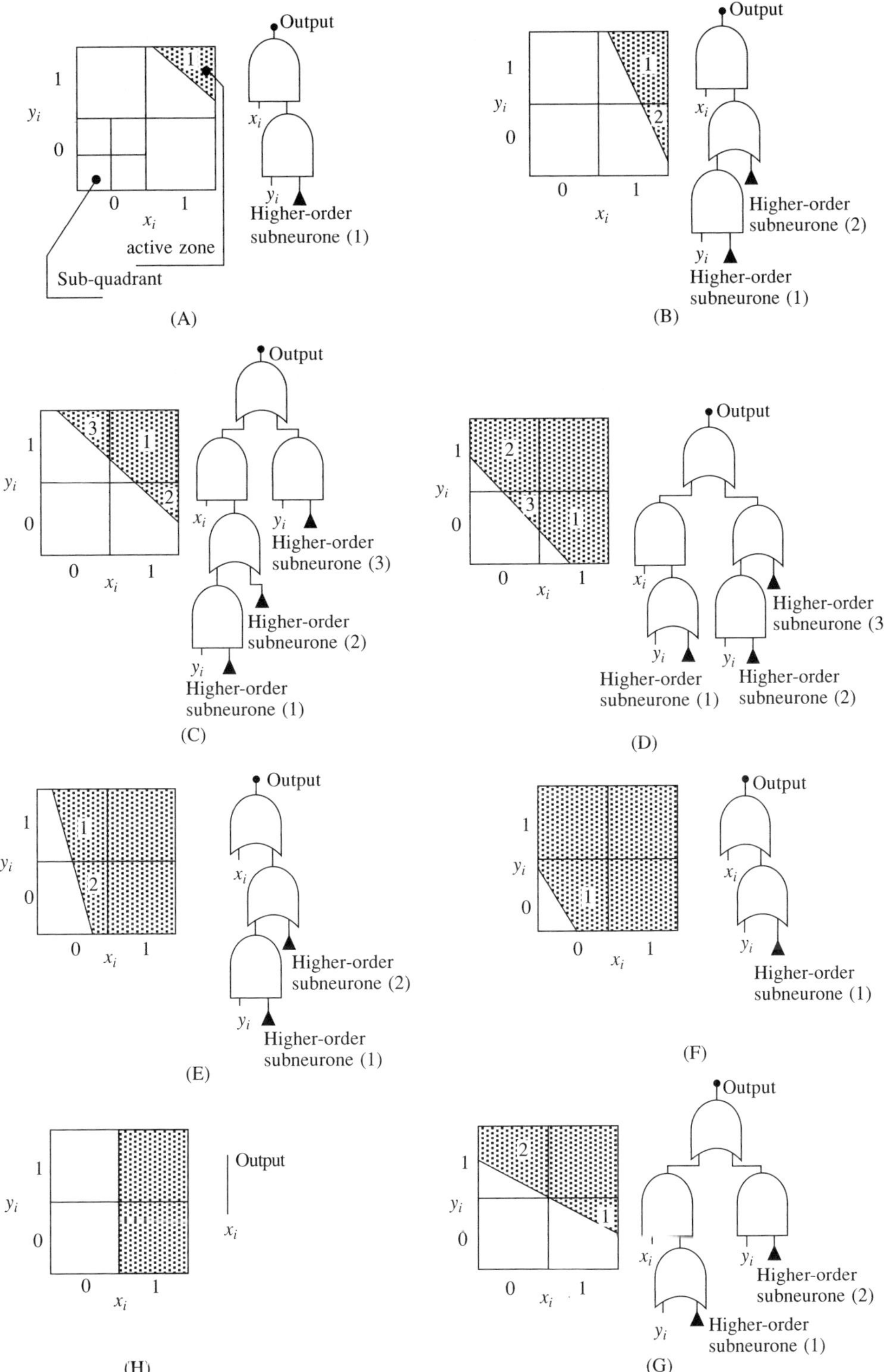


**Fig. 5.21** The division of a $(i-1)$-order quadrant in $i$-order quadrants and the corresponding subneurone implementations $(1 < i < n_b)$

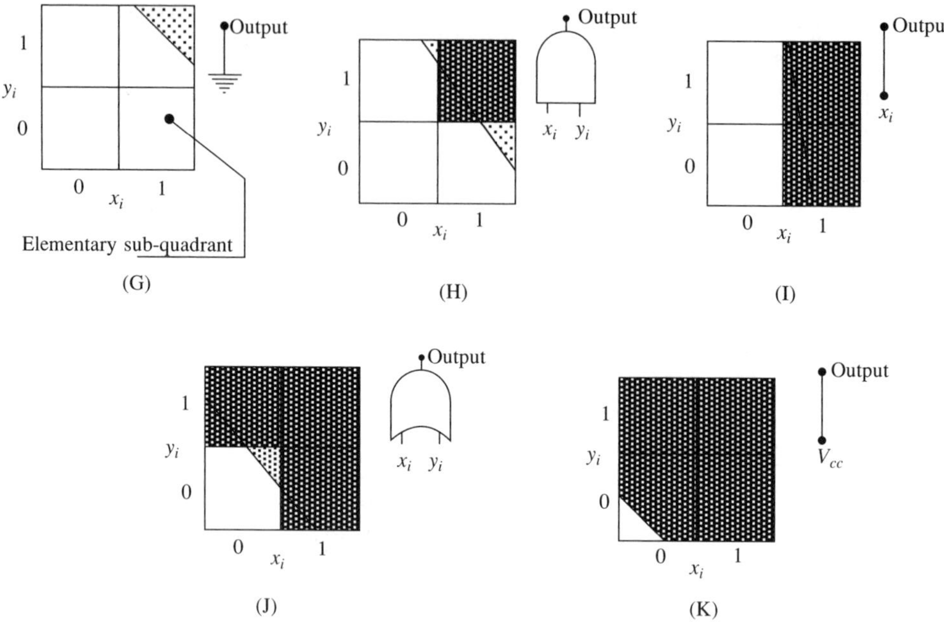


**Fig. 5.22** Types of $n_b$-order subneurones

inside the current quadrant. Some of the leaves use the signals generated by subneurones to produce the correct neurone output signals while others directly use the input signals of the original neurone. Figure 5.23 illustrates a three-dimensional subneurone ($n_a = 3$) that generates a maximal hardware implementation because the hyperplane crosses all subquadrants except $(x_i, y_i, z_i) = (1, 1, 1)$. In this subquadrant, the subneurone output is '1'. The maximal number of nodes in the binary tree, except the neurone output, is:

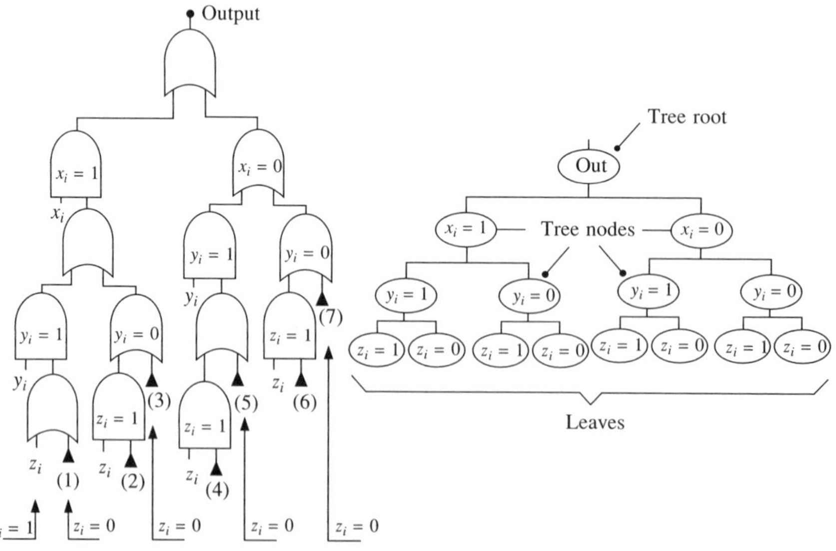


**Fig. 5.23** A three-dimensional subneurone implementation and the corresponding binary tree

$$\sum_{i=1}^{n_a} 2^i = 2^{n_a+1} - 2 \tag{5.50}$$

As illustrated in Fig. 5.23, each node of the binary tree generates one input in the logic gate structure if the related input bit ($x_i$, $y_i$, $z_i$) is '1' on that node. Otherwise, no input is necessary. Only half of the nodes correspond to input bits '1'. Thus, the total number of inputs in the logic gate structure is the sum between the number of subneurones and the number of nodes divided by two (5.51).

$$N_{\text{inputs}}(n_a) = 2^{n_a} - 1 + \frac{2^{n_a+1} - 2}{2} = 2^{n_a+1} - 2 \tag{5.51}$$

The maximal hardware complexity of one subneurone can be calculated as a function of $n_a$, according to (5.52).

$$N_{\text{GT2-max}}(n_a) = N_{\text{inputs}}(n_a) - 1 = 2^{n_a+1} - 3 \tag{5.52}$$

The total pyramidal structure of logic gates corresponding to one neurone has a complicated structure due to the variable number of higher-order subneurones feeding each lower-order subneurone. However, it is possible to determine the total number of subneurones of any order without analysing the detailed interconnections between them and the subneurones of other orders. The number of subneurones equals the number of subquadrants of corresponding size that are crossed by the hyperplane in the input data space. In a two-dimensional case, the subquadrants are square-shaped and there are $2^{i-1}$ subquadrants on each side of the square input data space. The maximal number of crossed subquadrants is $2 \cdot 2^{i-1} - 1$. In an $n_a$-dimensional data space the subquadrants are cubes ($n_a = 3$) or hypercubes ($n_a > 3$) and the previous result can be generalised to:

$$N_{\text{subn}}(i) = 2 \cdot [2^{(i-1)}]^{(n_a-1)} - 1 = 2^{(i-1)\cdot(n_a-1)+1} - 1 < 2^{(i-1)\cdot(n_a-1)+1} \tag{5.53}$$

The generalisation is based on the fact that the hyperplane can be projected on a base with $n_a - 1$ dimensions upon which lie a number of maximum $2^{(i-1)\cdot(na-1)}$ quadrants. Each of these quadrants is the bottom of an $n_a$-dimensional prism containing at most two subquadrants that are crossed by the hyperplane. Figure 5.24 illustrates the two-dimensional and the three-dimensional situations. Therefore, the total number of subneurones is given by equation (5.54).

$$N_{\text{total-subn}} = \sum_{i=1}^{n_b} 2 \cdot 2^{(n_a-1)\cdot(i-1)} - 1 < 2 \cdot \frac{2^{(n_a-1)\cdot n_b} - 1}{2^{n_a-1} - 1} \tag{5.54}$$

Based on the previous results, an absolute upper limit of the total implementation complexity of all the analogue neurones can be calculated as in (5.55). This calculation does not take into account the fact that order-$n_b$ subneurones have a smaller hardware complexity.

$$N_{\text{GT2-total}} = N_{\text{neur}} \cdot N_{\text{total-subn}} \cdot N_{\text{GT2-max}} < N_{\text{neur}}\left(2 \cdot \frac{2^{n_a n_b - n_b} - 1}{2^{n_a-1} - 1}\right) \cdot (2^{n_a+1} - 3) \tag{5.55}$$

The expression (5.55) can be replaced with the higher maximal limit that has a simpler expression, as in (5.56).

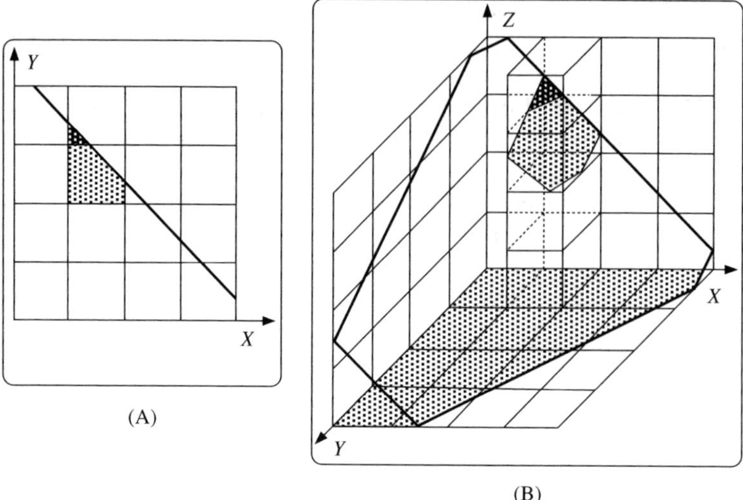


**Fig. 5.24** The intersection between the hyperplane and the subquadrants of second order ($i = 2$) in the two-dimensional case (A), and three-dimensional case (B)

$$N_{\text{GT2-total}} < N_{\text{neur}}\left(2 \cdot \frac{2^{n_a n_b - n_b} - 1}{2^{n_a - 2}}\right) \cdot (2^{n_a+1} - 3) < 2^4 \cdot N_{\text{neur}} \cdot (2^{n_a n_b - n_b} - 1) \tag{5.56}$$

The order of complexity generated by the adopted implementation method is given by (5.58).

$$O(N_{\text{neur}} \cdot 2^{n_a n_b - n_b}) \tag{5.57}$$

This result is superior to the one presented in (5.49) which corresponds to the method presented in [37] because the order of complexity (5.57) is smaller than the order of complexity (5.49). Therefore, the implementation of large neural networks designed with Voronoi diagrams is most efficient using the implementation strategy presented in section 5.1.3 even if no hardware optimisation is carried out.

### 5.3.2.2 Optimised implementations

The optimisation process presented in section 5.2 decreases even further the level of the initial hardware complexity. Very efficient optimisations are possible because, despite the large number of high-order subneurones, the number of different subneurone types is relatively small. This means that many of the subneurones are redundant. Given the correspondence between the hyperplanes in the input data space and neurones hardware implementation, the number of different subneurones of a certain order is estimated using geometrical considerations. An order-$i$ subneurone, in a two-dimensional data space, corresponds to a hyperplane (a straight line) in a quadrant with the side length $2^{n_b - i + 1}$ elementary subquadrants. The slope of the two-dimensional hyperplane (the straight line) is the same in all the subquadrants, regardless of their order. Thus, the subquadrants differ only by the intersection points between the hyperplane and the corresponding subquadrant sides (Fig. 5.25). For a certain order '$i$', there are a number of $N_c = 2 \times 2^{n_b-i+1}$ classes of two-dimensional subquadrants, depending on these intersection points. Similarly, in an $n_a$-dimensional data space the number of classes is $N_c = n_a \cdot 2^{n_b-i+1}$.

The subquadrants in the same class can differ by the exact shape of the boundary between the active region and the inactive region. The shape of this boundary depends on the slope and position of the hyperplane and it contains a precisely determined sequence of steps as in the example presented in Fig. 5.25. The maximal number of different step patterns for a given class of subquadrants can be calculated using algebraic and geometrical considerations. First, this upper limit is determined in a two-dimensional input data space and then the result is generalised for an $n$-dimensional situation.

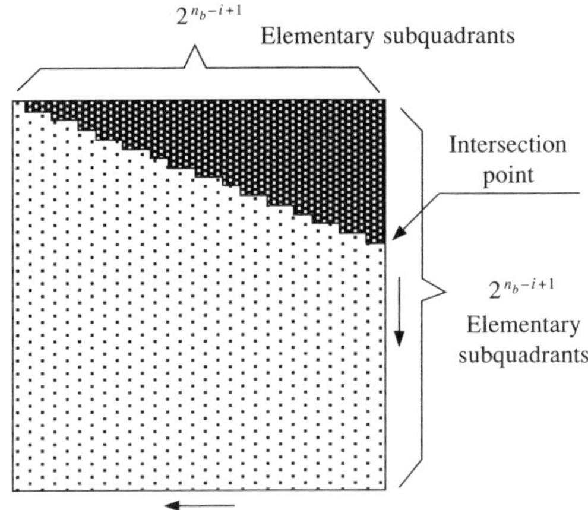


**Fig. 5.25** Example of order-$i$ subquadrant

The step pattern in each subquadrant of order '$i$' depends on the exact position of the straight line in the subquadrant. An elementary subquadrant is included in the active region of the two-dimensional input data space if more than half of its surface is situated on the active side of the hyperplane defining the neurone. In a two-dimensional case, the hyperplane is reduced to a straight line. The inclusion or the exclusion of each elementary quadrant can be determined by analysing the position of its centre. If the centre lies on the active side of the hyperplane then it has to be included in the active region of the input data space. Otherwise, it is excluded from the active region of the input data space. All the step patterns included in one of the $N_c$ classes contain a common elementary quadrant on one side of the current subquadrant. This elementary subquadrant is determined by the intersection point between the hyperplane and the side of the subquadrant, as shown in Fig. 5.26.

The exact position of the intersection point can vary by as much as $1/2^{nb+1}$ for one class of patterns, and it determines which other elementary subquadrants are included in the active region. This corresponds to two extreme hyperplane positions (a) and (b), presented in Fig. 5.26. When the hyperplane is in position (a) then only the elementary subquadrants marked by number '1' in Fig. 5.26 are included in the active region. If the hyperplane position changes continuously from position (a) to position (b) then new elementary quadrants are included in the active region in the order determined by the

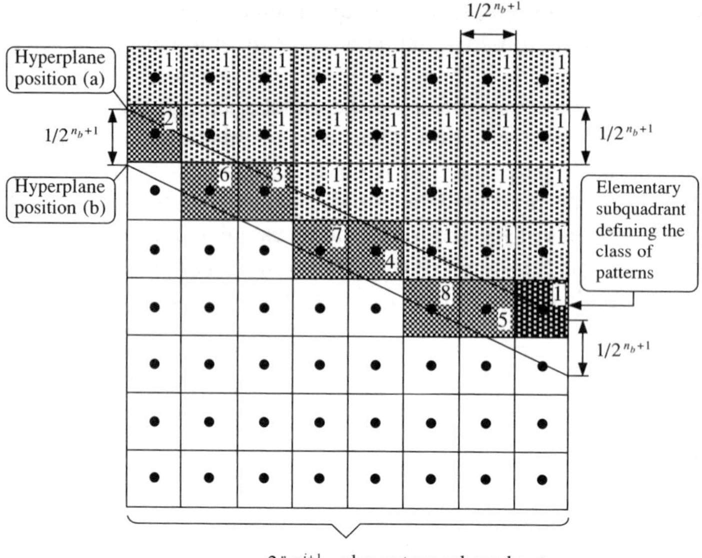


**Fig. 5.26** The step patterns included in one class of order '*i*' subquadrants

distance between the corresponding centres and the initial hyperplane (a). This order is indicated for the example in Fig. 5.26 by the numbers 2 through 7. The inclusion of each elementary quadrant generates a new pattern included in the current class. In the end, when the position (b) is attained, a maximal number of $2^{n_b-i+1}-1$ new elementary subquadrants have been added. This means that in a two-dimensional input data space the maximum number of patterns in a class is $N_p(i) = 2^{n_b-i+1}$.

In a three-dimensional situation, the hyperplane is immersed in a three-dimensional data space as initially presented in Fig. 5.24(B). The elementary subquadrants are in this case cubes grouped together into prisms with square bases included in the (*X*, *Y*) plane. Depending on its slope, the projection of the hyperplane can cover the entire (*X*, *Y*) square included in the current subquadrant or only part of it. If the entire (*X*, *Y*) square is covered then the two extreme positions of the boundary plane enclose a number of up to $(2^{n_b-i+1})^2-1$ centres of elementary quadrants and therefore the maximal number of patterns inside one class is $N_p(i) = 2^{2\cdot(n_b-i+1)}$. In an $n_a$-dimensional space, this can be generalised as $N_p(i) = 2^{(n_a-1)\cdot(n_b-i+1)}$. Therefore, the maximal number of different subquadrants, and implicitly the maximal number of subneurone types of the order '*i*', is given by (5.58).

$$N_{\text{sub-q}}(i) = N_c(i)\cdot N_p(i) = n_a\cdot 2^{n_a\cdot(n_b-i+1)} \tag{5.58}$$

The number of subneurones per layer is an increasing exponential as shown by equation (5.53), while the number of possible subneurone implementations is a decreasing exponential as demonstrated by (5.58). After the optimisation, the number of subneurones of each order is the minimum between the number of possible subneurones and the actual number of subneurones, as shown in (5.59). Therefore, due to the exponential variation of the two functions, the remaining neurones after the optimisation are very few as compared to the initial number of neurones. This situation is illustrated by the example in Fig. 5.27.

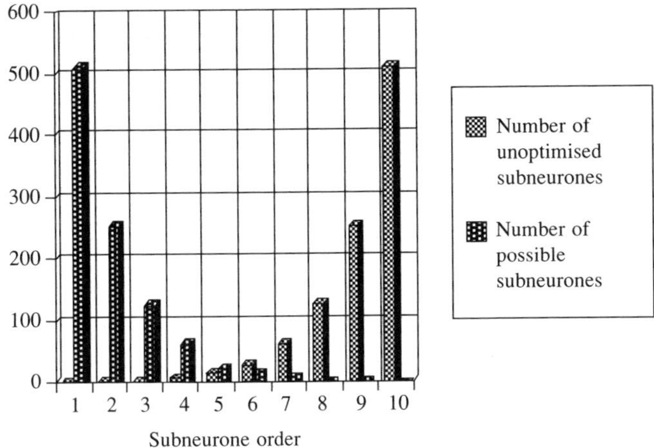


**Fig. 5.27** Graphical representation of the optimisation process ($n_a = 2$, $n_b = 10$)

$$N_{\text{subn-opt}}(i) = \min\{N_{\text{subn}}(i);\quad N_{\text{sub-q}}(i)\} = \min\{2^{(n_a-1)(i-1)+1};\quad n_a \cdot 2^{n_a(n_b-i+1)} \tag{5.59}$$

$$N_{\text{subn-opt}}(i) = \min\{N_{\text{subn}}(i); N_{\text{sub-q}}(i)\}=\min\{2^{(n_a-1)(i-1)+1}; 2^{n_a(n_b-i+1)+\log_2 n_a} \tag{5.60}$$

The intersection point between the two graphs illustrated by Fig. 5.27 is placed at the location corresponding to equal exponents in (5.60). This is given by the solution of the equation (5.61).

$$(n_a - 1) \cdot (i - 1) + 1 = n_a(n_b - i + 1) + \log_2 n_a \tag{5.61}$$

To compare the present algorithm with the algorithm presented in [37], the limit situation, with $n_a$ and $n_b$ having very large values, is analysed in (5.62).

$$\lim_{n_a;\, n_b \to \infty} \{i\} = \lim_{n_a;\, n_b \to \infty} \left\{ \frac{n_a \cdot n_b + 2n_a + \log_2 n_a - 2}{2n_a - 1} \right\} = \frac{n_b}{2} \tag{5.62}$$

Therefore, the number of optimised subneurones can be calculated as in equations (5.63), (5.64) and (5.65).

$$N_{\text{subn-opt}} < \sum_{i=1}^{n_b/2} 2 \cdot 2^{(n_a-1)\cdot(i-1)} + \sum_{i=n_b/2+1}^{n_b} n_a \cdot 2^{n_a(n_b-i+1)} \tag{5.63}$$

$$N_{\text{subn opt}} = 2 \cdot \frac{2^{(n_a-1)\cdot\frac{n_b}{2}} - 1}{2^{n_a-1} - 1} + n_a \cdot \left( \frac{2^{n_a \cdot \left(\frac{n_b}{2}+1\right)} - 1}{2^{n_a} - 1} - 1 \right) \tag{5.64}$$

$$N_{\text{subn-opt}} \approx 2^{(n_a-1)\cdot\frac{n_b}{2} - n_a + 2} + n_a \cdot 2^{n_a \frac{n_b}{2}} - n_a \tag{5.65}$$

The resulting total complexity is given by (5.66).

$$N_{\text{LG2-total}} = N_{\text{neur}} \cdot N_{\text{subn-opt}} \cdot (2^{n_a+1} - 3) \approx N_{\text{neur}} \cdot \left( 2^{\frac{n_a n_b}{2} - \frac{n_b}{2} + 3} + n_a \cdot 2^{n_a \frac{n_b}{2} + n_a + 1} \right) \tag{5.66}$$

Therefore, the resulting order of complexity is (5.67).

$$O\left( N_{\text{neur}} \cdot n_a \cdot 2^{\frac{n_a n_b}{2} + n_a} \right) \tag{5.67}$$

This result does not take into account the cross-neurone optimisations, which can bring further hardware reduction. Still, it shows that the increase of the network implementation with parameters $n_a$ and $n_b$ is much slower in the case of the new implementation method, as compared with the standard method presented in [37]. The result in (5.67) is slightly larger than the square root of the initial result (5.49). This implies a **substantial** gain in terms of hardware complexity reduction for large neural networks.

The method in [37] considers a particular case of Voronoi algorithm where all the hyperplanes are parallel to the axes in the input data space, in which case the generated order of complexity is reasonable. If a general Voronoi approach using oblique hyperplanes is necessary then method [37] generates much larger implementations than the one presented in this chapter. In conclusion, the present implementation method is the most adequate for ANN network implementation designed using Voronoi diagrams with oblique hyperplanes.

# 6

# Fuzzy logic fundamentals

Over the past few years, the use of fuzzy set theory, or fuzzy logic, in control systems has been gaining widespread popularity, especially in Japan. From as early as the mid-1970s, Japanese scientists have been instrumental in transforming the theory of fuzzy logic into a technological realisation. Today, fuzzy logic-based control systems, or simply *fuzzy logic controllers* (FLCs), can be found in a growing number of products, from washing machines to speedboats, from air condition units to hand-held autofocus cameras. In the present book, fuzzy logic is exemplified in the speed governing system of a synchronous generator set.

The success of fuzzy logic controllers is mainly due to their ability to cope with knowledge represented in a linguistic form instead of representation in the conventional mathematical framework. Control engineers have traditionally relied on mathematical models for their designs. However, the more complex a system, the less effective the mathematical model. This is the fundamental concept that provided the motivation for fuzzy logic and is formulated by Lofti Zadeh, the founder of fuzzy set theory, as the *Principle of Incompatibility*.

Zadeh stated that [240]:

*As the complexity of a system increases, our ability to make precise and yet significant statements about its behaviour diminishes until a threshold is reached beyond which precision and significance (or relevance) become almost mutually exclusive characteristics.*

Real-world problems can be extremely complex and complex systems are inherently fuzzy. The main advantage of fuzzy logic controllers is their ability to incorporate experience, intuition and heuristics into the system instead of relying on mathematical models. This makes them more effective in applications where existing models are ill-defined and not reliable enough.

## 6.1 Historical review

The term 'fuzzy' in fuzzy logic was first coined in 1965 by Professor Lofti Zadeh, then Chair of UC Berkeley's Electrical Engineering Department. He used the term to describe multivalued sets in the seminal paper, 'Fuzzy Sets' [239]. The work in his paper is derived from multivalued logic, a concept which emerged in the 1920s to deal with

Heisenberg's *Uncertainty Principle* in quantum mechanics. Multivalued logic was further developed by distinguished logicians such as Jan Lukasiewicz, Bertrand Russell and Max Black. At the time, multivalence was usually described by the term 'vagueness'. When Zadeh developed his theory, he introduced the term 'fuzzy'.

Zadeh applied Lukasiewicz's multivalued logic to set theory and created what he called *fuzzy sets* – sets whose elements belong to it in different degrees. According to the *fuzzy principle,* 'everything is a matter of degree'. While conventional logic is bivalence (TRUE or FALSE, 1 or 0), fuzzy logic is multivalence (*from* 0 to 1). It is a shift from conventional mathematics and number crunching to philosophy and language. At the beginning, fuzzy logic remained very much a theoretical concept with little practical applications. The work Zadeh was involved in consisted mainly of the computer simulation of mathematical ideas. In the 1970s, Professor Edrahim Mamdani of Queen Mary College, London, built the first fuzzy system, a steam engine controller, and later the first fuzzy traffic lights. This led to the extensive development of fuzzy control applications and products seen today.

## 6.2 Fuzzy sets and fuzzy logic

Classical set theory was founded by the German mathematician Georg Cantor (1845–1918). In the theory, a *universe of discourse, U*, is defined as a collection of objects all having the same characteristics. A classical set is then a collection of a number of those elements. The member elements of a classical set belong to the set 100 per cent. Other elements in the universe of discourse, which are non-member elements of the set, are not related to the set at all. A definitive boundary can be drawn for the set, as depicted in Fig. 6.1.

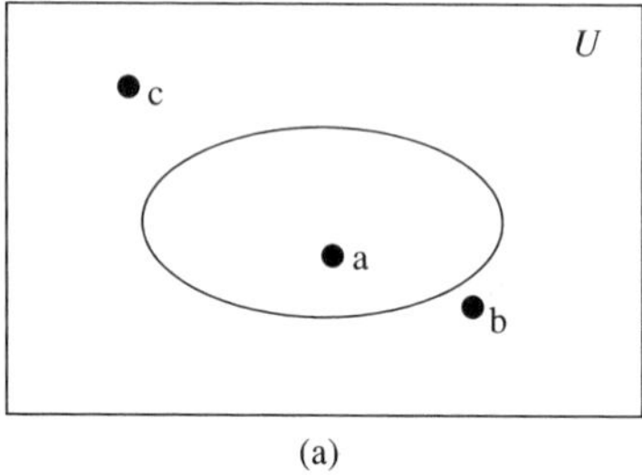


(a)

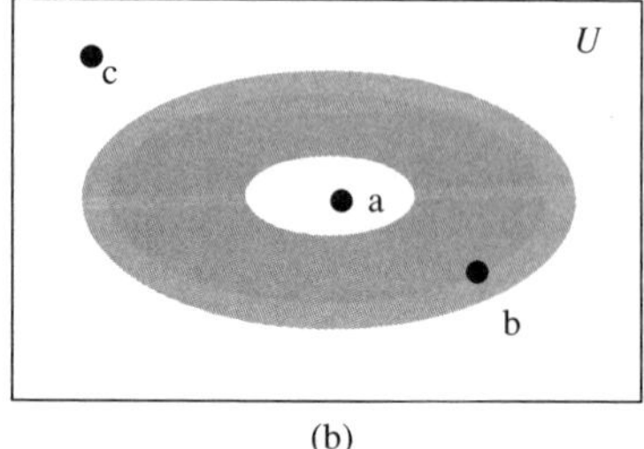


(b)

**Fig. 6.1** (a) Classical/crisp set boundary; (b) fuzzy set boundary

A *classical set* can be denoted by $A = \{x \in U \mid P(x)\}$ where the elements of $A$ have the property $P$, and $U$ is the universe of discourse. The *characteristic function* $\mu_A(x){:}U \rightarrow \{0, 1\}$ is defined as '0' if $x$ is not an element of $A$ and '1' if $x$ is an element of $A$. Here, $U$ contains only two elements, '1' and '0'. Therefore, an element $x$, in the universe of discourse is either a member of set $A$ or not a member of set $A$. There is ambiguity about membership. For example, consider the set ADULT, which contains elements classified by the variable AGE. It can be said that an element with AGE = '5' would not be a member of the set whereas an element with AGE = '45' would be. The question which arises is, where can a sharp and discrete line be drawn in order to separate members from non-members? At AGE = '18'? By doing so, it means that elements with AGE = '17.9' are not members of the set ADULT but those with AGE = '18.1' are. This system is obviously not realistic to model the definition of an adult human. Simple problems such as this one embody the notion behind Zadeh's Principle of Incompatibility.

In fuzzy set theory, the concept of characteristic function is extended into a more generalised form, known as *membership function*: $\mu_A(x):U \rightarrow [0, 1]$. While a characteristic function exists in a two-element set of $\{0, 1\}$, a membership function can take up any value between the unit interval [0, 1] (note that curly brackets are used to represent discrete membership while square brackets are used to represent continuous membership). The set which is defined by this extended membership function is called a *fuzzy set*. In contrast, a classical set which is defined by the two-element characteristic function, as described earlier, is called a *crisp set*. Fuzzy set theory essentially extends the concept of sets to encompass *vagueness*. Membership to a set is no longer a matter of 'true' or 'false', '1' or '0', but a matter of degree. The degree of membership becomes important. The boundary of a fuzzy set is shown in Fig. 6.1(b). While point a is a member of the fuzzy set and point c is not a member, the membership of point b is ambiguous as it falls on the boundary. The concept of *membership function* is used to define the extent to which a point on the boundary belongs to the set. A *fuzzy set F* can be defined by the set of tuples $F = \{(\mu_F(x), x) \mid x \in U\}$. Zadeh proposed a notation for describing fuzzy sets whereby '+' denotes enumeration and '/' denotes a tuple. Therefore, the fuzzy set $F$ becomes:

$$F = \int_U \mu_F(x)/x \quad \text{for a continuous universe } U$$

or $$F = \sum_{x \in U} \mu_F(x)/x \quad \text{for a discrete universe } U$$

Returning to the earlier example, an element with AGE = '18.1' may now be assigned with the membership degree to the set ADULT of, say, 1.0. An element of AGE = '17.9' may then have a membership degree of 0.8 instead of 0. Such gradual change in the degree of membership provides a better representation of the real world. However, the exact shape of the membership function is very subjective and depends on the designer and the context of the application. While set operations such as complement, union and intersection are straightforward definitions in classical set theory, their interpretation is more complicated in fuzzy set theory due to the graded attribute of membership functions. Zadeh [239] proposed the following *fuzzy set operation* definitions as an extension to the classical operations:

- Complement $\quad \forall x \in X: \mu_{A'}(x) = 1 - \mu_A(x)$
- Union $\quad \forall x \in X: \mu_{A \cup B}(x) = \max[\mu_A(x), \mu_B(y)]$
- Intersection $\quad \forall x \in X: \mu_{A \cap B}(x) = \min[\mu_A(x), \mu_B(y)]$

These definitions form the foundations of the basics of fuzzy logic thoery. The relationship between an element in the universe of discourse and a fuzzy set is defined by its membership function. The exact nature of the relation depends on the shape or the type of membership function used.

## 6.3 Types of membership functions

Figure 6.2 shows various types of membership functions which are commonly used in fuzzy set theory. The choice of shape depends on the individual application. In fuzzy

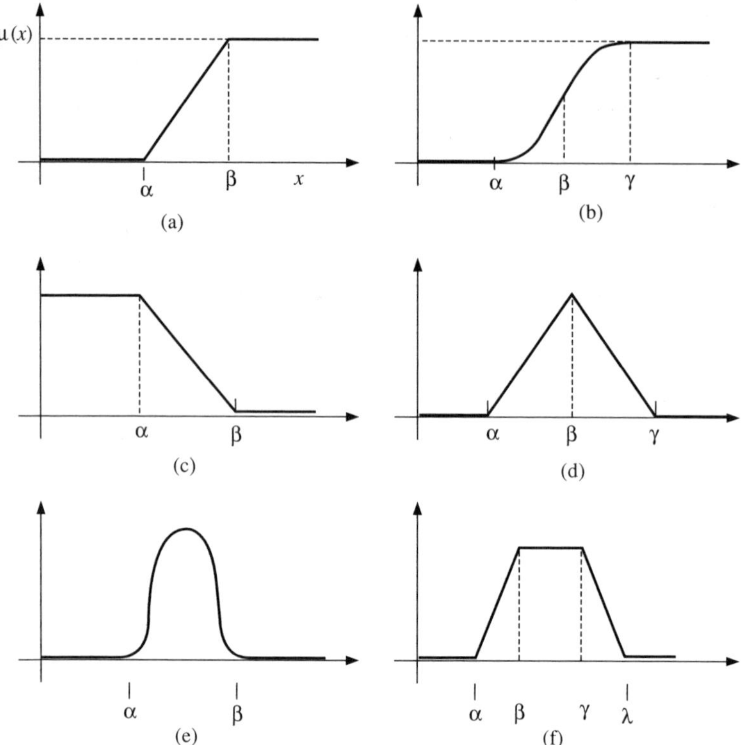


**Fig. 6.2** Types of membership functions: (a) Γ-function; (b) *S*-function; (c) *L*-function; (d) Λ-function; (e) Gaussian function; (f) Π-function

control applications, Gaussian or bell-shaped functions and *S*-functions are not normally used. Functions such as Γ-function, *L*-function and Λ-function are far more common.

The definitions of the membership functions chosen to be exemplified in this book are:

Γ-function, $\Gamma: U \rightarrow [0, 1]$

$$\Gamma(x; \alpha, \beta) = \begin{cases} 0 & x < \alpha \\ (x - \alpha)/(\beta - \alpha) & \alpha \leq x \leq \beta \\ 1 & x > \beta \end{cases}$$

*L*-function, $L: U \rightarrow [0, 1]$

$$L(x; \alpha, \beta) = \begin{cases} 1 & x < \alpha \\ (x - \beta)/(\alpha - \beta) & \alpha \leq x \leq \beta \\ 0 & x > \beta \end{cases}$$

$\Lambda$-function, $\Lambda{:}U \rightarrow [0, 1]$

$$\Lambda(x; \alpha, \beta, \gamma) = \begin{cases} 0 & x < \alpha \\ (x - \alpha)/(\beta - \alpha) & \alpha \leq x \leq \beta \\ (x - \gamma)/(\beta - \gamma) & \beta \leq x \leq \gamma \\ 0 & x > \gamma \end{cases}$$

## 6.4 Linguistic variables

The concept of a *linguistic variable*, a term which is later used to describe the inputs and outputs of the FLC, is the foundation of fuzzy logic control systems. A conventional variable is numerical and precise. It is not capable of supporting the vagueness in fuzzy set theory. By definition, a linguistic variable is made up of words, sentences or artificial language which are less precise than numbers. It provides the means of approximate characterisation of complex or ill-defined phenomena. For example, 'AGE' is a linguistic variable whose values may be the fuzzy sets 'YOUNG' and 'OLD'. A more common example in fuzzy control would be the linguistic variable 'ERROR', which may have *linguistic values* such as 'POSITIVE', 'ZERO' and 'NEGATIVE'. In this book, the following conventions are used to define linguistic variables. If $X_i$ is a linguistic variable defined over the universe of discourse $U$ where $x \in U$ then

| | |
|---|---|
| $LX_i^k$ | (for $k = 1, \ldots n$) are the linguistic values $X_i$ can take |
| $n$ | is the number of linguistic values $X_i$ have |
| $\mu_{LXi,k}(x)$ | is the $LX_i^k$ membership function for the value $x$ |
| $\underline{LX_i}$ | is the set containing $LX_i^k$, where $LX_i = \{LX_i^1, LX_i^2 \ldots Lx_i^n\}$. |

In the example above:

| | |
|---|---|
| $X_1$ | is 'ERROR' |
| $n = 3$ | is the number of linguistic values in $X_1$ |
| $LX_1^1$ | is 'POSITIVE' |
| $LX_1^2$ | is 'ZERO' |
| $LX_1^3$ | is 'NEGATIVE' |

and, for $x = \{-1, 0, 1\}$:

$$\mu_{LX1,1}(-1) = 0; \quad \mu_{LX1,1}(0) = 0; \quad \mu_{LX1,1}(1) = 1$$
$$\mu_{LX1,2}(-1) = 0; \quad \mu_{LX1,2}(0) = 1; \quad \mu_{LX1,2}(1) = 0$$
$$\mu_{LX1,3}(-1) = 1; \quad \mu_{LX1,3}(0) = 0; \quad \mu_{LX1,3}(1) = 0$$

## 6.5 Fuzzy logic operators

Logical connectives are also defined for fuzzy logic operations. They are closely related to Zadeh's definitions of fuzzy set operations. The following are four fuzzy operations which are significant for the second example presented in this book. $R$ denotes the relation between the fuzzy sets $A$ and $B$.

**Negation**

$$\mu_{\bar{A}}(x) = 1 - \mu_A(x)$$

**Disjunction**

$R$: $A$ OR $B$ $\qquad$ $\mu_R(x) = \max[\mu_A(x), \mu_B(x)]$

**Conjunction**

$R$: $A$ AND $B$ $\qquad$ $\mu_R(x) = \min[\mu_A(x), \mu_B(x)]$

**Implication**

$R$: $(x = A) \rightarrow (y = B)$ IF $x$ is $A$ THEN $y$ is $B$

Fuzzy implication is an important connective in fuzzy control systems because the control strategies are embodied by sets of IF-THEN rules. There are various different techniques in which fuzzy implication may be defined. These relationships are mostly derived from multivalued logic theory. The following are some of the common techniques of fuzzy implication found in literature.

**Zadeh's classical implication:**

$$\mu_R(x, y) = \max\{\min[\mu_A(x), \mu_B(y)], 1 - \mu_A(x)\}$$

**Mamdani's implication:**

$$\mu_R(x, y) = \min[\mu_A(x), \mu_B(y)]$$

Note that Mamdani's implication is equivalent to Zadeh's classical implication when $\mu_A(x) \geq 0.5$ and $\mu_B(y) \geq 0.5$.

**Godel's implication:**

$$\mu_R(x, y) = \begin{cases} 1 & \mu_A(x) \leq \mu_B(y) \\ \mu_B(y) & \text{otherwise} \end{cases}$$

**Lukasiewicz' implication:**

$$\mu_R(x, y) = \min\{1, [1 - \mu_A(x) + \mu_B(y)]\}$$

The differences in using the various implication techniques are described in [207]. It is fairly obvious by looking at the mathematical functions of the different implication techniques that Mamdani's technique is the most suitable for hardware implementation. It is also the most popular technique in control applications and is the technique used in the design example presented in the second part.

## 6.6 Fuzzy control systems

Figure 6.3 shows the block diagram of a typical fuzzy logic controller (FLC) and the system plant as described in [194]. There are five principal elements to a fuzzy logic controller:

- Fuzzification module (fuzzifier).
- Knowledge base.

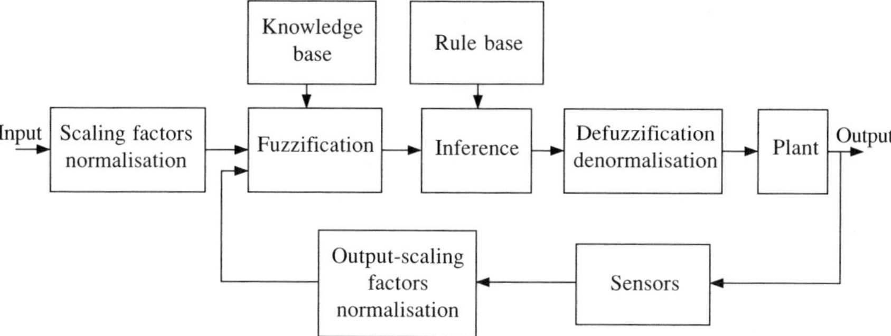


**Fig. 6.3** Block diagram of a typical fuzzy logic controller

- Rule base.
- Inference engine.
- Defuzzification module (defuzzifier).

Automatic changes in the design parameters of any of the five elements creates an adaptive fuzzy controller. Fuzzy control systems with fixed parameters are non-adaptive. Other non-fuzzy elements which are also part of the control system include the sensors, the analogue–digital converters, the digital–analogue converters and the normalisation circuits. There are usually two types of normalisation circuits: one maps the physical values of the control inputs onto a normalised universe of discourse and the other maps the normalised value of the control output variables back onto its physical domain.

### 6.6.1 Fuzzifier

The fuzzification module converts the crisp values of the control inputs into fuzzy values, so that they are compatible with the fuzzy set representation in the rule base. The choice of fuzzification strategy is dependent on the inference engine, i.e. whether it is composition based or individual-rule-firing based [88].

### 6.6.2 Knowledge base

The knowledge base consists of a database of the plant. It provides all the necessary definitions for the fuzzification process such as membership functions, fuzzy set representation of the input–output variables and the mapping functions between the physical and fuzzy domain.

### 6.6.3 Rule base

The rule base is essentially the control strategy of the system. It is usually obtained from expert knowledge or heuristics and expressed as a set of IF-THEN rules. The rules are based on the fuzzy inference concept and the antecedents and consequents are associated with linguistic variables. For example:

IF *error* ($e$) is *Positive Big* (PB) THEN *output* ($u$) is *Negative Big* (NB)

rule antecedent — rule consequent

*Error* ($e$) and *output* ($u$) are linguistic variables while *Positive Big* (PB) and *Negative Big* (NB) are the linguistic values. The rules are interpreted using a fuzzy implication technique. In fuzzy control theory, this is normally Mamdani's implication technique.

### 6.6.4 Defuzzifier

The diagram in Fig. 6.4 shows the membership functions related to a typical fuzzy controller's output variable defined over its universe of discourse. The FLC will process the input data and map the output to one or more of these linguistic values ($LU^1$ to $LU^5$). Depending on the conditions, the membership functions of the linguistic values may be clipped. Figure 6.5 shows an output condition with two significant (clipped above zero) output linguistic values. The union of the membership functions forms the fuzzy output value of the controller.

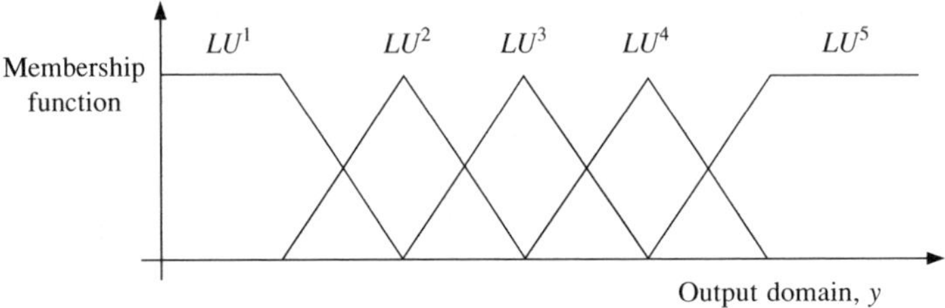


**Fig. 6.4** Membership function of the output linguistic values

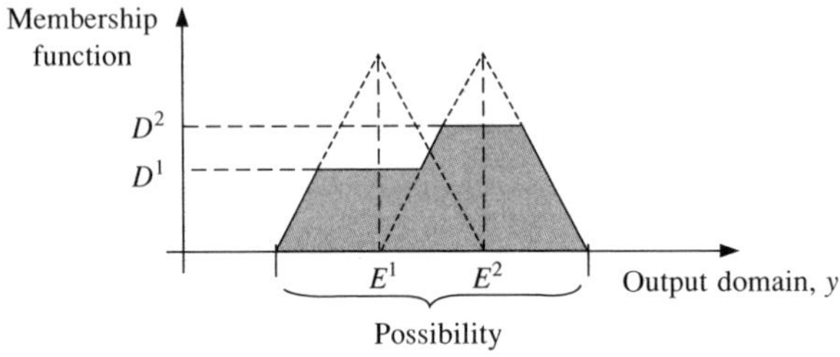


**Fig. 6.5** Possibility distribution of an output condition

This is represented by the shaded area in Fig. 6.5 and is expressed by the fuzzy set equation:

$$S = \bigcup_{i=1}^{k} S_i, \quad \mu_S(y) = \max_i [\mu_{S_i}(y)] \quad i = 1, 2, \ldots k$$

where:

$S$ is the union of all the output linguistic values
$S_i$ is an output linguistic value with clipped membership function
$k$ is the total number of output linguistic values defined in the universe of discourse.

In most cases, the fuzzy output value $S$ has very little practical use as most applications require non-fuzzy (crisp) control actions. Therefore, it is necessary to produce a crisp value to represent the possibility distribution of the output. The mathematical procedure of converting fuzzy values into crisp values is known as 'defuzzification'. A number of defuzzification methods have been suggested. The different methods produce similar but not always the same results for a given input condition. The choice of defuzzification methods usually depends on the application and the available processing power.

The defuzzification method used in the example presented in the second part is the weighted average method. This method requires relatively little processing power and is ideal for FPGA implementation where 'area space' is a major consideration. However, it is only valid for symmetrical membership functions. Each membership function is assigned with a weighting, which is the output point where the membership value is maximum. Based on the diagram in Fig. 6.5, the defuzzification process can be expressed by:

$$f(y) = \frac{\Sigma\, \mu(y) \cdot y}{\Sigma\, \mu(y)}$$

and using the weighted average method becomes:

$$f(y) = \frac{\Sigma_{m=1}^{n} E^{m} \cdot D^{m}}{\Sigma_{m=1}^{n} D^{m}}$$

where:

$f(y)$ is the crisp output value
$E^m$ is the crisp weighting for the linguistic value $LU^m$
$D^m$ is the membership value of $y$ with relation to the linguistic value $LU^m$.

The crisp defuzzifier output is used as it is or via an interfacing block, to control the plant.

## 6.7 Fuzzy logic in power and control applications

Over the past two decades, there has been a tremendous growth in the use of fuzzy logic controllers in power systems as well as power electronic applications. A recent series of tutorials in the *IEE Power Engineering Journal* [208], [209] which focused entirely on the applications of fuzzy logic in power systems is evidence of its growing significance in the field. Current applications in power systems include power system stability control, power system stability assessment, line fault detection and process optimisation for generation, transmission and distribution. Fuzzy logic is also used in motion control [45], [210], control of wind turbines [203], motor efficiency optimisation [211] and waveform estimation [202]. The advantages of using fuzzy logic in such applications include the following:

- Fuzzy logic controllers are not dependent on accurate mathematical models. This is particularly useful in power system applications where large systems are difficult to model. It is also relevant to smaller applications with significant non-linearities in the system.

- Fuzzy logic controllers are based on heuristics and therefore able to incorporate human intuition and experience.

There are numerous ways to build and implement a fuzzy logic system. It can either be based on a *fuzzy logic development shell* or built using software programming languages such as C++ or even Java. In the example presented in this book a different approach is made. The fuzzy system is developed using a hardware description language, VHDL.

# 7

# VHDL fundamentals

## 7.1 Introduction

Today electronic systems are so complex that traditional design methods are not able to keep pace with the increased demands of higher levels of integration coupled with a faster time-to-market cycle. As most of the large electronic systems typically involve years of research and development time prior to their actual production and with the exponential increase in complexity of integrated circuit design, it became clear that a standard design language that referenced a much higher level of design abstraction was required in order to facilitate the design process.

Therefore, modern design methodologies use a top-down approach in order to shorten the design cycle and to manage such an increased complexity. A key point of this approach is a Hardware Description Language (HDL) capable of offering support at all levels of description (behavioural, structural, physical). The standard represents a stand-alone specification that is not necessarily dependent on any specific tool developer or system subcontractor [15]. The verification of textual description correctness requires a test-bench that will ideally be able to exercise the modelled device under test at all levels of description.

The development of the VHDL (**V**ery High Speed Integrated Circuit **H**ardware **D**escription **L**anguage) began with a mandate set by the US Department of Defense (DoD) in the early 1980s as part of the Very High Speed Integrated Circuit (VHSIC) Program. This resulted in the adoption and release of the VHDL's IEEE Standard approved by the IEEE Standards Board in September 1993.

This chapter is a quick guide [61] to the main aspects of VHDL. The definition of the language is detailed in [10] and in-depth discussions can be found in all major VHDL text books [180], [189], [204], [15]. Many VHDL worked examples can be found in [61].

### 7.1.1 Top-down design methodology

The challenge for the developers of VHDL was to produce a hardware description language flexible enough to simulate conceptual designs but versatile enough to allow detailed timing simulation. The result is that VHDL is so robust that it can support a wide spectrum of modelling styles. For example, VHDL can represent a detailed description of the implementation aspects, or the most abstract ideas.

By supporting a wide array of modelling styles the language allows the designer to quickly develop and simulate ideas without being preocupied with the details of implementation. Simultaneously, as the design evolves to completion, the language is able to support a complex digital system. Top-down design is the term given to the design flow (Fig. 7.1) which begins with modelling an idea at an abstract level, and then proceeding through the iterative steps necessary to further refine this process into a detailed digital system.

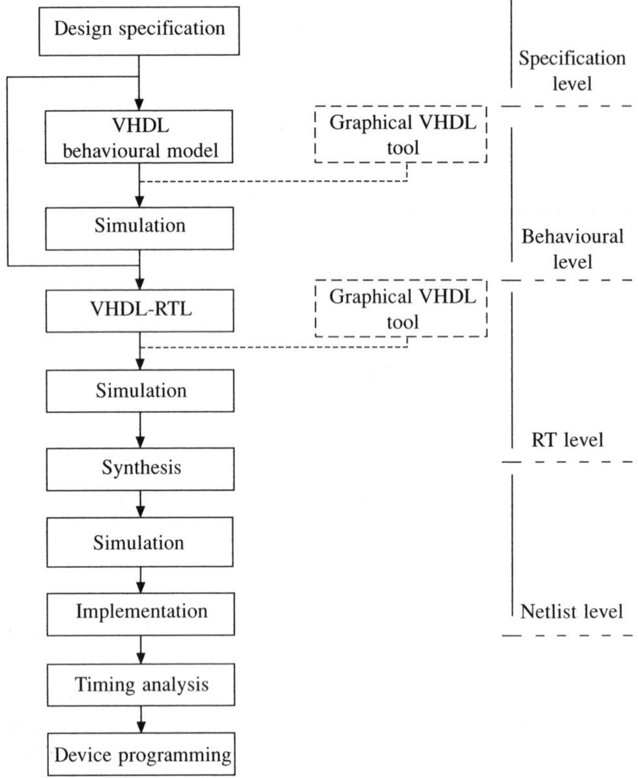


**Fig. 7.1** Top-down design flow

By producing a model at a very high level of abstraction, a test environment can be developed early in the design cycle. The development of a test environment allows verification that the concept is functionally correct within its intended scope. As the design evolves to each new level of detail, the test environment is used to test for ultimate compliance with the original specification.

VHDL's flexibility and choice of modelling styles enable a natural progression from idea to implementation and the top-down design process encourages the designer to develop and simulate ideas at their conception. A primary benefit of VHDL for the designer is the ability to quickly create, simulate and verify an abstract model. By simulating early in the process, design concepts can be tested long before investment is made in the implementation phase. A major feature of VHDL is its inherent ability to handle all levels of abstraction. This means that the designer requires the use of only a single language, as well as a single simulator for all phases of design.

### 7.1.2 Terminology

- **Behavioural model** – a simulation model that does not contain any implementation details and is described in abstract terms.
- **Structural model** – a netlist representation of a device consisting of instantiated components and their interconnections.
- **Register Transfer Level (RTL) model** or **dataflow model** – a model typically described with basic procedural constructs and functional operators and which may be targeted for automatic synthesis tools.
- **Mixed-level modelling** – a mixture of behavioural and structural modelling constructs.
- **Sequential** – describes procedural activity similar to most general-purpose programming languages, implying that ordering of statements is important to the meaning of the program and therefore statements are executed in the order in which they occur.
- **Concurrent** – describes statements operating in parallel with each other and initiated by signal value changes.
- **Driver** – signals within VHDL are controlled by drivers, through which the new signal values are propagated.
- **Event** – is a change in a signal value as a result of a new value being applied to that signal.
- **Transaction** – is said to have occurred on a signal whenever a value is assigned to the signal; the new value may or may not cause the signal value to change.
- **Sensitivity list** – a list of signals that activate the execution of concurrent elements when an event occurs on any one of them.
- **LRM** – the IEEE Standard 1076 Language Reference Manual.
- A modified **Bachus–Naur Format (BNF)** description for the VHDL statements discussed will be used in this book.
- **Square brackets '[]'** – items contained within the square brackets are optional items for that VHDL statement.
- **Squiggly brackets '{}'** – items contained within the squiggly brackets indicate items that can be included within the VHDL statement zero, one, or multiple times.

### 7.1.3 Behavioural and structural design

VHDL behavioural descriptions can be categorised as descriptions in which there is no reference to submodules within a specific VHDL architecture. This does not preclude the use of subprograms within VHDL descriptions but precludes the use of other VHDL components. Behavioural descriptions are generally procedural descriptions defining the design functionality. For the simple reason that one cannot necessarily relate this functional description to physical digital devices, this type of VHDL is classed as behavioural. The beauty of behavioural descriptions is that a designer can quickly model a digital system without being concerned with considering the details of the physical implementation.

Structural VHDL descriptions are categorised by the instantiation and interconnection of VHDL components. Structural VHDL descriptions can be viewed as VHDL netlists. Purely structural architectures are rare in practice because they do not exploit the abstraction capacity of VHDL language and the synthesis tool ability to deal with abstractions and efficiently generate hardware structures corresponding to complex mathematical models.

VHDL encourages a design partitioning approach, with detailed block development and structure implementation accomplished gradually.

VHDL supports behavioural as well as structural modelling. With behavioural modelling, few if any implementation details are revealed. The advantage to behavioural modelling is that one can quickly describe the functionality of a device or design using abstract concepts, such as signals of data types like integer. Structural modelling with component instantiation allows the designer to describe how the device will perform the function.

A designer should not assume that they must choose between a structural or behavioural modelling style since that assumption would place severe restrictions on the language. VHDL provides constructs that support a mix of both high level behavioural and structural modelling in all phases of design and throughout the entire top-down design process, allowing the designer to make a gradual transition from a behavioural model to a structural model. A behavioural (therefore abstract) VHDL description means a reduced number of program lines and shorter development time. Conversely, more structural descriptions may generate a lower number of hardware primitives because the performance of present synthesis tools is limited and they are not always capable of finding the optimal solution to certain applications. However, detailed knowledge of both the synthesis tool and the architecture of the target integrated circuit are required in order to write an optimal structural description of a complex entity. An inefficient structural description may often take more of the chip area than the solution automatically generated by the synthesis tool and based on a behavioural architecture. Therefore it is advisable for inexperienced engineers to keep the abstraction at a high level and rely on the synthesis tool capabilities.

## 7.2 VHDL design units

There are five design units in VHDL: entity, architecture, configuration, package and package body. Design units within a library (libraries are defined in the next section) are classified as either **primary** or **secondary** design units. Secondary design units are dependent upon the information contained within the primary design units. Therefore, secondary design units may only be analysed after the analysis of the primary unit. The primary design units within a design library are:

- the **entity** design unit which has as its secondary design unit the **architecture** (within a library an entity may have multiple architectures);
- the **package** design unit which has a **package body** as its secondary design unit (a package may have at most one package body within a design library);
- the **configuration** which has no secondary design units dependent upon it.

All primary design units (entity, package, and configuration) within a design library must have unique names, known as identifiers. They must be legal VHDL identifiers, which means that they must begin with an alphabetic character (a–z), all characters must be alphanumeric (a–z, 0–9) or underscores (_), and the identifier must not be a reserved word. Words like ENTITY, ARCHITECTURE, CONFIGURATION, PACKAGE, etc., are reserved by the language, and cannot be used outside the scope for which they were intended. Identifiers are not case sensitive and have no limitation to the number of characters used. Some examples of **illegal** identifiers for design units are:

**and** – reserved word,
**or** – reserved word,
**_xor_gate** – cannot begin with an underscore,
**5_entity** – cannot begin with a numeric character,
**this_is_pound£** – '£' is an illegal character in identifier.

The names of secondary design units must be unique for each primary design unit. For instance, for an entity called and2 there can be multiple architectures for that entity as long as they have unique names. There can be two different entities that have the same architecture name. It is legal to have an entity called and4 which has an architecture called behave that exists in the same library as an entity called or3 with an architecture called behave:

```
LIBRARY lib_1
ENTITY "and6"
   ARCHITECTURE "behave"
   ARCHITECTURE "stru"
ENTITY "or3"
   ARCHITECTURE "behave"
```

### 7.2.1 Entity

The entity design unit is the interface between the outside world and the design. The connection points (PORTs) to the design, the direction and type of information that flows through these connection points are specified in this unit. For example, a 2-input AND gate entity might look like:

```
ENTITY and2 IS
   PORT (in1,in2: IN bit;
      output: OUT bit);
END [ENTITY] [and2];
--comments in VHDLare written behind two dashes --
```

This design unit describes a device named and2. This device has three connection points, two inputs and one output. The type of information that flows through these connection points is a data type called bit that can take on the value '0' or '1'. The entity's name within an END statement is optional. Additional information can be placed to specify other parameters of the device such as speed or width. These are called generics. The example can be extended to include some delay parameters on the device that describe the propagation delay used in the behavioural description:

```
ENTITY and2 IS
      GENERIC(tplh: time := 3 ns;
            tphl: time := 5 ns);
      PORT (in1,in2: IN bit;
```

```
        output: OUT bit);
END and2;
```

Two generics (tplh and tphl) of type time have been placed. The default value of the generics (3 ns and 5 ns) are reprogrammable on an instance by instance basis. Generics are a very powerful and important feature of the language. The syntax of the entity statement requires the keywords ENTITY and IS to surround the user defined entity_name. Between this header and the keyword END is the entity's declarative region where a GENERIC statement and a PORT statement can be placed. An example of a more complicated entity, containing buses on the input and output ports, is:

```
ENTITY complex_with_buses IS
  GENERIC (width: integer := 8;
          tplh, tphl: time := 5 ns);
  PORT(input: IN bit_vector (width-1 DOWNTO 0);
          output: OUT bit_vector(0 TO width-1);
          clk: IN bit);
END complex_with_buses;
```

### 7.2.2 Architecture

The architecture defines the behaviour of an entity from a simulation point of view. It depends upon the information declared within an entity and has access to that information (ports and generics) within its body. An example of an architecture for the 2-input AND gate might look like:

```
ARCHITECTURE behave OF and2 IS
BEGIN
  output <= in1 and in2;
END behave;
```

Between the architecture's header and the BEGIN statement is the architecture's declarative region. In this region additional items can be declared which are available to the architecture. Possible declarations within an architecture include types, signals, subprograms and components. In the above example, the declarative region is empty. Between the BEGIN statement and the END statement is the architecture's statement region. Any concurrent statement may be placed within the architecture's statement region. In the example above, a signal assignment statement is being used as an illustration of a concurrent statement. This assignment statement uses the signal assignment operator "<=" to assign the logical AND of ports in1 and in2 to the port output.

Another example of an architecture can be a structural description of a 3-input AND gate. This example contains a COMPONENT and SIGNAL declaration; it uses the signal internal and the ports of the and3 entity (input1, input2, input3, output) to connect to the ports of the two instantiated components.

```
ARCHITECTURE struct OF and3 IS
  COMPONENT and2
      PORT(inl,in2: IN bit;
           out1: OUT bit);
```

```
    END COMPONENT;
    SIGNAL internal: bit;
BEGIN
    u1:and2 PORT MAP(in1 => input1, in2 => input2, out1 => internal);
    u2:and2 PORT MAP(in1 => input3, in2 => internal, out1 => output);
END;
```

### 7.2.3 Configuration

A configuration design unit is used to bind an architecture to an entity, so it defines which functionality (architecture) a particular entity will have. As there can be multiple different architectures for a particular entity, there can also be multiple different configurations for that entity. An example of a configuration written for the following entity and architecture might look like:

```
ENTITY and2 IS
    GENERIC(delay: time := 3 ns);
    PORT(in1, in2: IN bit; output: OUT bit);
END and2;
ARCHITECTURE behav OF and2 IS
BEGIN
    output <= in1 and in2 AFTER delay;
END behav;
CONFIGURATION config1 OF and2 IS
    FOR behav
    END FOR;
END config1;
```

An example of a configuration called conf which binds an entity called top_level to an architecture called struct is shown below. Inside the architecture structure are two instantiated components, U1 and U2, both of which are bound by individual FOR constructs which specify configurations for each of the subordinate components.

```
CONFIGURATION conf OF top_level IS
    FOR struct
            FOR u1:nand_gate USE CONFIGURATION WORK.config1;
            END FOR;
            FOR u2:or_gate USE CONFIGURATION WORK.config2;
            END FOR;
    END FOR;
END conf;
```

A configuration can become a detailed parts list for a design by reaching down through the hierarchy and binding an entity/architecture pair for every instantiated component in the design. A configuration that would bind the structural architecture to its and_or entity and bind entity/architectures to the instantiated components might look like:

```
LIBRARY design_1;
CONFIGURATION con_1 of and_or IS
```

```
    FOR structure
      FOR u1: and2 USE ENTITY WORK.and2(behave);
      END FOR;
      FOR u2: or2 USE CONFIGURATION design_1.or2_con;
      END FOR;
      END FOR;
  END con_1;
```

Each of the FOR statements above is a configuration specification. There are two basic types of configuration specifications and both are being used above. The first type uses an ENTITY configuration specification that indicates an instantiated component whose label is u1 and whose component declaration is and2. An entity and2 is bound to this component, found in the library whose symbolic name is WORK, and the architecture called behave is used for that and2 entity. The second type of configuration specification uses the CONFIGURATION keyword, and indicates an instantiated component whose label is u2 and whose component declaration is or2. A configuration called or2_con that can be found in the library with symbolic name design_1 is to be bound to this component.

Configuration specifications are not limited to configurations but can be placed directly into the architecture:

```
LIBRARY design_1;
ARCHITECTURE struct OF and_or IS
  COMPONENT and2
    PORT(and1,and2 : IN bit;
          and_out: OUT bit);
  END COMPONENT;
  --Configuration Specification
  FOR u1: and2 USE ENTITY WORK.and2(behave);
  COMPONENT or2
    PORT(or1,or2: IN bit;
          or_out: OUT bit);
  END COMPONENT;
  --Configuration Specification
  FOR u2: or2 USE CONFIGURATION design_1.or2_con;
  SIGNAL intern_1: bit;
BEGIN
  u1: and2 PORT MAP(in1, in2, intern_1);
  u2: or2  PORT MAP(in3, intern_1, out_1);
END structure;
```

In this last example the configuration specifications are shown within the declarative region of the architecture. This will produce the same results as shown in the first example. It seems simpler but in order to change this configuration specification the architecture has to be edited and recompiled, while in the first example only the configuration had to be edited and recompiled. So, by placing configuration specifications within a configuration, many configurations with different configuration specifications can be defined for the same entity/architecture pair. This allows the configuration of a design without editing it.

### 7.2.4 Package

Useful data type declarations and subprogram definitions are often placed into a package so that they can be accessed by a large number of entities, architectures or other packages. This allows a uniform coding style and produces easy to read and understandable code that, consequently, is easier to maintain and debug. Many useful VHDL constructs can be placed inside a package, such as: subprograms, types and/or subtypes, constants, signals (global), files, aliases, components, attributes. A simple example of a package is:

```
PACKAGE my_pack IS
   TYPE data_type is ('X','0','1','Z');
   FUNCTION data_to_bit (input: data_type) RETURN bit;
END my_pack;
```

### 7.2.5 Package body

The package body design unit contains the functionality of subprograms or the value of constants declared within a package. Consequently, the package body is dependent on the package declaration. Designers can use a package body to 'hide' the functionality of subprograms or the value of a constant from their user. An example of a package body written for the package my_pack might look like:

```
PACKAGE my_pack IS
   CONSTANT temp1: real;
   PROCEDURE do_nothing (input:IN bit);
END my_pack;
PACKAGE BODY my_pack IS
   CONSTANT temp1: real := 32.0;
   PROCEDURE do_nothing(input:IN bit) IS
   BEGIN
      RETURN;
   END do_nothing;
END my_pack;
```

## 7.3 Libraries, visibility and state system in VHDL

A design file is a VHDL text file that can be edited by the designer and will be the input to a VHDL analyser. The only requirement placed on a VHDL design file is that a design unit be completely contained within one file. It is therefore legal to create files that contain 1, 2, 3, ..., 100+ design units within them.

### 7.3.1 Libraries

The concept of a design library is fundamental to the understanding of design unit relationships, since a design unit resides within a design library. It is placed there by the VHDL analyse process. Therefore, a design unit within a design library is *always* syntactically correct. A VHDL design library is associated with a symbolic name that can be used by VHDL design units to access information contained within a library. This information may include the contents of packages, entities, architectures or configurations. Normally, a design library is implemented as a directory structure in the file system. Once a library (directory) has been established and associated with a symbolic name, design units may be placed into that library. This is accomplished through analysing (compiling) VHDL code. The symbolic name associated with a library can be thought of as an absolute address to that library structure.

There are two symbolic library names that are always available to designers. The first is a library designated by the identifier STD. The second is a library designated by the identifier WORK. The STD library is very special as it contains two packages that are basic and fundamental to any VHDL design. The first package is called STANDARD and is specified by the IEEE 1076 LRM. It contains the basic VHDL types such as integer, real, or Boolean. The second package is called TEXTIO and is also detailed by the IEEE 1076 standard. The TEXTIO package contains useful subprograms that enable the user to perform ASCII file manipulation.

When VHDL source files are analysed, the results are placed into the library associated with the symbolic name WORK. The WORK library may point to different libraries as each design library becomes the target of an analyse process. At any point in time, the design library to which one is compiling can be referenced in two ways: with its absolute address (the library's symbolic name) or with the relative address WORK.

### 7.3.2 Visibility

Visibility in VHDL refers to those declared constructs that are available to the current design unit. Formally, any information visible to an entity is visible within all the architectures for that entity. Also, any information visible to a package is visible to its package body. It is as though a secondary design unit adopts any information available to its parent, the associated primary design unit. To make information contained within a package available to another design unit, an entity, an architecture, or another package that package must be made 'visible' to that design unit. To do this, the LIBRARY and the USE statements must be used. A library can be referenced by the identifier, its symbolic name. The name of the library must be made visible. This is done by using the LIBRARY statement.

**Example:**

```
LIBRARY IEEE;
ENTITY test IS
END test;
```

This would make the symbolic name IEEE available for reference within the entity called test as well as any architecture written for this entity. VHDL implicitly provides two library statements before every design unit making the library STD and the library WORK available for all; they do not have to be stated explicitly. These are:

```
LIBRARY STD;
LIBRARY WORK;
```

In order to make a particular package within a library visible to a design unit, this package must be specified with a USE statement. The USE statement takes as arguments a symbolic name of a library followed by a period '.', the package name followed by a period '.', and a reference to a member element (type, constant, signal, function, procedure) in the package. The statement is ended with a semicolon ';'. A package called std_logic_1164 is considered as an example, which is contained in the library whose symbolic name is ieee and this package contains a function called 'and'. If this function is to be made visible to an entity called 'test' to be analysed into the 'ieee' library, then:

```
LIBRARY ieee;
USE ieee.std_logic_1164.and;
ENTITY test IS
END test;
```

If the designer wants to make all declarations within a package visible to a design unit, then 'ALL' is used:

```
LIBRARY ieee;
USE ieee.std_logic_1164.ALL;
ENTITY test IS
END test;
```

This code sequence makes everything in the package std_logic_1164 located in the library ieee visible to the entity test and all of its architectures. The library that is currently the target of the analysis can be referenced by the symbolic name WORK. VHDL implicitly gives all design units access to the library WORK. If the working library happened to be pointing to the same directory as the ieee library, then the above example could be written as:

```
USE WORK.std_logic_1164.ALL
   ENTITY test IS
   END test;
```

A USE statement is implicitly provided to make the STANDARD package available to all design units:

```
USE STD.STANDARD.ALL;
```

### 7.3.3 State system

Traditional simulators have a fixed palette of primitives and also have a fixed state/strength system, applicable to discrete digital system simulations. Therefore, like the limitations of fixed primitives, the designer has to be immediately concerned with implementation aspects when conceiving new ideas. Traditional simulators enforced the limitations of a fixed palette of primitives and a fixed state system by integrating a schematic capture system. This forced the engineer to think in terms of gates, flip-flops, multiplexers and registers instead of allowing concentration on the functionality of the design. With VHDL, the user defines the simulator's state system. One can duplicate a traditional simulator's fixed state system or define highly abstract states. For example:

```
TYPE week_day IS (Monday, Tuesday, Wednesday, Thursday, Friday, Saturday);
```

Within the context of VHDL, a TYPE is a set of values. This set of values encompasses all possible values held by a signal, variable, or constant of an assigned TYPE. Collectively, signals, variables and constants are called objects. When an object is declared, the set of values (type) that the object can hold is established. The simulator will not allow a value to be assigned to an object that is not within the set of the declared type. For example:

```
SIGNAL clock : std_ulogic;
VARIABLE crayon : colour := black;
CONSTANT clear : std_ulogic := '1';
```

In each case an object is declared and associated with a type. In the case of the variable, an initial value is declared. With the constant declaration, the constant value is specified. Types are usually declared within a package. This allows the type to be available to all design units which refer to that package. A typical type definition within a package might look like:

```
PACKAGE std_logic_1164 IS
   TYPE std_ulogic IS ('U','0','1','Z','X','W','L','H','-');
END std_logic_1164;
LIBRARY ieee;
USE ieee.std_logic_1164.ALL;
ENTITY mux IS
   GENERIC(tplh: time := 3 ns);
   PORT(in1,in2,sel: IN std_ulogic; output: OUT std_ulogic);
END mux;
```

Type std_ulogic is declared within the package std_logic_1164. Any design unit that references this package will be able to use std_ulogic. This is shown with the entity mux above. By making visible the package std_logic_1164, any object can be declared using the type std_ulogic. Types can also be defined within any declarative region of a design unit. For example, declaring the same type, std_ulogic, within an entity would look like:

```
ENTITY mux IS
   TYPE std_ulogic IS ('U','0','1','Z','X','W','L','H','-');
   GENERIC(tplh: time := 3 ns);
   PORT(in1,in2,sel: IN std_ulogic; output: OUT std_ulogic);
END mux;
```

In both the prior examples, the entity mux is identical in definition. The only difference is that the type std_ulogic, in the first example, is globally available while the type std_ulogic defined in the second example is only available within the entity mux and any associated architecture. Objects are said to be in one of three possible classes: signals, variables, or constants. Ports fall into the signals class while generics fall into the constants class.

### 7.3.4 VHDL simulation

A VHDL simulation is an event-driven simulation, which detects and reacts to changes on specific channels. In VHDL these channels are signals and the subsequent reaction to these changes is described by the concurrent VHDL statements that are sensitive to these signals. In a VHDL simulation cycle, signals are updated and events detected.

VHDL statements sensitive to these events are queued and then executed. These statements can also cause additional events without simulation time advancing. Therefore, a VHDL simulation cycle executes all statements on the present queue, updates all signals, detects new events and builds new execution lists. Since it is possible to have repeated executions of a VHDL simulation cycle without advancing time, a mechanism is needed for describing the forward motion of a VHDL simulation without regard to time. All signal assignments occur at least one delta in the future. This is called *delta ordered* simulation. Example:

```
ARCHITECTURE behave OF rom IS
  SIGNAL addr: integer;
BEGIN
  PROCESS (address)
  VARIABLE memory: mem_type;
  BEGIN
    addr <= vect_to_int(address);
    data <= memory(addr);
  END PROCESS
END behave;
```

In the example above a simple ROM is being modelled. The intent is to read in an 8-bit binary address (address), convert the address from binary to integer (addr), and drive the output signal (data) with the appropriate value selected by indexing memory (memory(addr)). The algorithm will not work since it depends on the signal addr to update its value immediately so that the memory can be indexed with the new value. Since addr will not update until the next delta point, the memory will be indexed with the wrong value. A correct implementation would be to declare addr as a variable. Variable values are assigned immediately. For example:

```
ARCHITECTURE behave OF rom IS
BEGIN
  PROCESS (address)
  VARIABLE memory: mem_type;
  VARIABLE addr: integer;
  BEGIN
    addr := vect_to_int(address);
    data <= memory(addr);
  END PROCESS
END behave;
```

## 7.4 Sequential statements

The process statement is the basic building block for behavioural modelling. A process statement is a concurrent shell in which sequential statements can be executed. All statements within a process are executed sequentially, when the process becomes active. The *process* as a concurrent statement is presented in 7.5.1.

The process statement can contain one and only one sensitivity list. A process with a sensitivity list can only be triggered by an event on a signal in this list. Once triggered, the process will execute all statements contained within the sequential statement section, and then suspend until another event is detected on one of those signals. If multiple signals are included in the sensitivity list, any one of those signals can trigger the process. Therefore, the use of a sensitivity list in a process is fairly limited. The WAIT sequential statement provides the user with more options than the process sensitivity list.

### 7.4.1 Wait

The first advantage to a WAIT statement is that it can be placed anywhere within the process body. With the process sensitivity list the process effectively suspends at the end of the process. With the WAIT statement, suspension occurs wherever a WAIT statement is encountered. The second advantage is that there is no limitation to the number of WAIT statements within a process. The third is that WAIT statements are more flexible. WAIT statements cannot be used in conjunction with a process sensitivity list. The process sensitivity list is equivalent to:

```
WAIT on signal1, signal2, ...
```

placed as the last sequential statement in a process. This WAIT statement simply means that if any of the listed signals has an event resume the process. There are four variations of WAIT statements and combinations of them. They are:

```
WAIT ON signal_list;
WAIT UNTIL condition;
WAIT FOR time;
WAIT;
```

**Examples:**

```
WAIT ON clock, clear, preset, d;
```

This statement will suspend a process and will resume if an event occurs on signals clock, clear, preset, or d.

```
WAIT UNTIL (clock = '1');
```

This statement will suspend a process and will resume when an event occurs on clock and the Boolean expression clock = '1' is true. All conditions following UNTIL must first have an event and the Boolean expression must be true.

```
WAIT FOR 10 ns;
```

Wait for simulation time to advance 10 ns, then resume the process. An effective way to model 'time-out'.

```
WAIT;
```

Suspend forever. There are number of instances where processes have been developed just to execute once.

```
WAIT ON clock UNTIL (clear = '0') FOR 10 ns;
```

This WAIT statement is an amalgamation of three types of WAIT statements. If a process does not have a sensitivity list and does not have a WAIT statement contained within it, the process will loop forever. During the initialisation phase of simulation all concurrent statements are executed once. The process statement is executed until it suspends during initialisation. If the process statement does not have a sensitivity list or at least one WAIT statement it will never suspend. Most VHDL compilers will check to ensure that a sensitivity list or WAIT statements are included in every process.

### 7.4.2 If-then-else

If-then-else is a basic mechanism for building decision structures. The VHDL syntax is:

```
IF condition THEN
    {sequence_of_statements}
[{ELSIF condition THEN}
    {sequence_of_statements }]
[ELSE
    {sequence_of_statements }]
  END IF;
```

A minimum if-then-else statement is:

```
IF condition THEN
  sequence_of_statements
END IF;
```

The execution of the if-then-else statement involves the sequential evaluation of the IF clause followed by any ELSIF clauses. The first clause to evaluate true will result in the associated sequence_of_statements to be executed. Once these sequence_of_statements have been executed, execution will end. If an ELSE clause is included and the IF and all ELSIF clauses have evaluated false, the ELSE will always evaluate true and execute. Note that the use of ELSIF is correct and that END IF is two words.

**Examples:**

```
IF (clock = '1') THEN
  q <= d AFTER 10 ns;
END IF;
```

This example is a simple representation of a D flip-flop having input and output, d and q respectively. If clock is a '1' then assign to the signal q the current value on signal d after 10 nanoseconds from the current simulation time. The next example shows the description of a clocked 4 to 1 mux:

```
PROCESS (clk)
  VARIABLE temp : bit;
  BEGIN
  IF clk = '1' THEN
    IF sel = "00" THEN
      temp := inputs(0);
    ELSIF sel = "01" THEN
      temp := inputs(1);
    ELSIF sel = "10" THEN
      temp := inputs(2);
```

```
            ELSE
               temp := inputs(3);
            END IF;
            output <= temp AFTER 5 ns;
         END IF;
      END PROCESS;
```

**Notes:**

- The syntax for the variable assignment operator ':=' is different from the signal assignment operator '<=.'
- Variables are assigned immediately, as opposed to signals, which are assigned at the next delta, if no delay is specified. Changes in values of variables do not cause simulation events.
- Variables are local to the process; their values are retained from one execution of a process to the next. Variables are used for local storage and signals are used as a vehicle to pass information among concurrent statements.

### 7.4.3 Case

The case statement can be thought of as a switch. An expression is evaluated and a sequence_of_statements associated with the matching choice is executed. If the expression evaluates to a choice, then the associated statements will be executed. Exit will occur when all statements associated with the first matching choice are executed. Syntax:

```
CASE expression IS
WHEN choice => sequence_of_statements
[WHEN choice => sequence_of_statements]
END CASE;
```

**Example:**

```
CASE vect IS
   WHEN "00" => int := 0;
   WHEN "01" => int := 1;
   WHEN "10" => int := 2;
   WHEN "11" => int := 3;
END CASE;
```

In this example vect is a two element bit_vector. By evaluating vect the matching WHEN choice will cause the variable int to be assigned the matching integer value. The clocked 4 to 1 mux example modelled before by using if-then-else statements can be rewritten using nested case statements.

```
PROCESS (clk)
   VARIABLE temp : bit;
BEGIN
   CASE clk IS
   WHEN '1' => CASE sel IS
      WHEN "00" => temp := inputs(0);
```

```
      WHEN "01" => temp := inputs(1);
      WHEN "10" => temp := inputs(2);
      WHEN "11" => temp := inputs(3);
      END CASE;
    output <= temp AFTER 5 ns;
    WHEN OTHERS => NULL;
    END CASE;
  END PROCESS;
```

### 7.4.4 Loop

Loop statements provide a mechanism to repeatedly execute a sequence of statements. The syntax is:

```
[label: ]   [iteration_scheme] LOOP
   sequence_of_statements
END LOOP [label];
```

Iteration_schemes can be:

```
WHILE condition
FOR loop_iteration_range
```

An iteration_scheme can be one of two types, WHILE or FOR. If the WHILE construct is used, the condition is evaluated and, if true, the sequence_of_statements within the loop statement will be executed. If the FOR construct is used, the sequence_of_statements within the loop will be repeatedly executed for the range specified.

Two statements can be used within the loop statement that will allow either the termination of a loop iteration or the complete termination of the loop statement. These statements are the NEXT and EXIT statements.

**Examples:**

```
FOR i IN 0 TO 3 LOOP
   IF (bus(i) = '1') THEN
      value := value + 2**i;
   END IF;
END LOOP;
```

This loop statement will examine each individual bit of the bit_vector bus. If the individual bit is a '1' then value will be increased by the exponential 2**i. After the fourth pass, the loop range will be exceeded and the loop will terminate. A feature of VHDL, unlike other programming languages, is that the range variable i was not declared. Any range variables used within the FOR construct do not have to be declared. The same range identifier can be used repeatedly from one loop statement to the next.

```
FOR i IN 3 DOWNTO 0 LOOP
   IF (bus(i) = '1') THEN
      value := value + 2**i;
   END IF;
END LOOP;
```

The above example shows that the range can be ascending or descending just by using the keywords TO or DOWNTO.

```
FUNCTION vect_to_int (bus: bit_vector(0 to 3))
RETURN INTEGER IS
VARIABLE value: integer := 0;
VARIABLE i: integer := 0;
BEGIN
    WHILE (i < 4) LOOP
    IF (bus(i) = '1') THEN
      value := value + 2**i;
    END IF;
    END LOOP;
  RETURN value;
END vector_to_integer;
```

This is the same example using different constructs. For the WHILE construct (unlike the FOR construct) the range identifier i must be declared.

### 7.4.5 Assert

The ASSERT statement is a mechanism that allows direct reporting to a VHDL simulator's standard output, usually displayed in the simulation window or console; some simulators allow the output to be redirected to a file. The ASSERT statement evaluates a Boolean expression and, **if false**, will report a string to the standard output. ASSERT messages are useful for reporting simulation progress, setup and hold violations, out of bounds addresses, overflow and underflow, etc. The syntax is:

```
ASSERT condition
[REPORT string]
[SEVERITY level]
```

The ASSERT condition is a Boolean expression. If the condition evaluates to false the assert statement will execute. According to the LRM, the error message reported by a VHDL simulator must contain a minimum of:

- ASSERT message identification.
- The SEVERITY level – defaults to ERROR if not specified.
- The REPORT string.
- The name of the design unit containing the assert statement.

The optional REPORT line allows for a user-defined message to be included with the message. This is unique in that no string formatting functions are required.

The SEVERITY level is a type defined in the package STD. The severity level can take on one of four values:

```
TYPE SEVERITY_LEVEL IS (NOTE, WARNING, ERROR, FAILURE);
```

**Examples:**

```
PROCESS (clk)
```

```
BEGIN
   IF clk = 'X' THEN
      ASSERT false REPORT "Clock is unknown"
      SEVERITY error;
   END IF;
END PROCESS;
```

In this example the assert statement will always execute if encountered. The Boolean condition is declared false. The if statement, clk = 'X', determines whether the assert statement should be executed.

```
PROCESS (clk)
BEGIN
   ASSERT (address < 1024)
   REPORT "Address out of range!" SEVERITY warning;
END PROCESS;
```

In this example, if address is less than 1024 the assert message will evaluate to true and not execute. When address is 1024 or greater the assert statement will execute.

```
ASSERT false  REPORT "All Done With Stimulus Application!"
              SEVERITY note;
```

This assert always evaluates false and will report to the simulator that the application of stimulus is complete.

## 7.5 Concurrent statements

### 7.5.1 Process

The process statement is a concurent statement and constitutes the basic building block for behavioural modelling. A process contains sequential statements which are executed sequentially when the process becomes active. The syntax for a process statement is:

```
[label] : PROCESS [(sensitivity_list)]
   process_declarative_part
BEGIN
   process_statement_part
END PROCESS [label];
```

The optional label allows for a user-defined name for the process. If the label is included with the END PROCESS clause, it must match the PROCESS label. A process statement may contain an optional sensitivity list. A sensitivity list contains signals that trigger the process statement. The process statement begins execution when any of the sensitivity list's signals contain an event. Once activated via a sensitivity list event, the process statement executes statements in a sequential manner. Upon reaching the end of the process, execution suspends until the next sensitivity list signal event. Without a sensitivity list a process will suspend only when a WAIT sequential statement is encountered. Example: a process sensitive to the signal clock. When an event occurs on clock the process will execute. Within the statement part of the process it is a signal assignment statement which inverts the value of clock 50 ns in the future. This will schedule an event on clock. This process toggles clock every 50 ns.

```
PROCESS (clock)
BEGIN
   clock <= not(clock) AFTER 50 ns;
END PROCESS;
```

### 7.5.2 Signal assignment

This is used quite frequently. The syntax for a signal assignment statement is:

```
target <= delay_options   straight_transforms or
   conditional_transforms or
   selected_transforms;
```

There are two delay_options, INERTIAL and TRANSPORT. Delay_options are an advanced topic and will not be discussed. If a delay_option is not specified the default is inertial. Signal assignments are self-contained VHDL processes. Depending on which transforms are used, straight transforms, conditional_transforms or selected_transforms, a concurrent signal assignment can replicate simple VHDL processes or VHDL processes with if-then-else statements or case statements, respectively.

```
ARCHITECTURE arch1 OF and2 IS
BEGIN
   output <= in1 and in2;
END arch1;
```

This architecture can be rewritten using a process statement that duplicates the functionality of the architecture above. Rewriting the signal assignment statement as a process statement provides insight into the operation of a concurrent signal assignment statement. All concurrent statements have signals to which they are sensitive. This sensitivity list of signals is what activates a concurrent statement. With the concurrent signal assignment statement, all signals to the right of the signal assignment operator, <=, comprise the sensitivity list of that statement. In this example, if an event occurs on signals in1 or in2 that statement will be executed.

```
ARCHITECTURE arch2 OF and2 IS
BEGIN
   PROCESS (in1, in2)
   BEGIN
      output <= in1 and in2;
   END PROCESS;
END arch2;
```

An example of a 2-input AND gate with a delay of 10 nanoseconds is:

```
ARCHITECTURE and2 OF and2 IS
      BEGIN
      output <= in1 and in2 AFTER 10 ns;
END and2;
```

In conditional transforms the keywords WHEN and ELSE indicate conditional transforms. This example replicates a VHDL process statement with if-then-else constructs. All signals to the right of the signal assignment operator are sensitive to events, and as such will activate this signal assignment statement.

```
............
BEGIN
    output <= in1 AFTER tplh WHEN sel = '0' and rising(in1) ELSE
              in1 AFTER tphl WHEN sel = '0' and falling(in1) ELSE
              in2 AFTER tplh WHEN sel = '1' and rising(in2) ELSE
              in2 AFTER tphl WHEN sel = '1' and falling(in2) ELSE
              '0' AFTER tphl;
END rtl;
```

The example below shows a 2 to 1 mux using selected_transforms within a concurrent signal assignment statement. Since the input and sel signals appear to the right of the signal assignment statement they make up the sensitivity list. If an event occurs on any of those signals, one of the four drivers will be selected. Selection is based on which WHEN condition matches the value on the sel signal.

```
.................
BEGIN
   WITH sel SELECT
      output <= input(0) AFTER 5 ns WHEN "00",
                input(1) AFTER 5 ns WHEN "01",
                input(2) AFTER 5 ns WHEN "10",
                input(3) AFTER 5 ns WHEN "11";
END simple;
```

The example below uses the sequential case statement within a process statement:

```
ENTITY mux IS
   PORT (input: IN bit_vector(0 to 3);
         sel: IN bit_vector(0 to 1);
         output: OUT bit);
END mux;
ARCHITECTURE simple OF mux IS
BEGIN
   PROCESS (input,sel)
   BEGIN
   CASE sel IS
      WHEN "00" => output <= input(0) AFTER 5 ns;
      WHEN "01" => output <= input(1) AFTER 5 ns;
      WHEN "10" => output <= input(2) AFTER 5 ns;
      WHEN "11" => output <= input(3) AFTER 5 ns;
   END CASE;
   END PROCESS;
END simple;
```

### 7.5.3 Component instantiation

The component instantiation statement is the vehicle used to provide structural design mechanisms to VHDL. It is with this construct that hierarchical designs may be developed. The component instantiation statement allows the placement of an instance of a VHDL entity, the connection of signals to the ports of the entity and, if necessary, the modification of generic values declared within the entity of the instantiated component. The syntax is:

```
label:  component_name
        [GENERIC MAP (association_list)]
        [PORT MAP (association_list)];
```

The label and component_name are the only required elements of the component instantiation statement. The GENERIC MAP is optional if there are no generics declared within the entity declaration of the instantiated component or there is no need to override the generics declared. The PORT MAP is optional if the component has no ports or if the designer does not have the intention to hook it up.

Before a component can be instantiated it must be declared. The declaration can be either in an architecture's declarative region or within a package. The package is referenced through the LIBRARY and the USE clauses. The component declaration clause only specifies an entity and not an entity/architecture pair. The intended architecture for the declared components is specified through configuration specifications which bind an architecture to an entity. The PORT MAP constructs wire up the devices and the GENERIC MAP constructs allow the overriding of the default generic values declared within the entity statements of the subcomponents. By default, the mapping of signals to the actual ports of the component is a positional relationship. The first signal listed in a port map construct will map to the first port defined in the component declaration.

With the remapping construct '=>' the designer can force the connection of signals to specific ports, despite the order in which they were declared. Each instantiated component executes concurrently with all other concurrent statements. If an event occurs on any ports of mode in or inout, the associated entity/architecture pair will execute.

### 7.5.4 Generate

A generate statement provides a mechanism for iterative or conditional elaboration of a portion of a description. Some schematic capture systems provide a mechanism that allows a designer to instantiate a single component and then attach a property that will cause the replication of the component a specific number of times. This expansion is handled by the schematic compiler. This provides a higher level of abstraction when designing at the gate level. The expansion of such a construct is iterative elaboration.

A conditional generate statement will be compiled regardless of the condition. During the elaboration phase of simulation, if the condition is false the behaviour within the generate statement will be discarded. The syntax is:

```
label: generation_scheme GENERATE
   {concurrent_statements}
   END GENERATE [label];
```

```
generation_scheme ::=
FOR generate_specification
IF condition
```

A label is required for a generate statement and is optional when used with the END GENERATE construct. If the optional label is used with the END GENERATE construct it must match the generate label. There are two generation_schemes available for the generate, which are the FOR generation scheme and the IF generation scheme. Depending on which generation scheme is chosen, the resulting behaviour will be either iterative or conditional elaboration. The FOR scheme will cause an iterative elaboration scheme and the IF scheme will cause a conditional elaboration scheme. Only the iterative elaboration scheme will be discussed. Example:

```
ENTITY reg16 IS
   PORT (input : IN bit_vector (0 to 15);
              clock : IN bit;
              output : OUT bit_vector (0 to 15));
END reg16;
ARCHITECTURE struct OF reg16 IS
   COMPONENT dff
      PORT (d, clk : IN bit; q : OUT bit);
   END COMPONENT;
BEGIN
   G1: FOR i IN 0 TO 15 GENERATE
      G1: dff PORT MAP (input(i),clock,output(i));
   END GENERATE G1;
END struct;
```

This example shows the VHDL equivalent of the description given using schematic capture to imply a 16-bit register without instantiating all 16 flip-flops. By using the generate statement and a FOR generation scheme the dff device can be easily replicated. As in the sequential loop statement, the counter i does not have to be declared and will increment by 1 for each loop through the generate statement. When 16 loops have been completed, generation stops.

### 7.5.5 Assert

The assert statement can be used both as a sequential statement and a concurrent statement. Instead of including the assert statement within a process as used in section 7.4, it will be stand-alone. Its functionality is identical but the signals within the assert condition constructs become the sensitivity list for the concurrent assert statement. Whenever an event occurs on one of those signals, the assert statement will become active and execute. Once active, the assert statement will only report to the simulator if the Boolean expression evaluates to false.

**Example:**

```
ASSERT (address < 5) REPORT "Address out of range!"
              SEVERITY warning;
```

## 7.6 Functions and procedures

### 7.6.1 Functions

A function can be thought of as a user-defined expression and, similar to an expression, returns a single value. A function can be called from anywhere within a VHDL statement that an expression can be placed. A function can have a parameter list, the same as a procedure but unlike a procedure, each parameter in the function's parameter list has a mode of input IN. The class of the parameter can be explicitly declared but, if absent, will default to CONSTANT. This means that a function can only read the value on any of its parameters and cannot make assignments to any of these parameters. A function cannot have an assignment to signals or variables that it up-level references and thus has no side-effects.

Functions have two parts, the declaration and the body. Function declarations are only allowed within a package. Functions declared within a package are available for use by any design unit that makes reference to it via the LIBRARY and USE statements. The simplified syntax for a function declaration is:

```
FUNCTION function_name [( parameter_list )] RETURN type_mark;
```

The keywords FUNCTION and RETURN signify the declaration of a function. The function_name is a user-chosen identifier for the function. Formal parameters of functions can be objects of class constant or signal and can only be of mode IN. If the object class of a formal parameter is not explicitly specified, the class will default to CONSTANT. The parameter_list is optional. Although the passing of parameters to a function is optional the returning of one value is mandatory. This is required since a function call is used as an expression and not as a VHDL statement. The function declaration includes the type of the returned value. Function examples:

```
PACKAGE utilities IS
   FUNCTION v_to_int ( vect : bit_vector(0 TO 3))
      RETURN integer;
   FUNCTION rising (SIGNAL sig : bit) RETURN boolean;
   FUNCTION falling(SIGNAL sig : bit) RETURN boolean;
END utilities;
```

The first function declared is a vector to integer conversion. A 4-element bit_vector is passed in and an integer value is returned. The class of the formal vector parameter is not specified and therefore becomes CONSTANT and the mode is unspecified and therefore defaults to IN. The mode of parameters in function parameter lists is rarely explicitly specified as the only acceptable mode is IN. The functions rising and falling will decide if a rising or falling edge occurred on the signal passed in. Edge detection routines commonly use information stored by the simulator in the attributes of a signal. To access this information the parameter must be declared to have a class of SIGNAL.

### 7.6.2 Function body

The body of a function contains the procedural instructions which encompass the behaviour of a function. The function declaration defines the calling interface of the function.

Since functions are used as expressions in VHDL, a function body defines an evaluation or conversion the function will carry out. The simplified syntax is:

```
FUNCTION function_name [( parameter_list )] RETURN type_mark IS
   [declarative_items ;          ]
BEGIN
   [sequential_statements ;    ]
END [function_name ];
```

The keywords FUNCTION, RETURN and IS represent the beginning delimiter of a function body and the keyword END marks the ending delimiter. The function_name used with the END construct is optional and, if used, must match the function_name used with the FUNCTION construct. The parameter_list requirements are identical to those discussed with the function declaration. A function body can exist without a corresponding function declaration.

Matching function declarations and function bodies are required if the function body is defined within a package and the function must be visible to design units that reference it via the library and use statements. If a function will only be used by other functions or procedures within the same package, the declaration can be omitted. The function body can be thought of as a function declaration and function body all wrapped into one. This capability is needed since functions can be defined within declarative regions of entity and architecture design units or within the declarative regions of a process, generate, block, function or procedure VHDL statements.

The function body has a declarative region that begins immediately after the keyword IS and ends right before the keyword BEGIN. Inside this region declarative_items can be declared. Declarative_items include: types, variables, constants, procedures, functions, and so forth. The keyword BEGIN delimits the beginning of the sequential statements that define the action the function will execute. A function call is only executed as a result of encountering its call within a calling expression. A function initialises any internal variables declared in the declarative region to their initial value and begins sequential execution of statements located after the BEGIN statement until it hits a RETURN statement. The RETURN statement must be accompanied by an expression that evaluates to the type declared in the function's RETURN type_mark. It is a run-time error if the end of a function is encountered prior to hitting an explicit RETURN statement.

**Examples:**

```
PACKAGE BODY utilities IS
   FUNCTION vector_to_integer ( vect : bit_vector(0 to 3))
      RETURN integer IS
   VARIABLE result: integer := 0;
   VARIABLE weight: integer := 1;
   BEGIN
   For i IN 0 TO 3 LOOP
      IF vect(i) = '1' THEN
         result := result+weight;
      END IF;
   weight := weight * 2;
```

```
        END LOOP;
        RETURN result;
        END vector_to_integer;
        FUNCTION rising(SIGNAL sig : bit) RETURN boolean IS
        BEGIN
        IF sig = '0' THEN
          RETURN false;
        ELSE
          RETURN true;
        END IF;
        END rising;
        FUNCTION falling(SIGNAL sig : bit) RETURN boolean IS
        BEGIN
          IF sig = '0' THEN
                  RETURN true;
          ELSE
                  RETURN false;
          END IF;
        END falling;
      END utilities;
```

### 7.6.3 Procedures

A procedure is a group of sequential VHDL statements that typically represent a single logical action. Therefore, a procedure call is a single stand-alone VHDL statement. Typical situations for a procedure could include:

```
request_bus(request_level,timeout);
initialise_memory(memory_id,data_file);
refresh_screen;
```

Modular design with subprograms results in clean and easily maintainable code. If there is a problem with the protocol or the functionality of one of the procedures, it is far easier to fix that anomaly within the affected procedure than to go to each of the instances where that procedure is used and make the changes. Modular coding also allows a designer to concentrate on the functionality of the model rather than the implementation.

A VHDL procedure can be called as a sequential statement or as a concurrent statement. If it is being called sequentially, the procedure begins execution when the call is encountered. As a concurrent statement, the signals in the procedure's parameter list (of mode IN or INOUT) comprise the sensitivity list for the procedure. That is, whenever a signal in a concurrent procedure call's parameter list has an event, the procedure begins execution.

Independent of whether the procedure is activated via sequential or concurrent activity, all statements within a procedure are executed sequentially from the top to the bottom; therefore, all statements within a procedure's body must be sequential. There is no limit to the number of input or output parameters that can be passed to or from a procedure. A function returns one value only and because of this property, a procedure is often a more appropriate choice of implementation than a function. For instance, when implementing a vector addition operation, it is often necessary to produce the carry output and overflow status along with the sum. Since three items must be returned, the appropriate choice for this implementation is a procedure.

Procedures can have side-effects. These are modifications of objects declared outside the procedure that were not passed through the procedure's argument list. This capability, commonly referred to as up-level referencing, is extremely powerful; however, these types of procedures can make the operation of a design difficult to understand.

Technically, a procedure has two parts: the declaration and the body. A procedure declaration is only allowed within a package. Procedures declared within a package are available for use by any design unit that refers to it via the LIBRARY and USE statements. The syntax for a procedure declaration within a package is:

```
PROCEDURE procedure_name [( parameter_list )];
```

The keyword PROCEDURE signifies the declaration of a procedure. The procedure_name is a user-chosen name for the procedure and the parameter_list is the argument list of objects that can be passed to or from the procedure. Formal parameters of procedures have an associated class and their class can be constant, variable, or signal. Formal parameters of procedures are also associated with a direction or mode; this mode can be in, out, or inout. Although an explicit definition for the class and mode of a parameter is optional, an implicit declaration for these exists. If neither the parameter's mode nor the class is explicitly defined, the resulting parameter will be treated as mode IN with class CONSTANT. If, however, the mode is explicitly specified but no class is given, then the parameter will be treated as having the VARIABLE class. For a procedure, the parameter_list is optional.

A procedure's name, the number of arguments passed, the object class and mode of the arguments all make up the signature of a procedure call. It is, therefore, possible to have a single procedure name that has many different signatures. The mode of the parameter determines what the procedure may do with the parameter. If the mode on a parameter is IN, then the procedure may only read the value on the parameter. If the mode on the parameter is OUT, then the procedure may only assign to this parameter. If the parameter has a mode of INOUT, then the procedure may read the value of the parameter and also make an assignment to the parameter.

The result of a variable assignment statement occurs immediately and the result is available to the next sequential statement. Whereas, in a signal assignment statement, all effects of that signal assignment statement occur in the future, and the result is not available to the next sequential statement. Therefore, a procedure must know whether it is dealing with variables or signals. Signals carry more information about themselves than simply their current value, this attribute information including the event status of the signal, the last value of the signal, and more. If a parameter has been given the explicit class of SIGNAL, then the procedure also has access to this important attribute information for the parameter. Procedure examples:

```
PACKAGE utilities IS
   PROCEDURE add_element(      element:IN real;
      VARIABLE filter_data: INOUT filter_data_type);
   PROCEDURE zero_out(input:INOUT filter_data_type);
   PROCEDURE still_busy;
END utilities;
```

The example shows procedure declarations within a PACKAGE. In the 'utilities' package three procedures are being declared. The first procedure declares add_element as having two parameters in its parameter_list, element and filter_data. The first parameter has no explicit class but is of mode IN. This parameter will default to a class of CONSTANT. The second parameter has an explicit class of VARIABLE and has an explicit mode of INOUT. Within the add_element procedure, no assignment to the parameter called element is allowed (mode IN) but both assignment and reading of the parameter called filter_data (mode INOUT) are permissible. The second procedure declared is called zero_out and has a single argument in its parameter list. Since the class is not explicitly specified and the mode is INOUT, the class of the parameter defaults to variable. Within this procedure, therefore, this parameter can be read and can also be assigned a new value by using the variable assignment operator (:=). The last procedure declaration shown is called still_busy and does not have a parameter list. Therefore, no arguments are passed in or returned. This can still perform a valuable function by utilising WAIT statements to delay simulation or by using up-level referencing to gather information or make assignments to signals that have not been explicitly passed through the parameter list.

### 7.6.4 Procedure body

The procedure body contains the procedural instructions that encompass the behaviour of a procedure. In contrast, the procedure declaration defines the calling interface of the procedure and identifies whether or not the procedure is available for global use. The procedure body uses the procedure's declaration for its interface and defines the action the procedure will carry out. The syntax for the procedure body is:

```
PROCEDURE procedure_name [( parameter_list )] IS [declarative_items; ]
BEGIN
   [sequential_statements; ]
END [ procedure_name ];
```

The keywords PROCEDURE and IS represent the beginning delimiter of a procedure body and the keyword END marks the ending delimiter. The procedure_name used with the END construct is optional and, if used, must match the procedure_name used with the procedure construct. The parameter_list requirements are identical to those discussed with the procedure declaration.

The procedure body can be thought of as a procedure declaration and procedure body all wrapped into one. This is needed since procedures can be defined within declarative regions of entity and architecture design units or within the declarative regions of a process, generate, block, function or procedure VHDL statements. This provides the ability to localise procedures. The procedure body has a declarative region that begins immediately after the keyword IS and ends right before the keyword BEGIN. Inside this region declarative_items can be declared. Declarative_items include: types, variables, constants, procedures, functions, and so forth. The keyword BEGIN delimits the beginning of the sequential statements that define the action a procedure will execute. The keyword END delimits the end of a procedure body with the optional procedure_name. If the procedure's label is used in conjunction with the END statement, it must match the procedure_name declared at the beginning of the procedure body.

Once activated, a procedure initialises any internal variables declared in the declarative region to their initial value and begins sequential execution of statements. It continues execution until it hits a WAIT statement, a RETURN statement, or the end of the procedure (implicit RETURN). Upon hitting a RETURN statement, the calling program resumes execution with any parameters passed out of the procedure updated with their new values.

**Examples**

```
PROCEDURE add_element(element:IN real;
   VARIABLE filter_data: INOUT filter_data_type) IS
BEGIN
   FOR i IN filter_data'high DOWNTO filter_data'low+1 LOOP
      filter_data(i) := filter_data(i-1);
   END LOOP;
   filter_data(filter_data'low) := element;
END add_element;
PROCEDURE zero_out(input:INOUT filter_data_type) is
BEGIN
   FOR i IN input'range LOOP
      input(i) := 0.0;
   END LOOP;
END zero_out;
```

## 7.7 Advanced features in VHDL

### 7.7.1 Attributes

There are two types of attributes: predefined attributes and user-defined attributes. The user-defined attribute is an advanced feature that allows the user to expand the language beyond what is available. It would be useful to maintain information about a signal other than its present value. Such information could include: the prior value of the signal, the time elapsed since the signal value changed or last event, the time elapsed since the signal value was driven or last transaction, if the signal is an array, what is the length of the array, etc.

```
ENTITY reg IS
   GENERIC (tsu:time := 3 ns);
   PORT( d: IN bit_vector(3 DOWNTO 0);
         clk: IN bit;
         q: OUT bit_vector(3 DOWNTO 0));
BEGIN
   PROCESS (clk)
   BEGIN
      IF clock = '1' THEN
         ASSERT (d'LAST_EVENT > tsu)
            REPORT "Setup Violation on the d Input!"
            SEVERITY error;
```

```
        END IF;
      END PROCESS;
    END reg;
```

The expression which the assert statement evaluates has the 'LAST_EVENT attribute on the d input port and this is a signal attribute. Port d is a signal and 'LAST_EVENT is one of the predefined attributes for a signal. The time since the last event occurred on the signal d is stored within the attribute d'LAST_EVENT. There are predefined attributes for signals, types and blocks. Attribute information is also maintained for objects declared as arrays allowing the determination of array length, dimension, indexing of an array. Predefined attributes include:

- 'EVENT Boolean value that is true when signal that it is attached to has had an event in the current delta.
- 'LEFT Value that indicates the leftmost bound of an array index.
- 'RIGHT Value that indicates the rightmost bound of an array index.
- 'HIGH Value that indicates the highest value of an array index.
- 'LOW Value that indicates the lowest value of an array index.
- 'RANGE The range of values that an array index can take on.
- 'LAST_VALUE The value that the signal this is attached to was prior to its current value.

### 7.7.2 Overloading

Overloading refers to using the same identifier to mean different things based on the context in which it is used. Overloading is a powerful concept which applies to three different areas: subprogram overloading, enumerated value overloading, operator overloading. With respect to subprogram overloading an example is given of an overloaded subprogram (function) called rotate_left which can be used to work on different input parameter types:

```
PACKAGE util IS
  TYPE mvl3 IS ('X','0','1');
  TYPE mvl4 IS ('X','0','1','Z');
  TYPE mvl3_vector IS ARRAY(natural RANGE <>) OF mvl3;
  TYPE mvl4_vector IS ARRAY(natural RANGE <>) OF mvl4;
  FUNCTION rotate_left(input:mvl3_vector) RETURN mvl3_vector;
  FUNCTION rotate_left(input:mvl4_vector) RETURN mvl4_vector;
END util;
```

Two overloaded functions are being declared. The functions are overloaded because the identifiers for the functions are the same but the signature of the functions is different. A subprogram's signature includes the subprogram name, the number of subprogram parameters, the types of each subprogram's parameters, the class of each subprogram's parameters and, in the case of functions, the subprogram return type. If any of these are different between two subprograms of the same name, they are said to be overloaded. In the example above there are two subprograms (functions) being declared that have the same name, the same number of parameters and each subprogram parameter has the same class. However, the type of the parameters and the type being returned by the subprogram are different. These two functions are overloaded. VHDL allows the end-user to overload the common operators for various data types. Other operators that may be overloaded include "+", "–", "*", "/", etc.

### 7.7.3 Passive processes

A passive PROCESS is any PROCESS which does not affect the simulation behaviour of any signal, therefore no signal assignments can occur. All PROCESSes communicate with the outside world by manipulating signal values. A passive PROCESS can be used to monitor simulation activity and report on erroneous behaviour through textual messages. A passive PROCESS can appear either within an architecture or within an entity's body. The only type of PROCESS that may appear within an entity's body is a passive PROCESS. Typically, timing checks are implemented as passive PROCESSes within the entity's body section. An example is given below showing how a single passive PROCESS is activated by an event on the clk port. If this is a rising edge, the PROCESS checks to make sure that the d input has been stable for at least tsu time units. If it is violated, a message will be reported:

```
ENTITY reg IS
  GENERIC (tsu:time := 3 ns);
  PORT(d: IN bit_vector(3 DOWNTO 0);
       clk: IN bit;
       q: OUT bit_vector(3 DOWNTO 0));
BEGIN
  PROCESS (clk)
  BEGIN
    IF clock = '1' THEN
      ASSERT (d'LAST_EVENT > tsu)
              REPORT "Setup Violation on the d Input!"
              SEVERITY error;
    END IF;
  END PROCESS;
END reg;
```

### 7.7.4 TEXTIO

Within the STD library TEXTIO contains file input/output routines. These allow the user to read and write ASCII files within a VHDL model. In order to access a text file from VHDL, the user must create a symbolic name, known as a file designator for the file, using the FILE statement. An example shows how a file designator called demo_file is being created. The type of the file is text and the file is read only (mode is IN). The only other mode of a file designator is OUT (write only). A line variable is an access type with each element being of type character. To read a line from a file into a line variable the readline routine from TEXTIO is engaged. The PROCESS reads a line out of the demo_file every time there is an event on the clk signal, it then places the line read from that file into a variable (l_var) that is of elastic array type called a line type:

```
USE STD.textio.ALL;
ARCHITECTURE behave OF t_bench IS
FILE demo_file: text IS IN "/home/stim_file.txt";
BEGIN
   PROCESS(clk)
   VARIABLE l_var:line;
   BEGIN
      readline(demo_file,l_var);
   END PROCESS;
END behave;
```

## 7.8 Summary

- VHDL supports behavioural as well as structural modelling. With behavioural modelling, few if any implementation details are revealed. The advantage to behavioural modelling is that the functionality of a device or design can easily be described using abstract concepts. Structural modelling with component instantiation allows the designer to describe how the device will perform the function.
- VHDL provides constructs that support a mix of both high level behavioural and structural modelling in all phases of design and throughout the entire top-down design process.
- Five different design units were discussed that comprise the fundamental building blocks in VHDL. The entity, the architecture and the configuration constitute the physical aspects of a design. The entity describes the interface information for a device. The architecture describes the behaviour for a device. There could be many different architectures for a given entity. The configuration design unit describes the binding of an architecture to an entity.
- The packages are a repository for common elements that may be used by many different designs. These common elements quite often include subprograms, components, constants and other objects that may need to be generally available. The package body design unit contains the functionality of subprograms or the value of constants declared within a package.
- The library is the physical repository in which design units (that have been successfully analysed by the VHDL analyser) are stored. There are two statements that make design units visible to other design units: LIBRARY and USE. The LIBRARY statement activates a library for potential use by a design unit. Depending on the syntax used, the USE statement engages other elements. Three statements are always implied, or made available to VHDL design units. These are:

```
LIBRARY std;
LIBRARY work;
   USE std.standard.ALL;
```

- VHDL provides multi-level design simulations, which can range from most abstract to details of a switch level simulation. A VHDL simulator is a concurrent event simulator. When an event is detected the VHDL statements that are sensitive to that event are queued for execution. This cycle is repeated. Signals are only updated at the beginning of a simulation cycle and are not updated immediately after execution. Delta ordering is a vehicle in which forward motion in a VHDL simulation can be expressed without advancing simulation time. Time in a VHDL simulation is represented by two axes, delta ordering and simulation time. Simulation time will never advance until all deltas scheduled for the present simulation time have been executed.

- Since VHDL is a concurrent programming language, all sequential statements are contained within a concurrent shell called a PROCESS. The most common used five concurrent statements were discussed: Process, Assert, Component Instantiation, Generate, Signal Assignment. Concurrent statements are miniature forms of describing a VHDL PROCESS. Each concurrent statement has a sensitivity list and each concurrent statement will execute asynchronously when an event occurs in its respective sensitivity list. Most concurrent statements could be modelled as PROCESSes and, as such, provide a mechanism for partitioning a large design.
- Through a combination of TEXTIO techniques, portable VHDL test-benches can be developed. It can be seen how a component could be stimulated by information contained within a file, the results of that component's behaviour can be compared against an expected results file and any errata could be reported to a file.

# Case Studies

# 8

# Neural current and speed control of induction motors

This chapter presents the mathematical principles and algorithms underlying a new sensorless control strategy for three-phase cage induction motors. The control method comprises two elements: the stator current control strategy and the sensorless speed control strategy. Both of them are based on an equivalent three-phase *R–L–e* circuit whose parameters are derived from the space vector model of the induction motor. An induction motor speed estimation strategy is presented. A new sensorless speed control method is formulated and tested by simulation. The VHDL design and FPGA implementation of the new controller are presented, including simulation and experimental results.

## 8.1 The induction motor equivalent circuit

The proposed motor control strategy uses the classical sensorless drive system structure with the motor supplied by a VSI-PWM inverter, which is controlled by a digital circuit based only on the stator current feedback information. As mentioned in Chapter 3, the predictive current control method uses an equivalent *R–L–e* circuit for the load modelled by the equation

$$\underline{u}(t) = R\underline{i}(t) + L\frac{\mathrm{d}\underline{i}(t)}{\mathrm{d}t} + \underline{e}(t) \tag{8.1}$$

The *R–L–e* equivalent circuit parameters for an induction motor can be derived from its general space vector model

$$\begin{cases} \underline{u}_s^{\theta} = R_s \underline{i}_s^{\theta} + \dfrac{\mathrm{d}\underline{\Psi}_s^{\theta}}{\mathrm{d}t} \\ \underline{u}_r^{\theta} = R_r \underline{i}_r^{\theta} + \dfrac{\mathrm{d}\underline{\Psi}_r^{\theta}}{\mathrm{d}t} - j(\omega_e - \omega_{er})\underline{\Psi}_r^{\theta} = 0 \\ \underline{\Psi}_s^{\theta} = L_s \underline{i}_s^{\theta} + L_m \underline{i}_r^{\theta} \\ \underline{\Psi}_r^{\theta} = L_r \underline{i}_r^{\theta} + L_m \underline{i}_s^{\theta} \end{cases} \tag{8.2}$$

particularised for the stator reference frame. Thus, the parameters defining the reference frame are $\theta = 0$ and $\omega = \omega_{es} = 0$, which yields the equation system

$$\begin{cases} \underline{u}_s^s = R_s \underline{i}_s^s + \dfrac{\mathrm{d}\underline{\Psi}_s^s}{\mathrm{d}t} \\ \underline{u}_r^s = R_r \underline{i}_r^s + \dfrac{\mathrm{d}\underline{\Psi}_r^s}{\mathrm{d}t} - j\omega_{er}\, \underline{\Psi}_r^s = 0 \\ \underline{\Psi}_s^s = L_s \underline{i}_s^s + L_m \underline{i}_r^s \\ \underline{\Psi}_r^s = L_r \underline{i}_r^s + L_m \underline{i}_s^s \end{cases} \tag{8.3}$$

The two fluxes can be eliminated from the equations giving:

$$\begin{cases} \underline{u}_s^s = R_s \underline{i}_s^s + L_s \dfrac{\mathrm{d}\underline{i}_s^s}{\mathrm{d}t} + L_m \dfrac{\mathrm{d}\underline{i}_r^s}{\mathrm{d}t} \\ 0 = R_r \underline{i}_r^s + L_m \dfrac{\mathrm{d}\underline{i}_s^s}{\mathrm{d}t} + L_r \dfrac{\mathrm{d}\underline{i}_r^s}{\mathrm{d}t} - j\omega_{er}(L_m \underline{i}_s^s + L_r \underline{i}_r^s) \end{cases} \tag{8.4}$$

Therefore the derivative of the rotor current vector is

$$\frac{\mathrm{d}\underline{i}_r^s}{\mathrm{d}t} = \frac{1}{L_r}\left[-R_r \underline{i}_r^s - L_m \frac{\mathrm{d}\underline{i}_s^s}{\mathrm{d}t} + j\omega_{er}(L_r \underline{i}_r^s + L_m \underline{i}_s^s)\right] \tag{8.5}$$

Substituting this in (8.4) gives

$$\underline{u}_s^s = R_s \underline{i}_s^s + \left(L_s - L_m \frac{L_m}{L_r}\right)\frac{\mathrm{d}\underline{i}_s^s}{\mathrm{d}t} + \frac{L_m}{L_r}[-R_r \underline{i}_r^s + j\omega_{er}(L_r \underline{i}_r^s + L_m \underline{i}_s^s)] \tag{8.6}$$

Identifying this result with the fundamental equation (8.1), the parameters of the equivalent circuit are determined as follows:

$$\begin{cases} L = \dfrac{L_s L_r - L_m^2}{L_r} \\ R = R_s \\ \underline{e} = \dfrac{L_m}{L_r}[-R_r \underline{i}_r^s + j\omega_{er}(L_r \underline{i}_r^s + L_m \underline{i}_s^s)] = \dfrac{L_m}{L_r}(-R_r \underline{i}_r^s + j\omega_{er} \underline{\Psi}_r^s) \\ \underline{u} = \underline{u}_s \\ \underline{i} = \underline{i}_s \end{cases} \tag{8.7}$$

Consequently, the voltage space vector $\underline{u}$ is the voltage supplying the motor, the current $\underline{i}$ is the stator current, while the internal voltage $\underline{e}$ is a quantity bearing information on the motor operation parameters (speed and rotor current). Table 8.1 presents the electrical parameters of five different three-phase cage induction motors. It reflects the influence of the motor power on other parameter values. Thus, the stator and the rotor resistances are larger at lower powers and smaller at higher powers. In a well-designed motor, the leakage inductances are always small compared to the mutual inductance and the total leakage inductance $L_{\sigma s} + L_{\sigma r}$ does not generally exceed 10 per cent of the mutual inductance $L_m$.

Under these conditions, the expression of the equivalent inductance $L$ can be transformed as shown in (8.8) and it can be approximated by the sum of the two leakage inductances. The result is the approximate equivalent circuit illustrated in Fig. 8.1.

**Table 8.1** Electrical parameters of three-phase induction motors

| Quantity | (a) | (b) | (c) | (d) | (e) |
|---|---|---|---|---|---|
| $R_s$ | 0.0114 Ω | 0.371 Ω | 0.79 Ω | 2.89 Ω | 5.9 Ω |
| $R_r$ | 0.011 Ω | 0.415 Ω | 0.76 Ω | 2.39 Ω | 4.62 Ω |
| $L_{s\sigma}$ | 0.32 mH | 2.72 mH | 1.57 mH | 11 mH | 22 mH |
| $L_{r\sigma}$ | 0.36 mH | 3.3 mH | 1.59 mH | 6 mH | 24 mH |
| $L_m$ | 11.68 mH | 84.33 mH | 65 mH | 214 mH | 809 mH |
| $p$ (pole pairs) | 2 | 1 | 2 | 1 | 1 |
| $P$ | 110 kW | 11.1 kW | 5 kW | 3 kW | 2 kW |

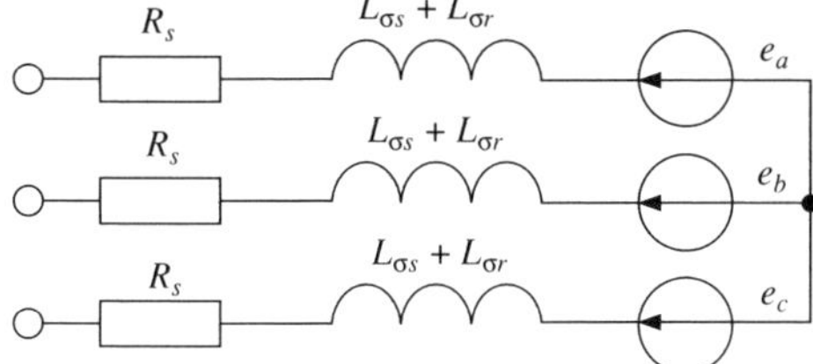


**Fig. 8.1** The approximate *R–L–e* equivalent circuit of a three-phase induction motor

$$L = \frac{(L_{\sigma s} + L_m) \cdot (L_{\sigma r} + L_m) - L_m^2}{L_m + L_{\sigma r}} = \frac{L_{\sigma s} L_{\sigma r} + L_m (L_{\sigma s} + L_{\sigma r})}{L_m + L_{\sigma r}} \approx L_{\sigma s} + L_{\sigma r} \quad (8.8)$$

Thus, despite the large number of turns in the motor windings, the equivalent inductance $L$ is relatively small due to the tight magnetic coupling between stator and rotor. However, the precise circuit contains a slightly larger equivalent inductance that can be calculated according to (8.7).

The parameters of the 11.1 kW motor presented on column (b) in Table 8.1 are used in the example to illustrate all the control principles formulated in this chapter. Using the same parameters for all simulations and calculations facilitates meaningful comparisons between alternative control strategies.

## 8.2 The current control algorithm

### 8.2.1 The switching strategy

The predictive current control strategy proposed in this section [81] involves the concept of non-inductive voltage, which is defined as the sum of the resistive voltage component $R\underline{i}$ and the internal voltage component $\underline{e}$. This quantity, denoted by $\underline{V}_{ni}$, excludes the inductive voltage component $L\mathrm{d}\underline{i}/\mathrm{d}t$ from the total voltage, hence the name of **non-inductive voltage**. This can be calculated using one of the two expressions:

$$\underline{V}_{ni} = R\underline{i} + \underline{e} = \underline{u} - L\frac{\mathrm{d}\underline{i}}{\mathrm{d}\underline{t}} \quad (8.9)$$

The second formulation is more profitable as it does not use the value of the internal voltage $\underline{e}$, which is difficult to calculate. The digital implementation of the control algorithm requires that all the quantities are sampled at equal time intervals. If the sampling process is taken into account then equation (8.1) becomes (8.10), where $T_s$ is the sampling period and $k$ is the index of the samples:

$$\underline{u}(kT_s) = R\underline{i}(kT_s) + \frac{L}{T_s}[\underline{i}(kT_s) - \underline{i}((k-1)T_s)] + \underline{e}(kT_s) + \mathrm{Err}(k, T_s) \quad (8.10)$$

The function $\mathrm{Err}(k, T_s)$ represents the calculation error generated by replacing the derivative in (8.1) with the approximation calculated based on the difference between two consecutive current values. The calculation error decreases with increasing frequency of the PWM and is negligible at the frequencies commonly used in induction motor drive systems (2 kHz to 20 kHz). Under these conditions, equation (8.10) can be written as:

$$\underline{u}(kT_s) \cong R\underline{i}(kT_s) + \frac{L}{T_s}[\underline{i}(kT_s) - \underline{i}((k-1)T_s)] + \underline{e}(kT_s) \quad (8.11)$$

The notation can be simplified by replacing the time argument '$kT_s$' with the sample index $k$, thereby transforming equation (8.11) into the equivalent form

$$\underline{u}(k) \cong R\underline{i}(k) + \frac{L}{T_s}[\underline{i}(k) - \underline{i}(k-1)] + \underline{e}(k) \quad (8.12)$$

Based on (8.9) and (8.12), the non-inductive voltage can be approximated as

$$\underline{V}_{ni}(k) \cong \underline{u}(k) - \frac{L}{T_s}[\underline{i}(k) - \underline{i}(k-1)] \quad (8.13)$$

The inverter output voltage is constant between two consecutive switching transients and if the PWM period is sufficiently short, the internal voltage $\underline{e}$ can be considered constant as well ($\underline{u}(t) = \underline{U}$), $\underline{e}(t) = \underline{E}$). This assumption substantially simplifies the current calculations. Thus, the relationship (8.1) becomes

$$\underline{U} = R\underline{i}(t) + L\frac{\mathrm{d}\underline{i}(t)}{\mathrm{d}t} + \underline{E} \quad (8.14)$$

and has the solution

$$\underline{i}(t) = \frac{\underline{U} - \underline{E}}{R} + \left(\underline{i}(0) - \frac{\underline{U} - \underline{E}}{R}\right) \cdot e^{-\frac{Rt}{L}} \quad (8.15)$$

The real and the imaginary part of the current space vector are

$$\begin{cases} x(t) = \mathrm{Re}\{\underline{i}(t)\} = \mathrm{Re}\left\{\frac{\underline{U} - \underline{E}}{R}\right\} + \mathrm{Re}\left\{\underline{i}(0) - \frac{\underline{U} - \underline{E}}{R}\right\} \cdot e^{-\frac{Rt}{L}} \\ y(t) = \mathrm{Im}\{\underline{i}(t)\} = \mathrm{Im}\left\{\frac{\underline{U} - \underline{E}}{R}\right\} + \mathrm{Im}\left\{\underline{i}(0) - \frac{\underline{U} - \underline{E}}{R}\right\} \cdot e^{-\frac{Rt}{L}} \end{cases} \quad (8.16)$$

Combining the two equations (8.16), the result is

$$y(t) = \mathrm{Im}\left\{\frac{\underline{U} - \underline{E}}{R}\right\} + \frac{\mathrm{Im}\left\{\underline{i}(0) - \frac{\underline{U} - \underline{E}}{R}\right\}}{\mathrm{Re}\left\{\underline{i}(0) - \frac{\underline{U} - \underline{E}}{R}\right\}} \cdot \left(x(t) - \mathrm{Re}\left\{\underline{i}(0) - \frac{\underline{U} - \underline{E}}{R}\right\}\right)$$

(8.17)

which can be reduced to

$$y(t) = \frac{\mathrm{Im}\left\{\underline{i}(0) - \frac{\underline{U} - \underline{E}}{R}\right\}}{\mathrm{Re}\left\{\underline{i}(0) - \frac{\underline{U} - \underline{E}}{R}\right\}} \cdot x(t) + \mathrm{Im}\{\underline{i}(0)\} = a \cdot x(t) + b \qquad (8.18)$$

The constants '$a$' and '$b$' in (8.18) are defined by

$$\begin{cases} a = \dfrac{\mathrm{Im}\left\{i(0) - \frac{\underline{U} - \underline{E}}{R}\right\}}{\mathrm{Re}\left\{i(0) - \frac{\underline{U} - \underline{E}}{R}\right\}} \\ b = \mathrm{Im}\{\underline{i}(0)\} \end{cases} \qquad (8.19)$$

Equation (8.18) is that of a straight line. This entails that the vertex of the current space vector $\underline{i}$ shifts with a variable speed along a straight trajectory. Due to the linear relationship between vectors $\underline{V}_{ni}$ and $\underline{i}$, the trajectory of $\underline{V}_{ni}$ is straight as well. Furthermore, according to (8.20), the vertices of the two space vectors shift along parallel trajectories whose direction is indicated by the argument $\varepsilon$ calculated according to (8.21).

$$\underline{V}_{ni}(t) = R \cdot \underline{i}(t) + \underline{E} \Rightarrow \mathrm{d}\underline{V}_{ni} = R \cdot \mathrm{d}\underline{i} \qquad (8.20)$$

$$\varepsilon = \arg(\mathrm{d}\underline{V}_{ni}) = \arg(\mathrm{d}\underline{i} = \arg(\underline{i}(+\infty) - \underline{i}(0)) = \arg(\underline{V}_{ni}(+\infty) - \underline{V}_{ni}(0)) \qquad (8.21)$$

The value $\underline{V}_{ni}(+\infty)$ is the non-inductive voltage after an infinitely long time and can be calculated as:

$$\underline{V}_{ni}(+\infty) = \lim_{t\to\infty}(R\underline{i}(t) + \underline{E}) = \lim_{t\to\infty}\left[R \cdot \frac{\underline{U} - \underline{E}}{R} + \left(\underline{i}(0) - \frac{\underline{U} - \underline{E}}{R}\right) \cdot e^{-\frac{Rt}{L}} + \underline{E}\right] = \underline{U} \qquad (8.22)$$

As a result, the angle $\varepsilon$ is

$$\varepsilon = \arg(\underline{U} - \underline{V}_{ni}(0)) \qquad (8.23)$$

Thus, as illustrated in Fig. 8.2, the trajectory of vector $\underline{V}_{ni}$ is a straight line oriented on the direction that links the point corresponding to the inverter output voltage with the initial value of the non-inductive voltage $\underline{V}_{ni}(0)$. These considerations can be used to determine the direction of the current trajectory in the complex plane without using the value of the internal voltage $\underline{e}$. The control voltages to the transistors in the PWM inverter need to be generated in such a way that the inverter output voltage maintains the required currents across the load. The required current modification during one sampling period is a complex quantity defined by the argument '$\arg\{\underline{i}_{ref}(k + 1) - \underline{i}(k)\}$' and the module $|\underline{i}_{ref}(k + 1) - \underline{i}(k)|$. These two parameters are often impossible to achieve simultaneously because only seven inverter output voltages are available, which means that only seven different current shifts can be performed at a given moment. Therefore, there are two alternative switching strategies:

- Minimising the module of the current error $|\underline{i}_{ref}(k + 1) - \underline{i}(k + 1)|$.

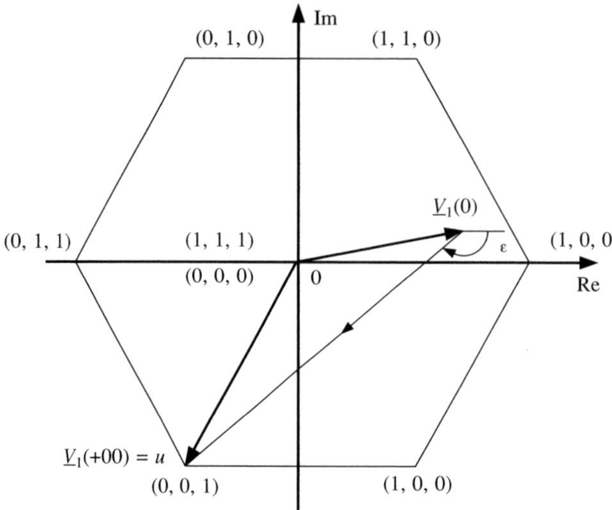


**Fig. 8.2** The trajectory of the vertex of $V_{ni}$ space vector

- Minimising the angle between the direction of the required current trajectory and the direction of the actual current trajectory $|\arg\{\underline{i}_{ref}(k+1) - \underline{i}(k)\} - \arg\{\underline{i}(k+1) - \underline{i}(k)\}|$.

The first alternative generates optimal control results but requires that equation (8.14) is solved for all seven possible output voltages and the results are compared. The most important disadvantage of this method is that the internal voltage $\underline{E}$ needs to be determined first. The calculation can be performed according to the equation

$$\underline{E}(k) = \underline{V}_{ni}(k) - R\underline{i}(k) \cong \underline{u}(k) - \frac{L}{T_s}[\underline{i}(k) - \underline{i}(k-1)] - R\underline{i}(k) \tag{8.24}$$

but this involves the value of the stator resistance $R$ that needs to be determined on-line due to temperature variation during the motor operation. Therefore, this method necessitates complicated calculations that make it impractical. The inverter switching strategy adopted in this example uses the second alternative. This approach yields good control results without requiring the value of the stator resistance, because it involves $\underline{V}_{ni}$ instead of $\underline{E}$ in the calculations. Thus, the directions $\varepsilon_j$ ($j = 1, 2, 3, \ldots 7$) of the possible current trajectories are first determined according to equation (8.23). Then the output voltage $\underline{u}_j$ that minimises the expression $|\varepsilon_j - \arg\{\underline{i}_{ref}(k+1) - \underline{i}(k)\}|$ is generated during the next sampling period. This current control strategy is illustrated by the example in Fig. 8.3. The current error vector $\Delta\underline{i}_{ref} = \underline{i}_{ref}(k+1) - \underline{i}(k)$ indicates a direction in the complex plane which is not identical to any of the directions that can be achieved using the available voltages. However, the voltage coded as (0, 1, 0) is capable of producing a current change in a direction that is much closer to the reference one than the other six possibilities (including the zero output voltage). As a result, this voltage is generated during the next sampling period $T_s$. A consequence of the fast voltage switching is that the non-inductive voltage vector $\underline{V}_{ni}$ never reaches the final value $\underline{V}_{ni}(+\infty) = \underline{u}$ during any of the sampling periods and therefore $\underline{V}_{ni}$ is always inside the hexagon in Fig. 8.3.

The switching strategy can include the null voltage generated by the inverter, thereby increasing the control flexibility, or it can exclude it improving the current response

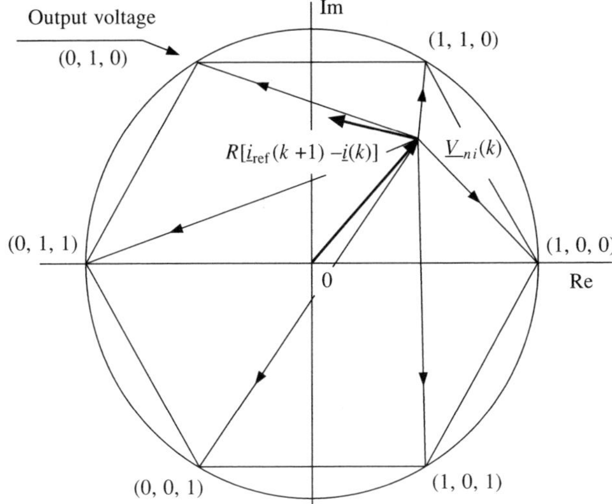


**Fig. 8.3** The graphic representation of the PWM current control principle

speed. Including the null output voltage in the switching strategy improves the current harmonic content [81], but presents the disadvantage that the current response is slow if the motor current is small. The current change rate $|\mathrm{d}\underline{i}/\mathrm{d}t|$ when $\underline{u} = 0$ is governed by equation

$$\left|\frac{\mathrm{d}\underline{i}}{\mathrm{d}t}\right| = -\frac{1}{L}|R\underline{i} + \underline{e}| \tag{8.25}$$

that is derived from (8.1). As demonstrated by (8.7), the internal voltage $\underline{e}$ is proportional to the motor currents, and therefore the module of vector $\underline{e}$ is approximately proportional with the module of the equivalent current vector (the stator current). In this case, the current change rate can be considered proportional to the module of the current vector, which means that the system response is infinitely slow when the motor currents tend to zero.

$$\left|\frac{\mathrm{d}\underline{i}}{\mathrm{d}t}\right| \cong -K\,|\underline{i}| \Rightarrow |\underline{i}| \cong i(0) \cdot e^{-Kt} \tag{8.26}$$

If the inverter generated voltage is not zero then the current change rate is given by

$$\left|\frac{\mathrm{d}\underline{i}}{\mathrm{d}t}\right| = -\frac{1}{L}|R\underline{i} + \underline{e} - \underline{u}| = -\frac{1}{L}|\underline{V}_{ni} - \underline{u}| \tag{8.27}$$

This ensures high response speed because the vector $\underline{V}_{ni}$ lies always inside the voltage hexagon and therefore $|V_{ni} - u|$ is always much larger than zero. As a result, the control method excluding the zero output voltage has to be adopted when the value of $|\underline{V}_{ni}|$ is below a critical limit $|\underline{V}_{ni}|_{\mathrm{crt}}$. The zero voltage can be involved in the switching strategy when $|\underline{V}_{ni}|$ is above this limit. The value of the critical limit is chosen based on the required current response and the parameters in the equivalent circuit. Consequently, the control method that always excludes the zero voltage can be considered a particular case of the general control strategy. This particular case is defined by $|\underline{V}_{ni}|_{\mathrm{crt}} = +\infty$ so that $|\underline{V}_{ni}| < |\underline{V}_{ni}|_{\mathrm{crt}}$ in all situations.

Therefore, the adopted current control strategy in this basic form is flexible as it allows adjusting the relationship between the response speed and the harmonic content. A small $|\underline{V}_{ni}|_{\mathrm{crt}}$ ensures a better harmonic content while a large $|\underline{V}_{ni}|_{\mathrm{crt}}$ generates faster transient response. The new current control strategy generates voltage pulses defined by widths that are integer multiples of the sampling period $T_s$. The motor currents are sampled before each switching process. Every time, the last two sets of current samples are used in the calculations for the next voltage. Therefore, the output voltage can change only at definite moments in time given by

$$t_k = kT_s + \delta t \quad k = 0, 1, 2, 3, \ldots \tag{8.28}$$

where $\delta t$ is the time required for the calculation process to be fulfilled (Fig. 8.4).

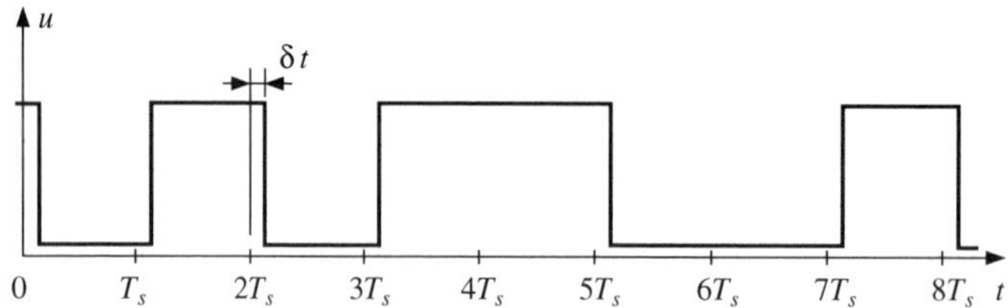


**Fig. 8.4** The PWM voltage signal generated by the basic version of the new control algorithm

The classical PWM signals are generated by comparing a sine wave (the modulator) with a triangular wave (the carrier). The widths of the generated voltage pulses vary continuously between zero and the period of the triangular carrier $T_c$. The frequency of the voltage pulses $f_{\mathrm{PWM}}$ equals the carrier frequency $f_c$. An alternative method is the space vector PWM. It changes the inverter voltage between the seven possible values in such a manner that the average voltage over several switching periods equals the reference voltage space vector. In both cases, the PWM frequency has to be vigorously controlled because it influences the quality of the generated voltage signal, but it is also approximately proportional to the losses in the inverter. Thus, it has to be limited to an acceptable value by adopting an appropriate carrier frequency. The PWM frequency used in common drive systems varies between 2 kHz and 20 kHz.

The PWM frequency generated by the new control method is not constant but is influenced by the motor operation conditions and it varies inside the interval $[0; 1/(2T_s)]$. The maximal number of switching processes per second is restricted by equation (8.28). This number needs to be large so that a voltage change can be generated at approximately the moment it is required. This consideration leads to the necessity of a short sampling period. Nonetheless, a short $T_s$ is equivalent to an increased upper limit of the PWM frequency. Therefore, an additional restriction needs to be imposed on the current control algorithm in order to limit the frequency of the PWM voltage while maintaining a short sampling period. A given PWM frequency $f_{\mathrm{PWM}}$ can be imposed if only two switching processes are allowed during a time interval of $T_{\mathrm{PWM}} = 1/f_{\mathrm{PWM}}$. If the voltage has already been switched twice during a certain time period, then the switching process is inhibited until a new period begins. Thus, $T_s$ and $T_{\mathrm{PWM}}$ are now independent quantities and $T_s$ can be set at much smaller values than $T_{\mathrm{PWM}}$. Typically, $T_{\mathrm{PWM}}$ is 10 to

40 times longer than $T_s$. As the PWM frequency can be as high as 20 kHz, it means that the sampling frequency can be up to 800 kHz, which requires fast A/D converters. This is the enhanced version of the current control algorithm. It is more flexible than the basic algorithm version because it allows a supplemental adjustment of the inverter losses beside control over the current harmonics.

### 8.2.2 The adopted neural architecture

The output voltages generated by a PWM inverter can be associated with a 3-bit code. The neural network has the task of generating the correct bit code related to each PWM rectangular voltage pulse. According to the current control strategy adopted, the neural network output signals depend on two factors: the vertex position of the non-inductive space vector $\underline{V}_{ni}(k)$ in the complex plane, and the direction indicated by error vector $[\underline{i}_{\text{ref}}(k+1) - \underline{i}(k)] = R\Delta\underline{i}_{\text{ref}}$. For each position in the complex plane and for each direction, one specific set of control signals is generated to the inverter.

The actual current $\underline{i}(k)$, the reference current $\underline{i}_{\text{ref}}(k+1)$ and the non-inductive voltage $\underline{V}_{ni}(k)$ are complex quantities treated as pairs of real values. This implies the construction of a four-dimensional Voronoi diagram: two dimensions correspond to current error vector, while the other two are the components of vector $\underline{V}_{ni}(k)$. To simplify the design process, the network has been decomposed into functional modules. Each module was then designed separately by means of two-dimensional Voronoi diagrams. The novel architecture is defined by three interconnected subnetworks (Fig. 8.5). The first neural component determines the position of the non-inductive voltage space vector $\underline{V}_{ni}(k)$ in the complex plane, while the second determines the direction of the current error vector $\Delta\underline{i}_{\text{ref}}$. The third subnetwork merges the two results and generates a 3-bit code associated with one of the output voltages of the PWM inverter.

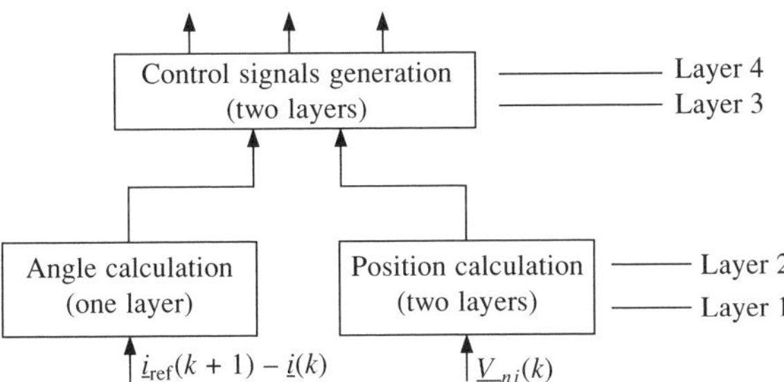


**Fig. 8.5** The architecture of the neural network

The first two subnetworks are designed by means of Voronoi diagrams. Two bi-dimensional Voronoi diagrams were used, each being the projection of a four-dimensional diagram on perpendicular planes. Each combination of four real inputs corresponds to two input-data points in two different diagrams. They are the projections of the unique input-data point in the four-dimensional input-data space.

The adopted implementation solution is: the angle subnetwork determines the argument of the error vector $\Delta\underline{i}_{\text{ref}}$ with a precision of $\pm\alpha°$, while the position subnetwork divides the complex plane into '$m$' polygonal cells and determines the cell which includes the vertex of $\underline{V}_{ni}(k)$. The generation of the control signals to be supplied to the PWM inverter considers that the vertex of $\underline{V}_{ni}(k)$ is located in the centre of the corresponding cell, and the calculation is performed accordingly.

#### *8.2.2.1 The angle subnetwork*

This subnetwork uses a number of '$n$' neurones placed within a single layer to divide the complex plane into '$2n$' sectors (Fig. 8.6). The calculation error '$\Delta\varepsilon$' is related to the number of neurones '$n$' according to equation (8.29):

$$\Delta\varepsilon = \frac{360°}{2 \cdot n} \tag{8.29}$$

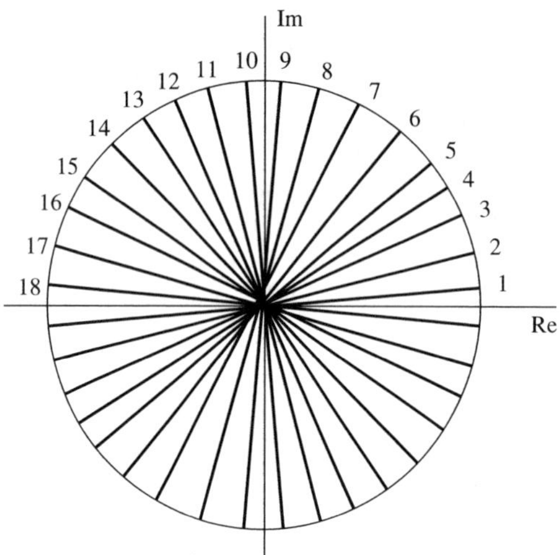


**Fig. 8.6** The division of the complex plane into sectors (angular Voronoi cells)

The neurone output is '1' if the vertex of the current error vector is located in the active region defined by the neurone in the complex plane, and it is '0' otherwise. Consequently, a different result is generated for each of the $n$ sectors. The obtained binary code has an important property: any group of consecutive sectors shares several identical bits on certain positions in the corresponding codes, as shown in Table 8.2.

The number of shared bits decreases when the width of the group increases. Thus, if the sectors are defined by $n$ neurones then the codes corresponding to a group of $m$ consecutive sectors share $n - m + 1$ identical bits. The positions of the shared bits depend on the position of the sector group inside the 360° interval. These properties are exploited by the control signal subnetwork described in section 8.2.2.3, and by the implementation of the on-line inductance estimator.

#### *8.2.2.2 The position subnetwork*

The position subnetwork divides the complex plane into polygonal cells. There is a large range of possible solutions for performing this division. According to the adopted current control principles, the argument of the difference vector $\underline{V}_{ni} - u_{\mathrm{PWM}}$, where $\underline{u}_{\mathrm{PWM}}$ is one

**Table 8.2** The codes generated by an angle subnetwork with $n = 6$ neurones

| Angle interval | Code | | | | | |
|---|---|---|---|---|---|---|
| [–15°; 15°) | 0 | 0 | 0 | 0 | 0 | 0 |
| [15°; 45°) | 1 | 0 | 0 | 0 | 0 | 0 |
| [45°; 75°) | 1 | 1 | 0 | 0 | 0 | 0 |
| [75°; 105°) | 1 | 1 | 1 | 0 | 0 | 0 |
| [105°; 135°) | 1 | 1 | 1 | 1 | 0 | 0 |
| [135°; 165°) | 1 | 1 | 1 | 1 | 1 | 0 |
| [165°; 195°) | 1 | 1 | 1 | 1 | 1 | 1 |
| [195°; 225°) | 0 | 1 | 1 | 1 | 1 | 1 |
| [225°; 255°) | 0 | 0 | 1 | 1 | 1 | 1 |
| [255°; 285°) | 0 | 0 | 0 | 1 | 1 | 1 |
| [285°; 315°) | 0 | 0 | 0 | 0 | 1 | 1 |
| [315°; 345°) | 0 | 0 | 0 | 0 | 0 | 1 |

of the inverter output voltages, equals to the argument of the corresponding current variation across the load. To ensure good operation accuracy, the arguments corresponding to the same voltage $\underline{u}_{\mathrm{PWM}}$ but to different vectors $\underline{V}_{ni}$ inside the same Voronoi cell, should have almost equal values. Using geometrical considerations, three conclusions can be drawn:

- Increasing the number of cells increases the operational accuracy of the neural network.
- The point in the complex plane corresponding to an inverter output voltage $\underline{u}_{\mathrm{PWM}}$ cannot be included in any Voronoi cell. If such a point were included in a cell then $\arg\{\underline{V}_{ni} - u_{\mathrm{PWM}}\}$ varies between 0° and 360° for different vectors $\underline{V}_{ni}$ in the same cell. Therefore, these points need to be part of the cell boundaries.
- As a consequence of the previous point, the vectors $\underline{V}_{ni}$ situated close to a voltage vector $\underline{u}_{\mathrm{PWM}}$ should be separated into a large number of cells, the criterion of separation being the value of $\arg\{\underline{V}_{ni} - u_{\mathrm{PWM}}\}$. This means that the complex plane division into cells should have a radial structure around each point corresponding to an inverter output voltage. Around such a point, the Voronoi diagram has to be similar to Fig. 8.6.

The adopted solution simultaneously takes into account the previous three points and the need to minimise the number of neurones. Therefore, the division into cells shown in Fig. 8.7 has been chosen. It has the advantage that one neurone can be involved in the radial configuration of two or three different inverter voltages thereby optimising the ratio between the current control quality and the hardware implementation complexity.

The position subnetwork contains two layers. The first layer models the boundaries of the triangular Voronoi cells. The second layer contains a number of neurones equal to the number of Voronoi cells. Each neurone is activated if the input-data point is situated inside the associated cell. The output data generated by this subnetwork is therefore a string of $N_V$ bits that contains a single bit '1' and $N_V - 1$ bits '0', where $N_V$ is the total number of Voronoi cells. The neurones in the second layer can be implemented as a combination of NOT gates and 3-input AND gates that are driven by the neurones in the first layer.

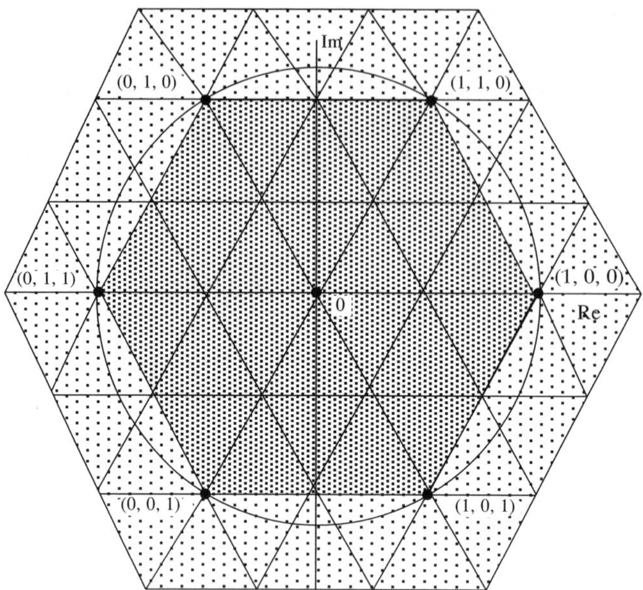


**Fig. 8.7** The partition of the interest area into Voronoi cells

### *8.2.2.3 The control signal subnetwork*

The control signal subnetwork has the task of generating the 3-bit output code related to the inverter voltage, using the information generated by the other two subnetworks. For each triangular Voronoi cell, the argument of the current error vector can have values between 0° and 360°. The interval [0°; 360°] corresponding to each cell is divided into sectors related to different inverter output voltages. The division into sectors is carried out considering that the vector $\underline{V}_{ni}$ is always situated in the centre of the corresponding triangular Voronoi cell. There are two alternative control strategies: the zero output voltage generated by the inverter is either included or excluded from the calculations. Therefore, the 360° interval is divided into either six or seven intervals, depending whether the zero voltage is used or not. The zero voltage is never used by the Voronoi cells in the immediate neighbourhood of the complex plane origin because in this case $|\underline{V}_{ni}|$ is small, and using the zero voltage would cause a very slow current response (Fig. 8.8).

The architecture of the control subnetwork contains two layers. The first layer includes six or seven neurones, one for each triangular Voronoi cell, depending on the adopted current control version and on the position of the cell with respect to the origin of the complex plane. Each neurone identifies a sector in the complex plane that is associated with a range of error vectors arguments $\arg\{\Delta\underline{i}_{\text{ref}}\}$. This argument information is coded by the angle subnetwork (Table 8.3, see page 222). Therefore, all the angle values inside a certain sector correspond to binary codes that share a given set of identical bits on certain consecutive positions inside these codes. As a result, the neurones in the first layer are implemented as AND connected to a certain number of NOT gates depending on the numbers of bits '0' and '1' to be tested. An additional input of the AND gate is connected to the output of one neurone in the position subnetwork. The output of this

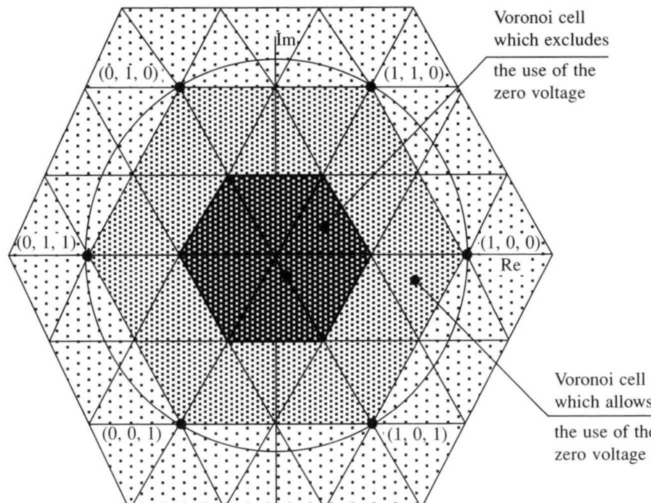


**Fig. 8.8** The triangular cell classification based on position with respect to the origin

neurone is activated when the vector $\underline{V}_{ni}$ is situated in the correct triangular Voronoi cell. The second layer consists of three neurones generating the three general output bits of the neural network. These neurones multiplex the information supplied by the first layer neurones, and they are implemented as OR logic gates.

### 8.2.3 The automated design process

A set of three programs has been developed in order to generate an adequate matrix description for the three neural subnetworks. These programs (see Appendix B) are used in conjunction with other three universal programs – CONV_NET, OPTIM and VHDL_TR – to obtain a complete automation of the neural network design and implementation. The conversion process is monitored by a master program (PWM_GEN) that controls the user interface and calls all the six specialised programs in the correct order. The logical connections between these programs are illustrated by Fig. 8.9. The master program allows the user to control the main parameters of the neural networks to be generated:

- The number of triangular Voronoi cells.
- The number '$n$' of sectors used to divide the 360° interval.
- The number of bits used to code the analogue inputs of the angle subnetwork.
- The number of bits used to code the analogue inputs of the position subnetwork.
- The maximum number of inputs allowed for one gate.
- Whether six or seven PWM output voltages are used.

**Note:** If the number of triangular Voronoi cells is only six, the present current control method becomes similar to the control algorithm presented in [179] where the complex plane is divided into six regions. The control algorithm in [179] uses very limited information on vector $\underline{e}$ because only the region in the complex plane that includes the vertex of $\underline{e}$ can be calculated. The information on $\underline{V}_{ni}$ used by the present control

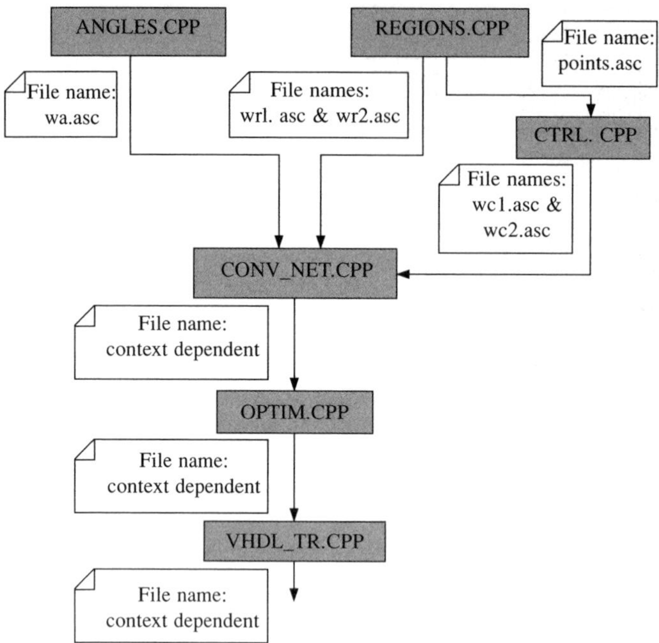


**Fig. 8.9** The neural PWM generator design programs and their interconnections

algorithm is more accurate because the inductance $L$ is estimated on-line in the manner described in 8.2.4.

Alternative architectures defined by different numerical parameters have been tested by means of computer simulation. The solution generating an optimal performance–complexity ratio has been adopted. The implementation solution uses $n_b = 5$ input bits to code each analogue input for both the angle subnetwork and the position subnetwork. The 360° interval is divided into 36 sectors while the complex plane is divided into 54 triangular Voronoi cells. The zero voltage is not used (the parameter $|\underline{V}_{ni}|_{\text{crt}}$ was considered infinite).

The initial netlist description of the angle subnetwork contained 660 logic gates arranged on 11 gate layers. The netlist was eventually optimised to 378 gates (representing 57.27 per cent of the initial gate count). The initial and the final number of gates for the position subnetwork are 567 and 242, which means compression to 42.68 per cent. This subnetwork has been implemented by a logic gate structure with 14 layers, while the control signal subnetwork has been optimised from 3026 to 709 gates resulting in a compression ratio of approx. 1:4. The corresponding hardware implementation contains only six layers of logic gates. Therefore, the total number of logic gates in the optimised neural implementation is 1329 logic gates. The overall circuit depth is 14 + 6 = 20 layers of logic gates. The VHDL descriptions of the angle subnetwork and of the position subnetwork are presented in Appendix C.

### 8.2.4 The on-line inductance estimation

For correct current control, the value of inductance $L$ needs to be either measured or estimated. An original on-line estimation method has been integrated into the switching algorithm to transform it into a universal control strategy that does not require any previous information about the motor parameters.

The on-line estimation starts with an initial inductance $\hat{L}(0)$. The inductance estimation $\hat{L}(0)$ is then incrementally updated and progressively more accurate estimations $\hat{L}(1), \hat{L}(2), \hat{L}(3)$ are calculated until the correct value is found. The algorithm convergence is guaranteed for any initial inductance value, but for reasons of simplicity, it is considered that $\hat{L}(0) = 0$. Each incremental correction is performed in parallel with one current control step (one output voltage being determined). The effect of using an estimated inductance instead of the exact value is that the non-inductive voltage $\underline{V}_{ni}$ cannot be calculated exactly, but an estimated value $\hat{\underline{V}}_{ni}$ is determined instead. Equation (8.13) can be therefore rewritten as

$$\hat{\underline{V}}_{ni}(k) = \underline{u}(k) - \frac{\hat{L}}{T_s}[\underline{i}(k) - \underline{i}(k-1)] = \underline{u}(k) - \frac{\hat{L}}{T_s} \cdot \Delta\underline{i}(k) \tag{8.30}$$

The estimated inductance is given by the relation $\hat{L} = L + \Delta L$ where $L$ is the real inductance and $\Delta L$ is the estimation error. As a result, equation (8.30) becomes

$$\hat{\underline{V}}_{ni}(k) = \underline{u}(k) - \frac{L + \Delta L}{T_s} \cdot \Delta\underline{i}(k) = \underline{V}_{ni}(k) - \frac{\Delta L}{T_s} \cdot \Delta\underline{i}(k) \tag{8.31}$$

During the next sampling period the current varies according to equation

$$\underline{u}(k+1) \cong R\underline{i}(k+1) + \underline{e}(k+1) + \frac{L}{T_s}\Delta\underline{i}(k+1) \tag{8.32}$$

Adding and subtracting $\underline{V}_{ni}(k) = R\underline{i}(k) + \underline{e}(k)$ gives

$$\underline{u}(k+1) \cong \underline{V}_{ni}(k) - R\underline{i}(k) - \underline{e}(k) + R\underline{i}(k+1) + \underline{e}(k+1) + \frac{L}{T_s}\Delta\underline{i}(k+1) \tag{8.33}$$

With the notation $\Delta\underline{e}(k+1) = \underline{e}(k+1) - \underline{e}(k)$, (8.33) can be written as

$$\underline{u}(k+1) \cong \underline{V}_{ni}(k) + R\Delta\underline{i}(k+1) + \Delta\underline{e}(k+1) + \frac{L}{T_s}\Delta\underline{i}(k+1) \tag{8.34}$$

and as

$$\underline{u}(k+1) - \underline{V}_{ni}(k) \cong \left(R + \frac{L}{T_s}\right) \cdot \Delta\underline{i}(k+1) + \Delta\underline{e}(k+1) \tag{8.35}$$

Substituting (8.31) in (8.35) gives:

$$\underline{u}(k+1) - \hat{\underline{V}}_{ni}(k) - \frac{\Delta L}{T_s} \cdot \Delta i(k) \cong \left(R + \frac{L}{T_s}\right) \cdot \Delta\underline{i}(k+1) + \Delta\underline{e}(k+1) \tag{8.36}$$

or

$$\hat{\underline{V}}_{\Delta}(k+1) \cong \left(R + \frac{L}{T_s}\right) \cdot \Delta\underline{i}(k+1) + \frac{\Delta L}{T_s} \cdot \Delta i(k) + \Delta\underline{e}(k+1) \tag{8.37}$$

where

$$\hat{\underline{V}}_{\Delta}(k+1) = \underline{u}(k+1) - \hat{\underline{V}}_{ni}(k) \tag{8.38}$$

The internal voltage $\underline{e}$ is shown to be a function of the motor currents, which has a rate of change limited by the motor inductance. Therefore, the change of the internal voltage $\underline{e}$ is similarly limited and $|\Delta\underline{e}(k+1)|$ decreases with the increase of the sampling frequency $f_s = 1/T_s$. As a result, in many practical applications $|\Delta\underline{e}(k+1)|$ is much smaller than the module of the other two terms in equation (8.37).
If conditions

$$\begin{cases} |\,\Delta\underline{e}(k+1)\,| \ll \left(R + \dfrac{L}{T_s}\right) \cdot |\,\Delta\underline{i}(k+1)\,| \\ |\,\Delta\underline{e}(k+1)\,| \ll \dfrac{|\,\Delta L \cdot \Delta\underline{i}(k)\,|}{T_s} \end{cases} \tag{8.39}$$

are fulfilled then equation (8.37) can be simplified as

$$\hat{\underline{V}}_\Delta(k+1) \cong \left(R + \frac{L}{T_s}\right) \cdot \Delta\underline{i}(k+1) + \frac{\Delta L}{T_s} \cdot \Delta\underline{i}(k) \tag{8.40}$$

The on-line induction estimation is based on the approximate equation (8.40) and on geometrical properties of the set of three space vectors involved in it. Thus, if the estimation error $\Delta L$ is positive then the space vector $\hat{\underline{V}}_\Delta(k+1)$ is situated between $\Delta\underline{i}(k+1)$ and $\Delta\underline{i}(k)$ as illustrated by Fig. 8.10(a). On the other hand, if $\Delta L$ is negative then $\Delta\underline{i}(k+1)$ lies between $\Delta\underline{i}(k)$ and $\hat{\underline{V}}_\Delta(k+1)$. In case $\Delta L = 0$, the direction of $\hat{\underline{V}}_\Delta(k+1)$ will be the same as the direction of $\Delta\underline{i}(k+1)$.

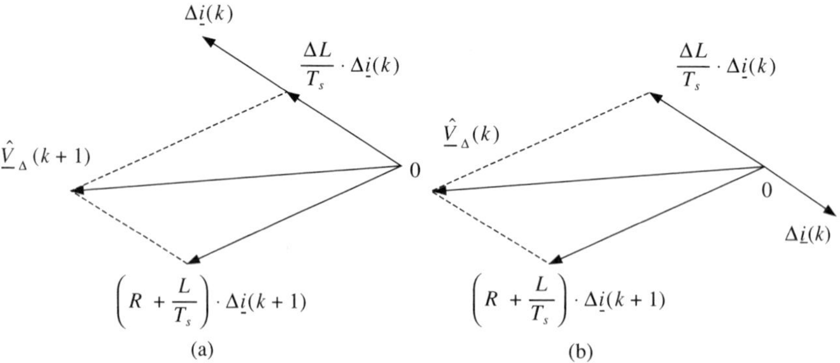


**Fig. 8.10** Inductance estimation principle

The estimated inductance $\hat{L} = L + \Delta L$ needs to be corrected by decreasing it whenever the situation in Fig. 8.10(a) occurs, and by increasing it in the situation illustrated by Fig. 8.10(b). The algorithm is concisely expressed as

$$\begin{cases} \hat{L}(k+1) = \hat{L}(k) + \delta L & \text{if } \Delta\underline{i}(k+1) \text{ is between } \hat{\underline{V}}_\Delta(k+1) \text{ and } \Delta\underline{i}(k) \\ \hat{L}(k+1) = \hat{L}(k) + \delta L & \text{if } \hat{\underline{V}}_\Delta(k+1) \text{ is between } \Delta\underline{i}(k) \text{ and } \Delta\underline{i}(k+1) \end{cases} \tag{8.41}$$

where the increment step $\delta L$ is a small positive quantity.

The presented algorithm operates correctly only if $|\Delta\underline{e}(k+1)|$ is negligible. The validity conditions (8.39) for induction estimation are a prerequisite for obtaining accurate estimation values. The larger the value of $|\Delta \underline{e}(k+1)|$ the larger the estimation errors. Other factors that influence the induction estimation process are the quantisation error of the A/D converters and the value of the step $\delta L$.

Due to the quantisation error, the vectors $\underline{\hat{V}}_{\Delta}(k+1)$ and $\Delta \underline{i}(k+1)$ are not always in the same direction even if the inductance estimation is correct. This causes small fluctuations of the estimated value after the approximate inductance has already been calculated. The amplitude of the fluctuations is proportional to the increment step size $\delta L$ and therefore it has to be small to ensure good estimation precision.

### 8.2.5 Conditions for accurate current control

As previously demonstrated, the PWM current controller operates correctly if the sampling frequency is high. Two conditions need to be fulfilled:

- The sampling frequency has to be sufficiently high to ensure that the approximate expression (8.13) of the non-inductive voltage $\underline{V}_{ni}$ is valid.
- The sampling frequency needs to be high enough to ensure that the variations of the internal voltage $|\Delta \underline{e}(k+1)|$ fulfil the conditions (8.39) for accurate inductance estimation.

The limitations imposed by these conditions in the particular case where the inverter load is an induction motor are investigated and presented in [81]. The calculations incorporate a series of approximations that do not diminish the generality of the conclusions. The values of the motor parameters in Table 8.1 are used as a guide to determining the validity of the approximations. The sampling frequency $f_s$ must comply [81] with the condition

$$f_s = \frac{1}{T_s} >> \omega_{max} = 2\pi \cdot 100 = 628 \text{ Hz} \tag{8.42}$$

It can be demonstrated [81] that the first condition for accurate induction estimation becomes

$$\left|\frac{d\underline{e}}{dt}\right| < (R_s + L\omega_{max}) \cdot \left|\frac{d\underline{i}_s^s}{dt}\right| << \left(R_s + \frac{L}{T_s}\right) \cdot \left|\frac{d\underline{i}_s^s}{dt}\right| \Leftrightarrow \frac{1}{T_s} >> \omega_{max} \tag{8.43}$$

which is equivalent to condition (8.42).

The second condition for accurate induction estimation is [81]:

$$\left(\frac{R_s}{L} + \omega_{max}\right) \cdot (1 + k_{PWM}) \cdot |\underline{u}_s^s| << \frac{\Delta L}{L \cdot T_s} \min \left|\frac{d\underline{i}_s^s}{dt}\right| = \frac{\Delta L}{L \cdot T_s}(1 - k_{PWM}) \cdot |\underline{u}_s^s| \tag{8.44}$$

The sampling frequency $f_s$ which fulfils this condition is given by

$$f_s = \frac{1}{T_s} >> \frac{1 + k_{PWM}}{1 - k_{PWM}} \cdot \frac{R_s + L\omega_{max}}{\Delta L} \tag{8.45}$$

The accepted inductance estimation error $\Delta L$ is an important factor that influences the minimal sampling frequency. Very accurate inductance estimations imply small $\Delta L$ that, according to (8.45), require high sampling frequencies. Less accurate inductance estimation can be performed at lower sampling frequency. For example, the inductance estimation with a precision of 0.5 mH for the 11.1 kW motor in Table 8.1, when $k_{PWM} = 0.7$, requires that the $f_s$ is much larger than 46.2 kHz. Therefore, an adequate sampling frequency is 450 kHz. Even larger sampling frequencies are required if either $k_{PWM}$ has larger values or if higher estimation precision is required. The practical solution to limit the sampling frequency $f_s$ while obtaining accurate induction estimation is to perform the estimation process at small stator angular frequency $\omega_{es}$.

In conclusion, the lower limit for the sampling frequency has been determined [81]. It is defined by conditions (8.42), (8.43) and (8.45) which must be simultaneously fulfilled. The first two conditions are less restrictive as they depend only on the maximal motor speed. The last condition is more restrictive and depends on the motor speed, on its electrical parameters and on the PWM modulation index. Consequently, condition (8.45) alone can be used for practical calculations as any sampling frequency determined based on (8.45) also fulfils conditions (8.42) and (8.43).

### 8.2.6 Current control implementation methods

The implementation of the two interrelated algorithms can be performed using DSPs or specialised digital architectures (ASICs or FPGAs). The DSPs approach is simple because the corresponding software development is straightforward. However, the two algorithms imply a large number of time-consuming mathematical operations to be performed for each PWM pulse and therefore they use most of the DSP resources, limiting its capability to perform the speed control task. The use of specialised digital structures ensures fast operation and allows the speed control and current control algorithms to be performed in parallel.

There are two possible software implementations for the current control algorithm: the direct implementation of calculating all the angles, and the indirect implementation that uses scalar products between vectors to find the optimal output voltage. The direct implementation implies calculating the six necessary angles using 'arctan' trigonometric function [81] as shown in (8.46), and then comparing the results.

$$\alpha = \begin{cases} \arctan\left(\dfrac{\mathrm{Im}\{\underline{u}_{\mathrm{INV}} - \underline{V}_1\}}{\mathrm{Re}\{\underline{u}_{\mathrm{INV}} - \underline{V}_1\}}\right) & \text{when } \mathrm{Re}\{\underline{u}_{\mathrm{INV}} - \underline{V}_1\} > 0 \\ \pi - \arctan\left(\dfrac{\mathrm{Im}\{\underline{u}_{\mathrm{INV}} - \underline{V}_1\}}{\mathrm{Re}\{\underline{u}_{\mathrm{INV}} - \underline{V}_1\}}\right) & \text{when } \mathrm{Re}\{\underline{u}_{\mathrm{INV}} - \underline{V}_1\} < 0 \\ \dfrac{\pi}{2} & \text{when } \begin{cases} \mathrm{Re}\{\underline{u}_{\mathrm{INV}} - \underline{V}_1\} = 0 \\ \mathrm{Im}\{\underline{u}_{\mathrm{INV}} - \underline{V}_1\} > 0 \end{cases} \\ -\dfrac{\pi}{2} & \text{when } \begin{cases} \mathrm{Re}\{\underline{u}_{\mathrm{INV}} - \underline{V}_1\} = 0 \\ \mathrm{Im}\{\underline{u}_{\mathrm{INV}} - \underline{V}_1\} < 0 \end{cases} \end{cases} \tag{8.46}$$

The 'arctan' function can be implemented as a look-up table thereby accelerating the calculations but the sequentially performed subtractions, divisions and comparisons required by (8.46) considerably slow down the calculation process. A realistic estimate of the computational effort can be obtained considering that the first two possibilities in (8.46) are equally probable while the last two are unlikely to be fulfilled due to the exact equalities which are involved. In this case between 12 and 18 subtractions (depending how many times case (a) and case (b) occur in (8.46)) are requested. Additionally, a further six divisions and between 6 + 6 = 12 and 9 + 6 = 15 comparisons have to be performed for each sampling period.

There are DSPs containing on-chip RAM and on-chip maskable ROM (for instance, TMS320C5x) [9]. Consequently, both the control program and the look-up table can reside either on-chip or off-chip in an external EPROM. On-chip configuration is advantageous because it is compact, reliable and simplifies the PCB design. Unfortunately, the on-chip ROM memory space is limited. A total of $8.2^{10}$ memory words are available for TMS320C51 and twice as much for TMS320C53 [9]. If the complete motor control program is large then space for a look-up table might not be available. In these circumstances the table would have to be placed in an external EPROM.

For a simple and compact hardware implementation, alternative algorithms must be used to eliminate the need for the external EPROM memory chip. As a result, the trigonometric function calculations must be avoided because any specialised routine for calculating such functions is a huge time consumer. The indirect implementation avoids trigonometric functions by utilising scalar products, which are performed between each of the six vectors and the reference vector. By definition the scalar product between two $n$-dimensional vectors $\bar{a}$ and $\bar{b}$ is given by

$$\bar{a} \cdot \bar{b} = |\bar{a}| \cdot |\bar{b}| \cdot \cos\varphi \tag{8.47}$$

The smaller the angle $\varphi$ between the two vectors, the larger the result. Thus, the correct output voltage is chosen by maximising the corresponding scalar product [81]. The calculations are relevant only if the six vectors have the same module. To fulfil this condition, the six vectors need to be normalised. Consequently the six scalar products $p_j$ (where $j = 1, 2, \ldots, 6$) have to be calculated.

The combined software implementation of the current control and inductance estimation algorithms requires a very large number of mathematical operations to be performed. In the case where the direct current control implementation is employed, then up to 12 + 24 = 36 algebraic calculations and 15 + 2 = 17 comparisons are requested for each algorithm step. The use of the indirect current control implementation in order to eliminate the need for look-up tables requires up to 12 + 54 = 66 algebraic operations and 5 + 2 = 7 comparisons for each algorithm step. In a typical situation where the PWM frequency of 20 kHz and $T_{\text{PWM}} = 10 \cdot T_s$, the algorithm requires that 200 000 calculation steps are performed each second. This amounts to a total of 7 200 000 algebraic operations plus 3 400 000 comparisons per second in the case of the direct implementation. The indirect implementation requires a computational effort of 13 200 000 algebraic operations and 1 400 000 comparisons per second.

A DSP program created to perform all these operations contains supplementary instructions: reading operands from memory, writing results to memory, program control instructions (jumps), and so on. An optimistic estimate is that the calculation and comparison instruction number is approximately equal to the number of all other instructions in the program. This means that a speed of at least 30 000 000 instructions per second (30 MIPS) is required for on-line operation. The DSPs commonly used in drive system applications belong to the TMS320C3x and TMS320C5x series and are fast and inexpensive processors. For example, the 16-bit, fixed-point DSPs TMS320C5x generation performs up to 50 MIPS while some of the TMS320C3x perform 30 MIPS but have a separate 60 MFLOPS floating point arithmetic unit, which increases the total computation power tremendously. The latest DSP devices offer much larger calculation speeds (TMS320C6x generation offers 1600 MIPS) but their price is still too high for inexpensive drive control applications. Thus, the computation effort required by the adopted current control strategy can be handled by an inexpensive DSP. However, the algorithm consumes a significant part of its total resources (up to 33 per cent if a DSP from the TMS320C3x generation is used and up to 60 per cent if a TMS320C5x is used). Therefore, a complex induction motor control strategy including the presented current control method can be difficult to implement using a single common inexpensive DSP. The resistance estimation algorithms plus the flux and speed control procedures involve a large number of calculations and the total requirements can easily surpass the calculation power of such a chip.

Multiprocessor DSP-based control systems are therefore not a practical solution, and hardware implementation using ASIC or FPGA technologies proves to be an adequate alternative strategy for a fast and efficient control system capable of providing high performance. The high speed is achieved by adapting the hardware architecture to the algorithm specific data flow requirements. In addition, pipelining and parallel processing can be used on a large scale to exploit all the opportunities offered by the specific calculation algorithms. The more parallelisms that can be found in one algorithm the faster the operation of its hardware implementation can be. Calculation parallelism is best exploited by hardware implemented neural networks containing tens or hundreds of elementary processing units cooperating to solve a particular problem. The neural approach is flexible as the neurone number can be increased or decreased and the calculation precision varied accordingly. The neural network size and architecture is determined based on the necessary calculation precision and the available hardware resources. The motor controller structure developed in this research work uses FPGA implemented neural networks alongside pipelined digital structures to carry out the computationally intensive task of controlling the stator current. This solution is fast, inexpensive and eliminates the timing problems related to the sequential operation of a DSP processor. Using this approach, the complexity of the control tasks performed is not significantly limited by the hardware operation speed. The only important limitation is given by the available amount of hardware resources.

### 8.2.7 Current control simulation

The system simulation approach combines the modelling flexibility of the VHDL software tools with the graphical capabilities of MATLAB. Thus, the simulation results have been generated in a numerical format using Workview Office 7.31 produced by Viewlogic, and then imported in MATLAB to generate the corresponding graphs.

A VHDL model of a three-phase induction motor has been created using the mathematical space vector model of the motor. A separate simplified model of the PWM inverter has been developed considering all power transistors as ideal switches. The two modules were combined with an abstract VHDL description of the adopted control strategy to generate a model of the entire drive system. This has been used to analyse the current control principles presented in section 8.2. The system operation has been simulated with different parameter values and the simulation results validated the current control principles previously presented.

In the VHDL Code Fragment 8.1, the motor model is an entity having two input ports (the stator voltage and load torque) and two output ports (the stator currents and the rotor angular frequency). All data regarding the motor operation during the simulation is stored in an output ASCII file (motor.txt). The file contains numerical data in matrix format, which is compatible with MATLAB. Each line in the matrix contains the set of quantities that characterise the motor operation at a certain moment in time: currents, voltages, speed and torque.

```
-- Code Fragment 8.1
LIBRARY math;
USE math.complex_basic.all;
USE math.mathtyx.all;
USE std.textio.all;

ENTITY motor IS
  PORT(us: IN COMPLEX;
  Tload: IN REAL;
  ist: OUT COMPLEX;
  omegar: OUT REAL);
END motor;

ARCHITECTURE arch_motor OF motor IS
  CONSTANT Rs: REAL :=0.371;
  CONSTANT Rr: REAL :=0.415;
  CONSTANT Ls: REAL :=0.08705;
  CONSTANT Lr: REAL :=0.08763;
  CONSTANT Lm: REAL :=0.08433;
  CONSTANT Jr: REAL :=0.1;
  CONSTANT p: REAL :=2.0;
  CONSTANT deltat: TIME :=50 ns;
  CONSTANT dt: REAL := 5.0e-8;
  SIGNAL next_step: INTEGER :=1;
  FILE outf : TEXT IS OUT "c:\andrei\motor.txt";
BEGIN
  PROCESS(next_step)
  VARIABLE my_line: LINE;
  VARIABLE ist1,ist2,ir1,ir2,Fist1,Fist2,Fir1,Fir2,z:
  COMPLEX :=(0.0,0.0);
  VARIABLE T,omegar1,omegar2: REAL :=0.0;
  CONSTANT d_space: STRING :=" ";
  BEGIN
  IF next_step=1 THEN
  WRITE(my_line,us.re);
  WRITE (my_line,d_space);
  WRITE(my_line,us.im);
  WRITE (my_line,d_space);
  WRITE(my_line,ist1.re);
  WRITE (my_line,d_space);
  WRITE(my_line,ist1.im);
  WRITE (my_line,d_space);
  WRITE(my_line,ir1.re);
```

```
  WRITE (my_line,d_space);
  WRITE(my_line,ir1.im);
  WRITE (my_line,d_space);
  WRITE(my_line,T);
  WRITE (my_line,d_space);
  WRITE(my_line,omegar1);
  WRITELINE(outf,my_line);
  END IF;
  ist1:=ist2;
  ir1:=ir2;
  Fist1:=Fist2;
  Fir1:=Fir2;
  omegar1:=omegar2;
  Fist2:=Fist1+(us-Rs*ist1)*dt;
  Fir2:=Fir1+(j*omegar1*Fir1-Rr*ir1)*dt;
  ist2:=(Lr*Fist2-Lm*Fir2)/(Lr*Ls-Lm*Lm);
  ir2:=(Ls*Fir2-Lm*Fist2)/(Lr*Ls-Lm*Lm);
  z:=ist1*conj(ir1);
  T:=3.0/4.0*p*Lm*(z.im);
  omegar2:=omegar1+(T-Tload)/Jr;
  IF next_step<1000 THEN
  next_step<=next_step+1 AFTER deltat;
  ELSE
  next_step<=1 AFTER deltat;
  END IF;
  ist<=ist1;
  omegar<=omegar1;
  END PROCESS;
END arch_motor;

CONFIGURATION conf_motor OF motor IS
  FOR arch_motor
  END FOR;
END conf_motor;
```

The PWM inverter is modelled as a simple VHDL process. The sensitivity list of the process contains the 6-bit vector ‘abcdef’ containing the control signals to the six power transistors. The first three bits uniquely define the inverter output voltage. They are used as the selection criterion in the CASE statement that generates the corresponding voltage space vector ‘us’.

```
-- Code Fragment 8.2
process(abcdef(5 downto 3))
  constant U0: REAL :=
  begin
  case abcdef(5 downto 3) is
  when "100"=> us<=U0*(1.0,0.0);
  when "110"=> us<=U0*(0.5,0.866);
  when "010"=> us<=U0*(-0.5,0.866);
  when "011"=> us<=U0*(-1.0,0.0);
  when "001"=> us<=U0*(-0.5,-0.866);
  when "101"=> us<=U0*(0.5,-0.866);
```

```
when others=> us<=(0.0,0.0);
end case;
end process;
```

The motor parameters used for the simulations presented in Figs. 8.11, 8.12 and 8.13 are given in column (b) of Table 8.1. Thus, according to equation (8.8) the inductance in the equivalent circuit is $L = 5.9$ mH. The parameters defining the operation of the current controller are as follows:

- The module of the reference stator current space vector: 10 A.
- The reference frequency: 50 Hz.
- The inductance updating step $\delta L$: 0.05 mH.
- The PWM frequency: 20 kHz.
- The sampling frequency: 450 kHz.
- $|\underline{V}_{ni}|_{\text{crt}} = \infty$.

In Fig. 8.11 it is shown that the correct inductance value is obtained in a short time interval (about 200 ms). The initial induction estimation error is very large causing very large errors in the calculation of the non-inductive voltage vector $\underline{V}_{ni}$. The increasing accuracy of the inductance estimate is reflected in the decreasing ripples of estimated $\underline{V}_{ni}$ shown in Fig. 8.12.

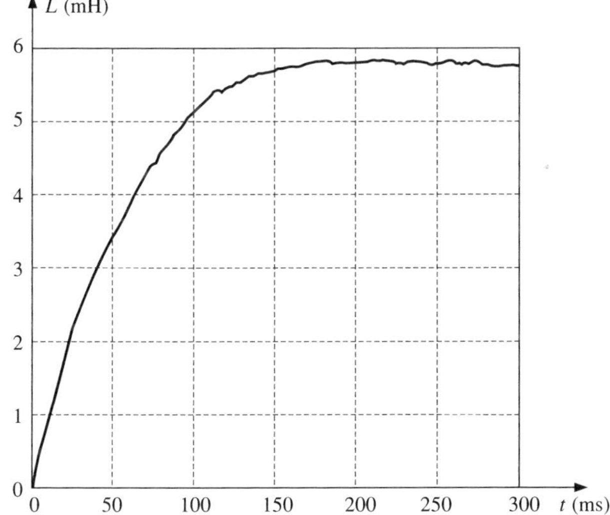


**Fig. 8.11** Inductance estimation values

Figure 8.13 presents the trajectory in the complex plane of the actual current space vector across the stator winding. It demonstrates that the current ripples are maintained at low levels.

Similar simulations have been performed for a current control algorithm version that always uses all the seven output voltages ($|\underline{V}_{ni}|_{\text{crt}} = 0$). Similar results are obtained but the inductance estimation process is much slower (about ten times slower). Moreover, the variations of the inductance estimation value in steady state operation are twice the

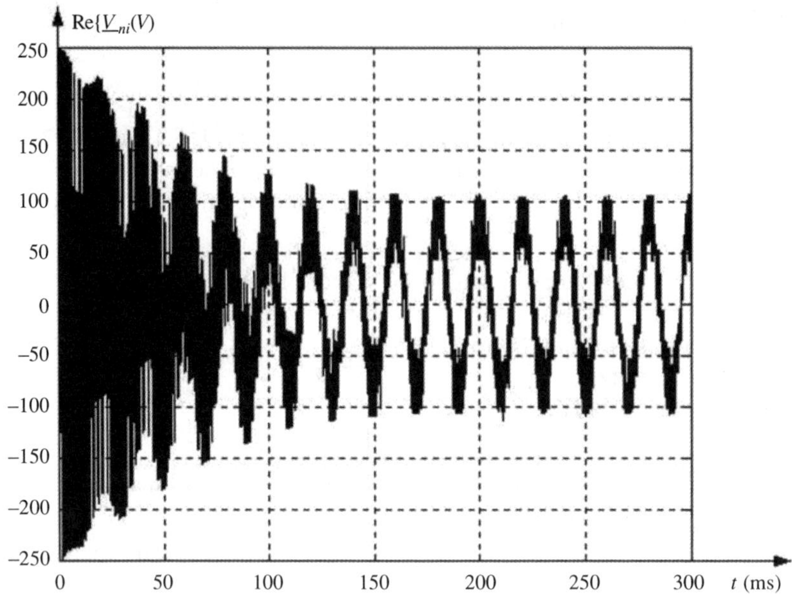


**Fig. 8.12** The non-inductive voltage estimation

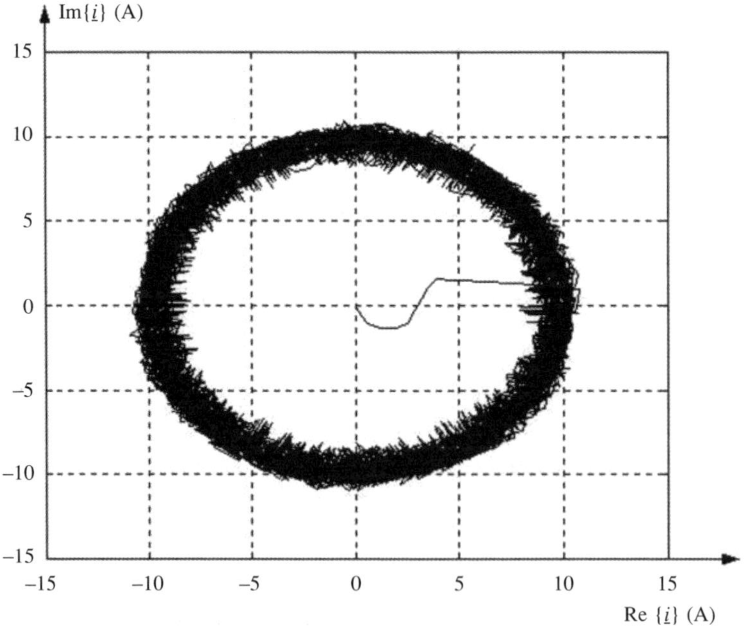


**Fig. 8.13** The load current space vector

amplitude of the result of Fig. 8.11. However, the current ripple amplitude decreased to approximately 70 per cent compared to Fig. 8.13 after the inductance estimation process was finished. Thus, although using the zero voltage decreases the current ripples, it also increases the inductance estimation errors, thereby affecting the accuracy of the information on motor operation. Consequently, this control method version decreases the precision of any sensorless speed control strategy based on the equivalent *R–L–e* circuit.

## 8.3 The new sensorless motor control strategy

The common sensorless induction motor control strategies are derived from the sensor-based field-oriented control methods that have been extended to include speed estimation algorithms. All field orientation methods require several transformations of the electromagnetic quantities from the stator reference frame into the flux reference frame, and back from the flux reference frame into stator reference frame. Thus, a general stator quantity '$\underline{A}$' is transformed from fixed stator coordinates into mobile flux coordinates, using equation (8.48). The inverse transformation is carried out according to (8.49) where $\theta(t)$ indicates the angle of the flux vector (the rotor flux, the stator flux or the airgap flux) and is a function of time. The complete control algorithms require several other mathematical calculations to be carried out: integrations, divisions, multiplications and square roots.

$$\begin{pmatrix} A_d^\theta \\ A_q^\theta \end{pmatrix} = \begin{pmatrix} \cos\theta(t) & \sin\theta(t) \\ -\sin\theta(t) & \cos\theta(t) \end{pmatrix} \cdot \begin{pmatrix} A_d^s \\ A_q^s \end{pmatrix} \tag{8.48}$$

$$\begin{pmatrix} A_d^s \\ A_q^s \end{pmatrix} = \begin{pmatrix} \cos\theta(t) & -\sin\theta(t) \\ \sin\theta(t) & \cos\theta(t) \end{pmatrix} \cdot \begin{pmatrix} A_d^\theta \\ A_q^\theta \end{pmatrix} \tag{8.49}$$

The sensorless speed control strategies are usually software implemented using DSPs or microcontrollers, because the hardware implementation is difficult to achieve due to the large number of different mathematical operations to be implemented. Attempts have been made to combine the software and the hardware approaches [98]. This strategy requires the implementation of a custom mathematical processor alongside specialised control modules, RAM memory and ROM memory (to store the control programs). The specialised modules implement routine tasks such as the PWM signal generation or the A/D converter control, while the mathematical processor carries out all the complex mathematical calculations and updates the operation parameters for the specialised modules. Such an approach combines the flexibility of software implementation and the speed of the hardware implementation. It requires large integrated circuits and complex design procedures including simulation tools capable of checking the operation of the mixed software–hardware control block. An inexpensive, simple and compact hardware implementation requires the calculation overhead to be minimised so that the software component of the control algorithm can be eliminated. Therefore, simpler speed estimation methods and control strategies need to be devised.

The calculation complexity can be much reduced if quantities that are invariant at reference frame transformations are used so that matrix equations like (8.48) and (8.49) can be eliminated. The two types of quantities having such a property are the space vector modules and the phase shifts between the space vectors. A control algorithm that operates with these quantities is more suitable to using polar coordinates than the classical rectangular coordinates. Consequently, new speed estimation algorithms and speed control algorithms need to be developed and expressed as simple equations in polar coordinates. The novel speed control strategy proposed can operate in conjunction with the current control method described in section 8.2, or it can be implemented independently. In both situations the speed control strategy is based on two principles:

- The speed information is extracted by analysing the magnitude and/or the phase shift between two space vectors $\underline{A}$ and $\underline{B}$, chosen from the electromagnetic variables in the equivalent *R–L–e* circuit ($\underline{u}$, $\underline{V}_{ni}$, $\underline{e}$, $\underline{i}$).
- The motor speed is controlled by modifying the amplitude and the angular speed of the stator current vector.

Using only quantities that are invariant at reference frame transformations (phase shifts and amplitudes) implies that the choice over the reference frame does not change the form of the speed estimation method or the form of the speed control algorithm. All reference frames are equivalent. However, the mathematical demonstration of the principles underlying the new control strategy is simpler in rectangular coordinates than in polar coordinates. For simplification the most appropriate approach is to define the reference frame orientation using vector $\underline{A}$ involved in the motor speed estimation (the real axis of the rectangular coordinates is maintained parallel to this vector as illustrated in Fig. 8.14). In this situation, the phase shift $\alpha_{BA}$ between $\underline{B}$ and $\underline{A}$ is calculated using only the rectangular components of vector $\underline{B}$:

$$\begin{cases} \mathrm{Re}\{\underline{B}_\theta\} = B \cdot \cos \alpha_{BA} \\ \mathrm{Im}\{\underline{B}_\theta\} = B \cdot \sin \alpha_{BA} \end{cases} \Rightarrow \alpha_{BA} = \arctan\left(\frac{\mathrm{Im}\{\underline{B}_\theta\}}{\mathrm{Re}\{\underline{B}_\theta\}}\right) \tag{8.50}$$

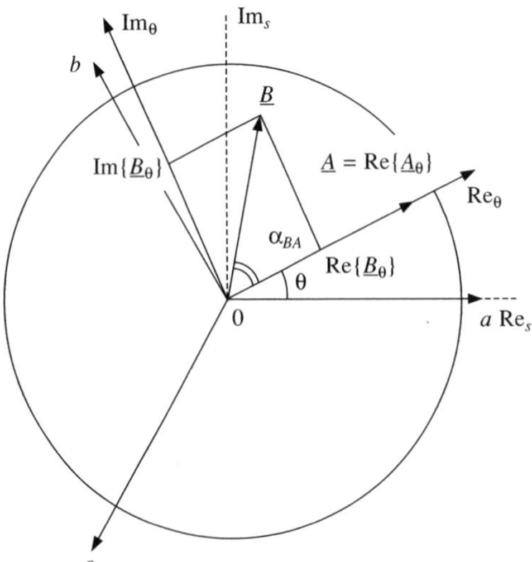


**Fig. 8.14** The reference frame oriented on vector *A*

Section 8.2.2.1 demonstrates that the calculation of the space vector arguments can be efficiently carried out by hardware implemented neural networks. The result is that the phase shift calculation is reduced to subtracting the space vector arguments, thereby avoiding trigonometric calculations and reducing the total chip area of the controller.

Therefore, equations like (8.50) are used only for theoretical analysis but do not have to be implemented directly into hardware.

### 8.3.1 Speed estimation algorithms

Several estimation methods can be developed depending on the vectors $\underline{A}$ and $\underline{B}$ that are chosen from the quantities available in the *R–L–e* circuit ($\underline{u}$, $\underline{V}_{ni}$, $\underline{e}$, $\underline{i}$). The methods have different degrees of accuracy and imply different calculation complexity levels. The most straightforward solution is operating with the voltage $\underline{u}$ and the current $\underline{i}$ because they are directly measurable quantities. The non-inductive voltage $\underline{V}_{ni}$ is a good option if the speed estimation is performed by a controller that uses the current control strategy presented in section 8.2. The vector $\underline{V}_{ni}$ is calculated for the current control algorithm but the information can also be transferred to the speed estimator thereby decreasing the computation effort. The use of the internal voltage $\underline{e}$ requires the largest number of calculations because its value needs to be derived from the space vectors $\underline{u}$ or $\underline{V}_{ni}$.

The class of estimation methods defined by $\underline{A} = \underline{i}_s$ is analysed in a stator current-oriented reference frame. Due to the stator current orientation, the imaginary part of the stator current vector is zero and the reference frame rotates with the angular speed $\omega_{es}$, which corresponds to the synchronism speed in steady-state operation. Throughout this section, the superscript 'syn' is attached to space vectors expressed in the synchronous stator current-oriented reference frame. The conditions defining the chosen reference frame are mathematically described by

$$\begin{cases} \theta = \omega_{es} \cdot t \\ \underline{i}_s = I_s \end{cases} \tag{8.51}$$

where $I_s$ designates a real quantity. The induction motor space vector model in a stator current-oriented reference frame is expressed as

$$\begin{cases} \underline{u}_s^{\text{syn}} = R_s I_s + \dfrac{\mathrm{d}\underline{\Psi}_s^{\text{syn}}}{\mathrm{d}t} + j\omega_{es}\,\underline{\Psi}_s^{\text{syn}} \\ \underline{u}_r^{\text{syn}} = R_r\,\underline{i}_r^{\text{syn}} + \dfrac{\mathrm{d}\underline{\Psi}_r^{\text{syn}}}{\mathrm{d}t} + j(\omega_{es} - \omega_{er}) \cdot \underline{\Psi}_r^{\text{syn}} \\ \underline{\Psi}_s^{\text{syn}} = L_s I_s + L_m\,\underline{i}_r^{\text{syn}} \\ \underline{\Psi}_s^{\text{syn}} = L_r\underline{i}_r^{\text{syn}} + L_m I_s \end{cases} \tag{8.52}$$

If the last two equations in (8.52) are substituted in the first two, the result is

$$\begin{cases} \underline{u}_s^{\text{syn}} = R_s I_s + L_m\dfrac{\mathrm{d}\underline{i}_r^{\text{syn}}}{\mathrm{d}t} + L_s\dfrac{\mathrm{d}I_s}{\mathrm{d}t} + j\omega_{es}(L_s I_s + L_m\,\underline{i}_r^{\text{syn}}) \\ 0 = R_r\,\underline{i}_r^{\text{syn}} + L_r\dfrac{\mathrm{d}\underline{i}_r^{\text{syn}}}{\mathrm{d}t} + L_m\dfrac{\mathrm{d}I_s}{\mathrm{d}t} + j\omega_{\text{slp}}(L_m I_s + L_r\,\underline{i}_r^{\text{syn}}) \end{cases} \tag{8.53}$$

The quantity '$\omega_{\text{slp}}$' in (8.53) is the 'slip angular frequency' and represents the difference between the stator and the rotor electrical angular frequencies.

$$\omega_{\text{slp}} = \omega_{es} \quad \omega_{er} \tag{8.54}$$

The slip angular frequency is related to the motor slip '*s*' as follows

$$\omega_{\mathrm{slp}} = s \cdot \omega_{es} \tag{8.55}$$

The calculation of $\omega_{\mathrm{slp}}$ is a prerequisite for the rotor speed estimation. The relation between $\omega_{\mathrm{slp}}$ and the rotor speed is described by

$$\omega_r = \frac{\omega_{es} - \omega_{\mathrm{slp}}}{p} = \omega_s - \frac{\omega_{\mathrm{slp}}}{p} \tag{8.56}$$

Therefore, the speed estimation methods can be reduced to methods of estimating the slip angular frequency. Due to their mathematical simplicity and hardware implementation advantages, the estimators based on the phase-shift information are adopted in this work as the optimal solution to the speed calculation problem [81]. To achieve the motor speed estimation, the information contained in the value of the angle $\alpha$ between the stator current vector $\underline{i}$ and one of the other vectors: $\underline{e}$, $\underline{u}_s$ or $\underline{V}_{ni}$ is processed. The calculations [81] yield the following results:

$$\tan^{-1}\alpha_{ui} = \tan^{-1}(\arg\{\underline{u}_r^{\mathrm{syn}}\}) = \frac{R_s L_r (R_r^2 + \omega_{\mathrm{slp}}^2 L_r^2) + \omega_{es}\omega_{\mathrm{slp}} L_m^2 L_r R_r}{(L_r L_s - L_m^2)\cdot(R_r^2 + \omega_{\mathrm{slp}}^2 L_r^2)\cdot \omega_{es} + L_m^2 R_r^2 \omega_{es}} \tag{8.57}$$

$$\tan^{-1}\alpha_{V_{ni}i} = \frac{\mathrm{Re}\{\underline{V}_{ni}^{\mathrm{syn}}\}}{\mathrm{Im}\{\underline{V}_{ni}^{\mathrm{syn}}\}} = \frac{\omega_{\mathrm{slp}} L_r}{R_r} + \frac{L_r R_s (R_r^2 + \omega_{\mathrm{slp}}^2 L_r^2}{\omega_{es} L_m^2 R_r^2} \tag{8.58}$$

$$\tan^{-1}\alpha_{ei} = \frac{\mathrm{Re}\{\underline{e}^{\mathrm{syn}}\}}{\mathrm{Im}\{\underline{e}^{\mathrm{syn}}\}} = \frac{\omega_{\mathrm{slp}} L_r}{R_r} \tag{8.59}$$

Relation (8.57) is very complicated and non-linear. It is not suitable to hardware implemented estimation of the slip angular speed. The result in (8.58) is simpler than (8.57), it contains one term that is proportional with $\omega_{\mathrm{slp}}$ and another term proportional with the slip angular frequency squared. If the stator angular frequency is high ($\omega_{es} \cong$ 314 rad/s), and the slip is small (it normally is during typical motor operation) then the last term in (8.58) can be neglected and an almost linear relationship between $\omega_{\mathrm{slp}}$ and $\tan^{-1}(\alpha_{V_{ni}i})$ is obtained:

$$\frac{\mathrm{Re}\{\underline{V}_{ni}^{\mathrm{syn}}\}}{\mathrm{Im}\{\underline{V}_{ni}^{\mathrm{syn}}\}} \approx \frac{\omega_{\mathrm{slp}} L_r}{R_r} \Rightarrow \omega_{\mathrm{slp}} \approx \frac{R_r}{L_r} \cdot \tan^{-1}\alpha_{V_{ni}i} \tag{8.60}$$

Figure 8.15 presents the numerical calculation results obtained for the 11.1 kW motor at a stator angular frequency $\omega_s$ = 314 rad/s. It can be seen that the $\mathrm{Re}(\underline{V}_{ni})/\mathrm{Im}(\underline{V}_{ni})$ characteristics are almost straight lines. Unfortunately, their curvature increases with the decrease of speed so that the relationship (8.60) is not valid for speeds much below the rated speed.

The approximate relationship (8.60) is appropriate for hardware implementation together with the current control strategy previously presented in section 8.2, because it provides the value of $\underline{V}_{ni}$. This version of the slip estimator is based on the equation

$$\omega_{\mathrm{slp}} = \tan^{-1}(\alpha_{V_{ni}i}) \cdot \frac{R_r}{L_r} = \tan^{-1}[\arg(\underline{V}_{ni}^s) - \arg(\underline{i}_s^s)] \cdot \frac{R_r}{L_r} \tag{8.61}$$

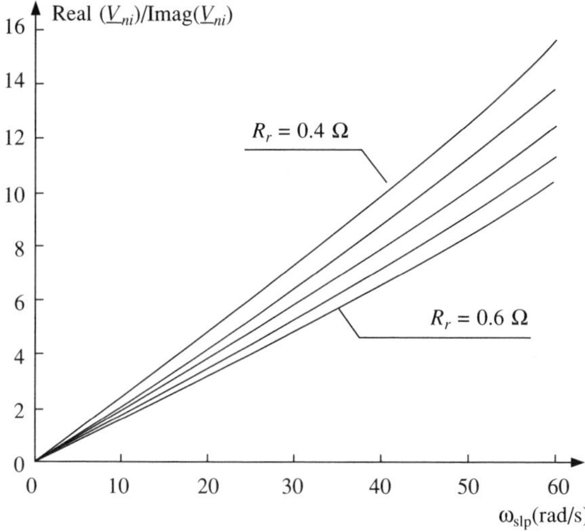


**Fig. 8.15** Quasi-linear dependency between $\omega_{\text{slp}}$ and Re($V_{ni}$)/Im($V_{ni}$) ratio ($\omega_{es}$ = 314 rad/s)

However, such an estimation method can be used only in a limited number of practical applications, where the motor speed is variable but always high. The correct slip estimation at any speed can only be performed using the equation

$$\omega_{\text{slp}} = \tan^{-1}(\alpha_{ei}) \cdot \frac{R_r}{L_r} = \tan^{-1}[\arg(\underline{e}^s) - \arg(\underline{i}_s^s)] \cdot \frac{R_r}{L_r} \tag{8.62}$$

that is derived from (8.59).

The internal voltage vector $\underline{e}$ can be calculated either as a function of $\underline{V}_{ni}$, $\underline{i}_s$ and $R_s$, or based on $\underline{u}_s$, $R_s$, $\omega_{es}$ and $\underline{i}_s$. The two alternatives are:

$$\begin{cases} \underline{e}_s^s = \underline{V}_1^s - R_s \underline{i}_s^s \\ \underline{e}_s^s = \underline{u}_s^s - R_s \underline{i}_s^s - \dfrac{L_s L_r - L_m^2}{L_r} \cdot \dfrac{d\underline{i}_s^s}{dt} \end{cases} \tag{8.63}$$

The choice made depends upon the electrical quantities available. If the new current control method is used, then the $\underline{V}_{ni}$-based estimator is optimal in terms of hardware implementation. Otherwise, the $\underline{u}$-based estimator is the better option. Thus, the two alternative estimators operate based on the equivalent equations

$$\begin{cases} \omega_{\text{slp}} = \tan^{-1}[\arg(\underline{V}_1^s - R_s \underline{i}_s^s) - \arg(\underline{i}_s^s)] \cdot \dfrac{R_r}{L_r} \\ \omega_{\text{slp}} = \tan^{-1}\left[\arg\left(\underline{u}_s^s - R_s \underline{i}_s^s - \dfrac{L_s L_r - L_m^2}{L_r} \cdot \dfrac{d\underline{i}_s^s}{dt}\right) - \arg(\underline{i}_s^s)\right] \cdot \dfrac{R_r}{L_r} \end{cases} \tag{8.64}$$

These two estimators are superior to the amplitude-based estimators because the division and the square root calculation are eliminated. The $\underline{V}_{ni}$-based estimator is particularly simple, as it requires only two multiplications and one subtraction. The $\underline{u}$-based estimator is slightly more complicated because it requires one additional multiplication and two additional subtractions (the current derivative being approximated by the difference of the last two current samples). The vector argument calculations in the stator reference frame can be performed by a hardware implemented neural network. Alternatively, the function $\tan^{-1}$ can be easily implemented as a small look-up table because it is a periodical and symmetrical function and only its values between 0° and 90° need to be stored. Such a small table is implementable into the same chip as the rest of the speed controller.

### 8.3.2 The novel speed control algorithm

In accordance with the general principles exposed at the beginning of section 8.3, a novel speed control algorithm is proposed which can be expressed as a set of simple mathematical equations written in polar coordinates. The proposed speed control strategy incorporates the slip estimator based on the phase-shift between the vectors $\underline{e}$ and $\underline{i}_s$. The new method simultaneously carries out two interrelated tasks:

- Controlling the rotor speed $\omega_r$ so that it follows the reference speed $\omega_{\text{ref}}$.
- Maintaining the slip angular frequency at a constant value: $\omega_{\text{slp}} = \Omega_{\text{slp}}$.

The two tasks are performed by controlling the angular frequency and the amplitude of the stator current. Thus, the speed controller contains two control loops. The slip control loop determines the stator current amplitude $I_s$ in such a manner that $\omega_{\text{slp}}$ is maintained as close as possible to the reference value $\Omega_{\text{slp}}$, while the speed control loop calculates the stator angular frequency $\omega_{es}$.

#### 8.3.2.1 The slip control loop

The slip control loop implements a non-linear control strategy to keep $\omega_{\text{slp}}$ constant by modifying the stator current amplitude $I_s$. The stator current controls the rotor current and the interaction of the two generates the motor torque, which in turn affects the slip angular frequency. The induction motor torque is given by the general equation

$$T = \frac{2}{3} L_m \cdot \text{Im}\,\{\underline{i}_s \cdot \underline{i}_r^*\} \tag{8.65}$$

If the expression of the rotor current for steady-state operation [81] is substituted in (8.65) then the steady-state motor torque is obtained as a function of current amplitude and slip angular frequency:

$$T(I_s, \omega_{\text{slp}}) = L_m \cdot \text{Im}\left\{\frac{j\omega_{\text{slp}} L_m I_s^2}{R_r - j\omega_{\text{slp}} L_r}\right\} \tag{8.66}$$

This expression can be further simplified as follows:

$$T(I_s, \omega_{\text{slp}}) = L_m \cdot \text{Im}\left\{\frac{j\omega_{\text{slp}} L_m I_s^2 (R_r + j\omega_{\text{slp}} L_r)}{R_r^2 + \omega_{\text{slp}}^2 L_r^2}\right\} \tag{8.67}$$

$$T(I_s, \omega_{\text{slp}}) = \frac{\omega_{\text{slp}} L_m^2 I_s^2 R_r}{R_r^2 + \omega_{\text{slp}}^2 L_r^2} \tag{8.68}$$

Thus, the motor torque increases as the stator current squared and has a non-linear variation against the slip speed. Figure 8.16 illustrates the torque-slip characteristics for the steady-state operation of the 11.1 kW induction motor. The critical slip angular frequency at which the torque attains its maximum corresponds to the null torque derivative:

$$\frac{\partial T(I_s, \omega_{\text{slp}})}{\partial \omega_{\text{slp}}} = \frac{L_m^2 I_s^2 R_r (R_r^2 - \omega_{\text{slp}}^2 L_r^2)}{(R_r^2 + \omega_{\text{slp}}^2 L_r^2)^2} = 0 \tag{8.69}$$

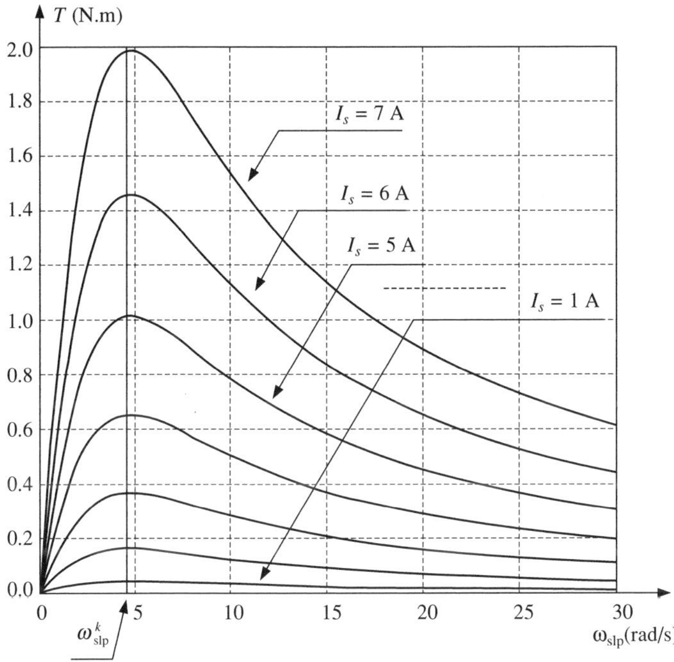


**Fig. 8.16** Steady-state torque variation for stator currents between 1 A and 7 A

From (8.69), the critical slip angular frequency $\omega_{\text{slp}}^k$ is calculated as

$$\omega_{\text{slp}}^k = \frac{R_r}{L_r} = \frac{1}{T_{er}} \tag{8.70}$$

The motor windings heat up during the operation. The result is a progressive increase of the stator and rotor resistances, which entails an increase of the rotor electrical time constant $T_{er}$, and therefore an increase of the critical slip angular frequency. Thus, $\omega_{\text{slp}}^k$ is independent of the stator current amplitude $I_s$ but depends on the rotor temperature. The actual variation of $\omega_{\text{slp}}^k$ during the motor operation depends on the construction details of the motor and on its operation mode. In practical applications, the load torque $T_l$ decreases with the decrease of the motor speed ($\partial T_l/\partial \omega_r > 0$). The stability of the motor operation is ensured only if the motor torque $T$ and the load torque $T_l$ comply with the condition

$$\text{sign}\left\{\frac{\partial T}{\partial \omega_r}\right\} = -\text{sign}\left\{\frac{\partial T_l}{\partial \omega_r}\right\} \Leftrightarrow \text{sign}\left\{\frac{\partial T}{\partial \omega_{\text{slp}}}\right\} = -\text{sign}\left\{\frac{\partial T_l}{\partial \omega_{\text{slp}}}\right\} \tag{8.71}$$

As a result, the motor speed is stable only if the slip angular frequency is in the interval $[0;\ \omega_{\text{slp}}^k)$. Therefore, the reference slip angular frequency $\Omega_{\text{slp}}$ has to be set to a value situated inside this interval. According to Fig. 8.16, the motor slip can be varied at constant load torque by controlling the stator current amplitude. The slip control loop needs to increase the stator current amplitude $I_s$ when the slip angular frequency $\omega_{\text{slp}}$ is larger than the reference $\Omega_{\text{slp}}$, and to decrease it when the slip angular frequency is smaller than $\Omega_{\text{slp}}$. The process requires information on the actual motor slip. To calculate this information, the control loop incorporates the slip estimation principles based on the phase shift between $\underline{e}^{\text{syn}}$ and $\underline{i}^{\text{syn}}$. Thus, maintaining a constant slip angular frequency during the steady-state operation is equivalent to maintaining a constant angle between $\underline{e}^{\text{syn}}$ and $\underline{i}^{\text{syn}}$.

The locus of $\underline{e}^{\text{syn}}$ is a set of circles tangential to the real axis of the rectangular synchronous reference frame [81]. The stator current amplitude $I_s$ is proportional to the circle radius so that for a given stator current amplitude the locus is made up of the two circles illustrated in Fig. 8.17. As demonstrated by (8.72), the quadrant where the internal voltage $\underline{e}^{\text{syn}}$ is situated depends on the sign of the stator angular frequency $\omega_{es}$ and on the sign of the slip angular frequency $\omega_{\text{slp}}$:

$$\begin{cases} \text{sign}\,\{\text{Re}\{\underline{e}^{\text{syn}}\}\} = \text{sign}\,\{\omega_{es}\} \cdot \text{sign}\,\{\omega_{\text{slp}} \\ \text{sign}\,\{\text{Im}\{\underline{e}^{\text{syn}}\}\} = \text{sign}\,\{\omega_{es}\} \end{cases} \tag{8.72}$$

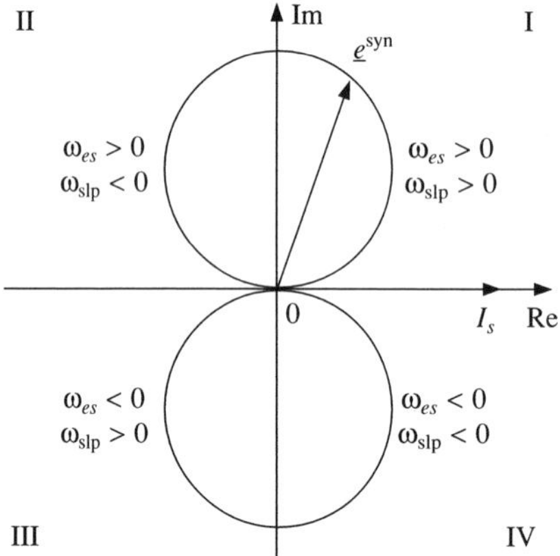


**Fig. 8.17** Internal load voltage locus in the complex plane

In most practical applications, the load torque opposes the motor shaft rotation. In this situation, the absolute value of the rotor speed is smaller than the absolute value of the magnetic field speed. Therefore $\omega_{es}$ and $\omega_{\text{slp}}$ have the same sign and $\underline{e}^{\text{syn}}$ is situated either in quadrant *I* or in quadrant IV of Fig. 8.17.

$$|\,\omega_{es}\,| > |\,\omega_{er}\,| \Rightarrow \text{sign}\,(\omega_{\text{slp}}) = \text{sign}\,(\omega_{es}) \tag{8.73}$$

There are applications where the torque may not be opposed to the shaft rotation, for example cranes and elevators. When an elevator is moving down, its weight creates a torque that tends to accelerate the shaft rotation. As a result, the rotor moves faster than the motor magnetic field, and the slip angular frequency sign is the opposite of the stator angular frequency sign. In this situation, the vector $\underline{e}^{syn}$ is situated in either quadrant II or quadrant III.

$$| \omega_{es} | < | \omega_{er} | \Rightarrow \text{sign}\ (\omega_{slp}) = - \text{sign}\ (\omega_{es}) \tag{8.74}$$

In conclusion, the sign of the reference slip angular frequency $\Omega_{slp}$ depends on the stator angular frequency sign and on the nature of the load. It is positive for motor operation in quadrants I and III and negative otherwise. The motor operation in the four quadrants corresponds to four different internal voltage vectors for steady-state operation: $\underline{e}^{syn}(\Omega_{slp1})$, $\underline{e}^{syn}(\Omega_{slp2})$, $\underline{e}^{syn}(\Omega_{slp3})$, $\underline{e}^{syn}(\Omega_{slp4})$ where $| \Omega_{slp1} | = | \Omega_{slp2} | = | \Omega_{slp3} | = | \Omega_{slp4} |$. The values of the four reference slip values $\Omega_{slp}$ have to be smaller in absolute value than the module of the critical slip angular frequency $|\omega^k_{slp}|$. If equation (8.70) is substituted in (8.59), the result is:

$$\tan^{-1}\alpha_{ei} = \frac{\omega_{slp}}{\omega^k_{slp}} \tag{8.75}$$

Consequently, the internal voltage vectors corresponding to the motor operation at critical slip are placed at 45° with regard to the reference frame axes. At slip values smaller in absolute values than $|\omega^k_{slp}|$ $\tan^{-1}\alpha_{ei}$ decreases and if $\omega_{slp}$ is null then $\tan^{-1}\alpha_{ei}$ is null as well. As shown in Fig. 8.18, the vectors $\underline{e}^{syn}(\Omega_{slp1})$, $\underline{e}^{syn}(\Omega_{slp2})$, $\underline{e}^{syn}(\Omega_{slp3})$, $\underline{e}^{syn}(\Omega_{slp4})$ need to be situated in the sectors limited by the imaginary axis of the synchronous reference system and by the vectors $\underline{e}^{syn}(\omega_{slp1})$, $\underline{e}^{syn}(\omega_{slp2})$, $\underline{e}^{syn}(\omega_{slp3})$, $\underline{e}^{syn}(\omega_{slp4})$. The slip control principle is formulated as follows:

- When the internal voltage vector $\underline{e}^{syn}$ lies in one of the shaded areas in Fig. 8.18, the controller decreases the stator current amplitude in order to increase the absolute value of the slip angular frequency $| \omega_{slp} |$.

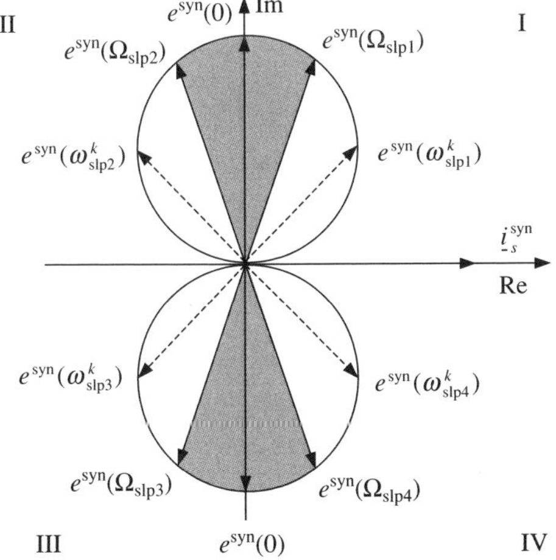


**Fig. 8.18** Characteristic points on $\underline{e}_{syn}$ locus and corresponding slip angular frequencies

- When the internal voltage vector lies outside the shaded sectors the speed controller needs to decrease the stator current amplitude in order to increase the $|\omega_{slp}|$.

Due to the symmetry in Fig. 8.18 the calculations referring to four quadrants can be reduced to equivalent calculations in only one quadrant. This transformation is carried out by replacing the real and imaginary parts of vector $\underline{e}^{syn}$ by absolute values. The result (Fig. 8.19) is an equivalent vector $\underline{E}^{syn}_{eqv}$ given by

$$\underline{E}^{syn}_{eqv} = |\mathrm{Re}\{\underline{e}^{syn}\}| + j \cdot |\mathrm{Im}\{\underline{e}^{syn}\}| = |\underline{E}^{syn}_{eqv}| \cdot [\cos(\alpha^{ref}_{eqv}) + j \cdot \sin(\alpha^{ref}_{eqv})] \quad (8.76)$$

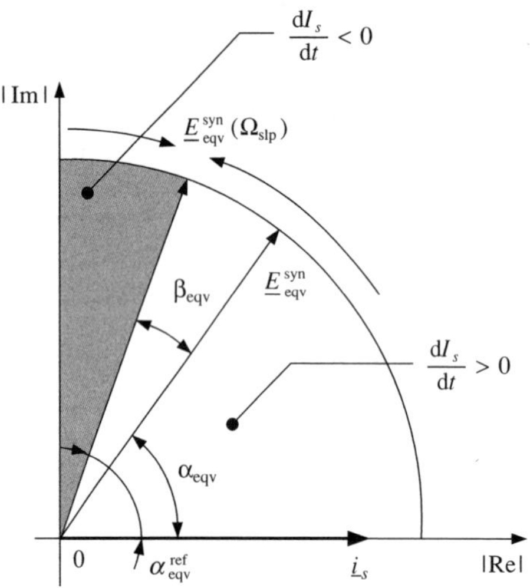


**Fig. 8.19** The reduction of the four quadrants to one

The rules governing the control of the stator current amplitude can therefore be expressed as the differential equation

$$\begin{cases} \dfrac{dI_s}{dt} = F_I(I_s, \beta_{eqv}) \\ \beta_{eqv} = \arg\{\underline{E}^{syn}_{eqv}(\Omega_{slp})\} - \arg\{\underline{E}^{syn}_{eqv}(\omega_{slp})\} = \arctan\left(\dfrac{\omega^k_{slp}}{\Omega_{slp}}\right) - \arg\{\underline{E}^{syn}_{eqv}(\omega_{slp}) \end{cases} \quad (8.77)$$

where the angle $\beta_{eqv}$ is the difference between the reference argument $\alpha^{ref}_{eqv} = \arg\{\underline{E}^{syn}_{eqv}(\Omega_{slp})$ and the argument $\alpha_{eqv}$ of the actual equivalent vector $\underline{E}^{syn}_{eqv}$.

There are several alternative expressions for the function $F_I$ but all of them have to limit the current amplitude within an interval of acceptable values $[I_{smin}; I_{smax}]$. The maximum limit is imposed by safety considerations as the motor and the power electronics circuitry has to be protected against overheating. The minimal stator current is imposed so that the internal voltage amplitude $|\underline{e}^{syn}|$ does not decrease under the limit where its argument cannot be calculated. Several versions of function $F_I$ are analysed in [81] and the corresponding motor control performance is assessed.

#### 8.3.2.2 The speed control loop

If the slip angular frequency $\omega_{\text{slp}}$ is maintained constant then the steady-state relationship between the rotor mechanical speed and the stator electrical angular frequency is linear:

$$\omega_{es} = \Omega_{\text{slp}} + \omega_{er} = \Omega_{\text{slp}} + p \cdot \omega_r \tag{8.78}$$

However, the slip angular frequency $\omega_{\text{slp}}$ cannot be kept constant at the reference value $\Omega_{\text{slp}}$ in transient operation. During transient operation all the quantities describing the motor operation undergo complicated changes that are difficult to control due to the non-linearities. Furthermore, transient operation causes the slip estimation errors, thereby increasing the difficulty of the control task. The speed control loop must simultaneously compensate the errors of the motor slip estimation and control the rotor speed. The control strategy can be expressed by the general differential equations

$$\begin{cases} \omega_{es}^{\text{ref}} = \text{sign}\,(\omega_r^{\text{ref}}\} \cdot \Omega_{\text{slp}} + \omega_{er}^{\text{ref}} = \text{sign}\,\{\omega_r^{\text{ref}}\} \cdot \Omega_{\text{slp}} + p \cdot \omega_r^{\text{ref}} \\ \dfrac{d\omega_{es}}{dt} = F_\omega(\omega_{es}^{\text{ref}}, \omega_{es}, I_s, \beta_{\text{eqv}}) \end{cases} \tag{8.79}$$

where the function 'sign' is defined by

$$\text{sign}\,(x) = \begin{cases} +1 \text{ when } x > 0 \\ \;\;0 \text{ when } x = 0 \\ -1 \text{ when } x < 0 \end{cases} \tag{8.80}$$

Any speed control strategy can be defined by the two functions $F_\omega$ and $F_I$ involved in equations (8.77) and (8.79). The simplest control version is defined by the functions

$$\begin{cases} F_I(I_s, \beta_{\text{eqv}}) = \begin{cases} K_I \cdot \beta_{\text{eqv}} & \text{when } I_s \in (I_{s-\min}\,; I_{s-\max}) \\ 0 & \text{when } I_s \notin (I_{s-\min}\,; I_{s-\max} \end{cases} \\ F_\omega(\omega_{es}^{\text{ref}}, \omega_{es}, I_s, \beta_{\text{eqv}}) = K_\omega \cdot \text{sign}\,(\omega_{es}^{\text{ref}} - \omega_{es}) \end{cases} \tag{8.81}$$

where $K_I$ and $K_\omega$ are proportionality constants. In this case, the derivative of the current amplitude is proportional to the angular error $\beta_{\text{eqv}}$, while the derivative of the angular frequency depends on the sign of the stator frequency error. Therefore, the motor control is linear and uses two P controllers operating in an independent manner, as the two functions $F_I$ and $F_\omega$ are calculated based on different parameters. This type of control is appropriate when the application requirements do not include fast transient response.

Using this control strategy, the motor behaves similarly to a synchronous machine with a start-up cage rotor:

- It is able to generate a constant speed for a certain range of load torque values.
- The rotor speed follows the variations of the stator frequency if this variation is slow.
- If the variations of the stator frequency are too fast, they take the rotor out of synchronism and the speed response becomes relatively slow.

The simulation results in [81] prove that the transient slip estimation errors do not affect the stability of the drive system operation. The errors cause oscillations of the angles $\alpha_{\text{eqv}}$ and $\beta_{\text{eqv}}$ but the stator current amplitude is given by

$$I_s(t) = K_I \cdot \int \beta_{\text{eqv}}(t) \cdot \mathrm{d}t \tag{8.82}$$

so that the effect of these oscillations is filtered out by integration. However, control strategies of increased complexity are required to obtain a fast response of the system to step changes of $\omega_{\text{ref}}$. Very fast induction motor transient responses are typically obtained using the rotor field-oriented control strategy. The new speed control strategy can be improved by finding two functions $F_I$ and $F_\omega$ that emulate the behaviour of a rotor field-oriented controller. Thus, the field generating current component $i_{sd}$ needs to be maintained constant while modifying the torque generating current component according to the speed error. This requires the calculation of the position $\theta_\psi(t)$ of the rotor flux vector $\underline{\Psi}_r$ and the equations (8.83) to be solved.

$$\begin{cases} \underline{i}_{sd} = I_s(t) \cdot \cos\left(\int_0^t \omega_{es}(t) \cdot t \cdot \mathrm{d}t - \theta_\psi(t)\right) = \text{const.} \\ \underline{i}_{sd} = I_s(t) \cdot \sin\left(\int_0^t \omega_{es}(t) \cdot t \cdot \mathrm{d}t - \theta_\psi(t)\right) = f(\omega_{\text{ref}} - \omega_r) \end{cases} \tag{8.83}$$

The rotor flux vector is not calculated by the new speed control strategy in order to minimise the calculation amount. On the other hand, solving the equation system (8.83) would increase the hardware implementation complexity to an unacceptable level. However, this strategy can approximate the position of $\underline{\Psi}_r$ using the position of $\underline{e}^{\text{syn}}$ for large and medium power motors. If the speed is larger than a few revolutions per second, then $\underline{e}^{\text{syn}}$ is approximately perpendicular on the rotor flux vector $\underline{\Psi}_r$ because the rotor resistance can be neglected as compared to motor reactance. Under these conditions, $\underline{e}^{\text{syn}}$ can be used to determine the position of vector $\underline{\Psi}_r$ in the complex plane.

Figure 8.20 indicates the typical positions of the vectors $\underline{e}$, $i$ and $\underline{\Psi}_r$ in the synchronous reference frame. Modifying the motor speed requires a modification of the motor torque. The field orientation solution is to alter the stator current component $i_{sq}$ while keeping

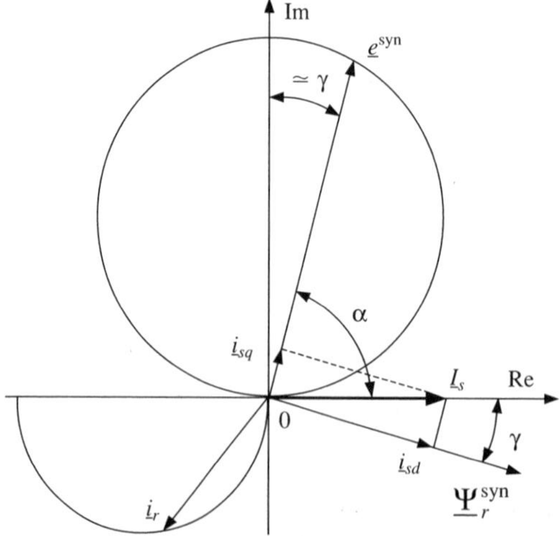


**Fig. 8.20** Relative position of vectors $e$, $i$ and $\Psi_r$ in the synchronous reference frame

$i_{sd}$ constant. According to the new control method, the task is achieved by simultaneously changing the stator angular frequency $\omega_{es}$ and the stator current amplitude $I_s$. The two stator current components are:

$$\begin{cases} i_{sd} = I_s \cdot \cos\gamma \\ i_{sq} = I_s \cdot \sin\gamma \end{cases} \tag{8.84}$$

where the angle $\gamma$ is indicated in Fig 8.20, while the derivatives of the two components are

$$\begin{cases} \dfrac{\mathrm{d}i_{sd}}{\mathrm{d}t} = \dfrac{\mathrm{d}I_s}{\mathrm{d}t}\cos\gamma - I_s \sin\gamma \cdot \dfrac{\mathrm{d}\gamma}{\mathrm{d}t} = 0 \\ \dfrac{\mathrm{d}i_{sq}}{\mathrm{d}t} = \dfrac{\mathrm{d}I_s}{\mathrm{d}t}\sin\gamma + I_s \cos\gamma \cdot \dfrac{\mathrm{d}\gamma}{\mathrm{d}t} \end{cases} \tag{8.85}$$

The derivative $\mathrm{d}i_{sd}/\mathrm{d}t$ is ideally null during the motor speed change and therefore the variation of the stator current amplitude $I_s$ depends on $\gamma$ according to

$$\frac{\mathrm{d}I_s}{\mathrm{d}t} = I_s \cdot \mathrm{tg}\,\gamma \cdot \frac{\mathrm{d}\gamma}{\mathrm{d}t} \tag{8.86}$$

which is derived from the first equation (8.85). Substituting (8.86) into the second equation (8.85), the result is

$$\frac{\mathrm{d}I_s}{\mathrm{d}t} = \frac{\mathrm{d}i_{qs}}{\mathrm{d}t} \cdot \frac{\mathrm{tg}\,\gamma}{\mathrm{tg}\,\gamma \cdot \sin\gamma + \cos\gamma} \tag{8.87}$$

which demonstrates that the stator current component $i_{sq}$ increases with the increase of $I_s$. At the same time, any variation of $I_s$ has to comply with the condition (8.86). Consequently, any increase of the stator current amplitude $I_s$ has to be simultaneous with an increase of the angle $\gamma = \pi/2 - \arg\{\underline{e}^{\mathrm{syn}}\}$. The variation of $I_s$ generates an initial increase of $\arg\{\underline{e}^{\mathrm{syn}}\}$ followed by a decrease [81]. This variation is reflected in the opposite alteration of the slip estimation results and of the angle $\gamma$ (unwanted result). To maintain the relationship between $I_s$ and $\gamma$ in accordance with (8.86), the stator angular frequency $\omega_{es}$ must be simultaneously altered with $I_s$ so that the effects of $\omega_{es}$ variations compensate the unwanted effects over angle $\gamma$. It was proven in [81] that increasing the slip angular frequency $\omega_{\mathrm{slp}} = \omega_{es} - \omega_{er}$ leads to oscillations starting with an initial decrease of $\arg\{\underline{e}^{\mathrm{syn}}\}$. This can cancel out the unwanted increase caused by the modification of $I_s$. The subsequent oscillations of angle $\gamma$ resulting from the modification of $I_s$ need to be cancelled out by suitable variations of $\omega_{es}$. The exact analytical non-linear solution to this problem is difficult to find. However, simplified solutions, equivalent to **quasi-field-oriented control methods**, can be investigated based on a few principles derived from the previous considerations and from the rules governing the slip estimation process. The principles are:

- The value of $\arg\{\underline{e}^{\mathrm{syn}}\}$ needs to be maintained at values close to 90° ($\Omega_{\mathrm{slp}}$ has to be small) to maintain the slip estimation errors at acceptable levels.
- The rotor speed changes are always initiated by the speed control loop according to equations (8.78) and (8.79). The stator current variations compensate for the unwanted oscillations of angle $\gamma$, which can be calculated as a function of $\beta_{\mathrm{eqv}}$ and $\alpha_{\mathrm{cqv}}^{\mathrm{ref}}$.

- The angle $\gamma$ must be allowed to increase during the speed changes simultaneously with the increase of $I_s$. This is equivalent with a simultaneous increase of $\beta_{eqv}$ and $I_s$, which can be easily achieved if $\partial F_I/\partial \beta_{eqv} > 0$.
- If the angles $\gamma$ and $\beta_{eqv}$ become too large, the speed of variation of the stator frequency has to be limited in order to reduce the motor slip and the error slip estimations. However, provided the motor slip has small values, the stator frequency can be allowed to undergo fast speed changes.

One of the simplest solutions that complies with the principles outlined is

$$\begin{cases} F_I(I_s, \beta_{eqv}) = \begin{cases} K_I \cdot \beta_{eqv} & \text{if } I_s \in (I_{s-\min}; I_{s-\max}) \\ 0 & \text{if } I_s \notin (I_{s-\min}; I_{s-\max}) \end{cases} \\ F_\omega(\omega_{es}^{ref}, \omega_{es}, I_s, \beta_{eqv}) = \begin{cases} \text{sign}(\omega_{es}^{ref} - \omega_{es}) \cdot (K_{\omega 1} - \beta_{eqv} \cdot K_{\omega 2}) & \text{if } \beta_{eqv} < \beta_{max} \\ \text{sign}(\omega_{es}^{ref} - \omega_{es}) \cdot (K_{\omega 1} - \beta_{max} \cdot K_{\omega 2}) & \text{if } \beta_{eqv} \geq \beta_{max} \end{cases} \end{cases} \tag{8.88}$$

where $F_\omega$ is a piecewise linear function (Fig. 8.21) defined by the constants $K_I$, $K_{\omega 1}$, $K_{\omega 2}$, $\beta$ whose optimal values depend on the motor parameters. The functions (8.88) represent

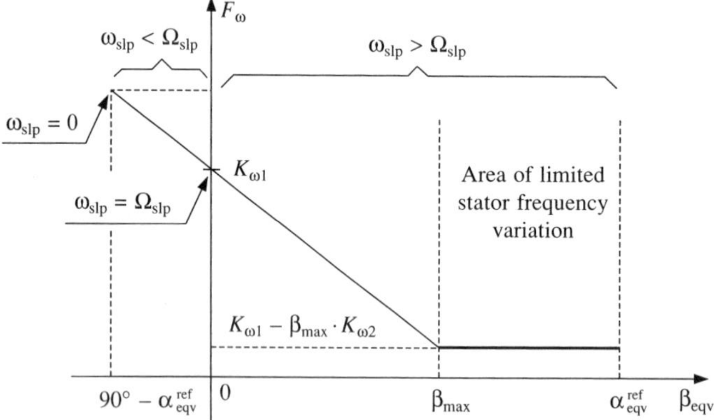


**Fig. 8.21** The variation of $F_\omega$ with $\beta$ when $\omega_{es}^{ref} > \omega_{es}$

the basic version of the new sensorless speed control algorithm proposed in this book. This control solution has been tested by simulations on the 11.1 kW motor using different values for the constants in (8.88). The MATLAB simulation results demonstrate substantially improved dynamic response, as exemplified in Fig. 8.22. In the same time, the method is capable of maintaining the rotor speed constant despite load torque variations (Fig. 8.23). The system response speed is approximately proportional to $K_{\omega 1}$ but increasing $K_{\omega 1}$ over a certain limit actually deteriorates the system response. This phenomenon is illustrated in Fig. 8.24, which can be compared with Fig. 8.22.

As shown by these simulation results, the stator angular frequency undergoes non-linear variation caused by the non-linear mathematical model (8.88) of the control strategy. Again, the motor behaves in a similar manner to a synchronous motor: the rotor speed follows the stator frequency changes only if the speed of these changes is below a critical limit depending on the rotor inertia and on the load torque.

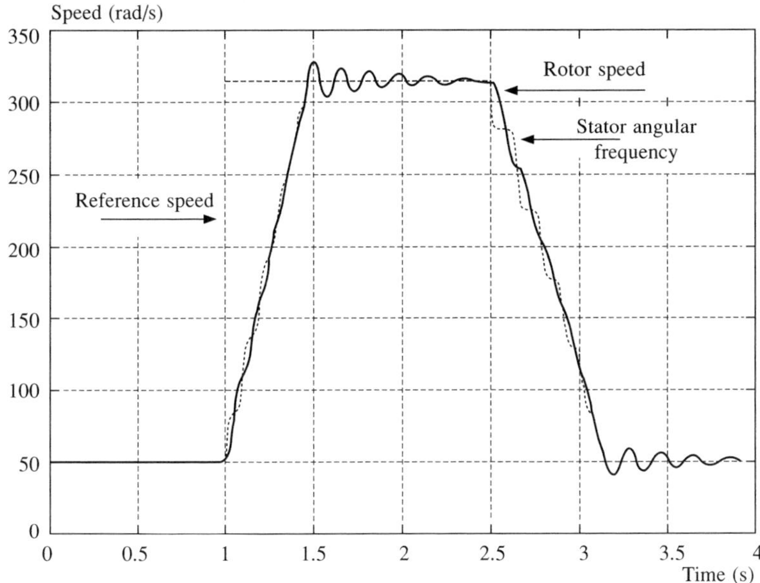


**Fig. 8.22** Quasi-field oriented control method results ($K_{\omega 1} = 1000\ \mathrm{s}^{-1}$)

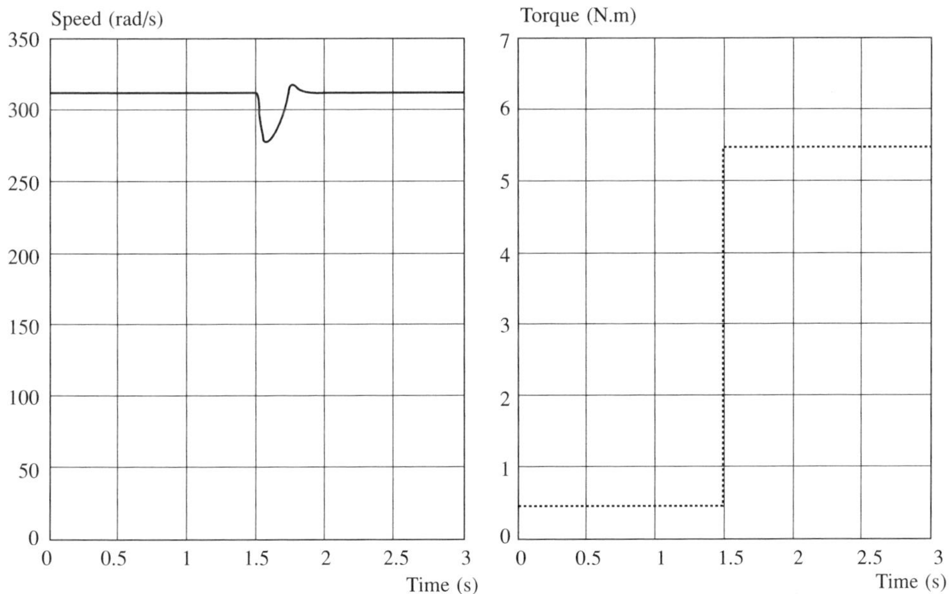


**Fig. 8.23** The motor speed variation during a step increase of the load torque ($K_{\omega 1} = 1000\ \mathrm{s}^{-1}$)

The speed control strategy can be further refined by using different values for the constant $K_I$ depending on the stator current amplitude $I_s$. A single value $K_I$ cannot be optimal for all the motor currents in the range ($I_{s\text{-min}}$; $I_{s\text{-max}}$) because, as demonstrated by equation (8.68), the motor torque is proportional to $I_s$ squared, and the same derivative

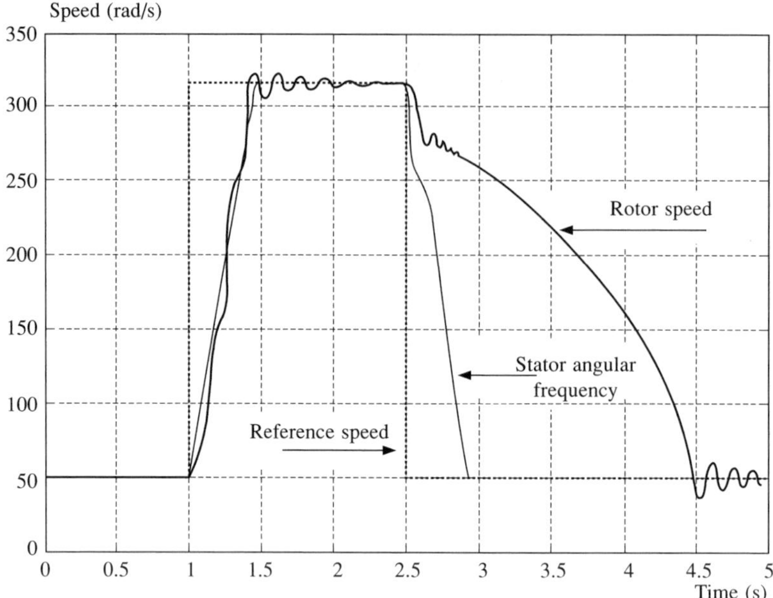


**Fig. 8.24** Quasi-field-oriented control method results ($K_{\omega 1} = 3000\ \mathrm{s}^{-1}$)

$\mathrm{d}I_s/\mathrm{d}t$ produces different torque variations at different stator current amplitudes. The effect is a slow dynamic response of the motor when the current is close to $I_{s\text{-min}}$ and a very fast one when the current is close to $I_{s\text{-max}}$. Thus, to optimise the motor response, an improved function $F_I$ needs to be found, which ensures the same dynamic parameters both at small stator currents and at large stator currents. This requires that the motor derivative does not depend on the stator current amplitude. The time derivative of the torque is:

$$\frac{\mathrm{d}T}{\mathrm{d}t} = \frac{\partial T(I_s, \alpha_{\mathrm{slp}})}{\partial I_s} \cdot \frac{\mathrm{d}I_s}{\mathrm{d}t} = \frac{2\omega_{\mathrm{slp}} L_m^2 I_s R_r}{R_r^2 + \omega_{\mathrm{slp}}^2 L_r^2} \cdot \frac{\mathrm{d}I_s}{\mathrm{d}t} \tag{8.89}$$

Consequently, the torque derivative $\mathrm{d}T/\mathrm{d}t$ is independent of the current amplitude $I_s$ if $\mathrm{d}I_s/\mathrm{d}t$ is inversely proportional to $I_s$. To include this improvement, the quasi-field-oriented control strategy initially formulated in (8.88) can be transformed into

$$\begin{cases} \dfrac{\mathrm{d}I_s}{\mathrm{d}t} = F_I(I_s, \beta_{\mathrm{eqv}}) = \begin{cases} K_I \cdot \beta_{\mathrm{eqv}} / I_s & \text{if } I_s \in (I_{s-\min}; I_{s-\max}) \\ 0 & \text{if } I_s \notin (I_{s-\min}; I_{s-\max}) \end{cases} \\ \dfrac{\mathrm{d}\omega_{es}}{\mathrm{d}t} = F_\omega(\omega_{es}^{\mathrm{ref}}, \omega_{es}, I_s, \beta_{\mathrm{eqv}}) = \begin{cases} \mathrm{sign}\,(\omega_{es}^{\mathrm{ref}} - \omega_{\mathrm{es}}) \cdot (K_{\omega 1} - \beta_{\mathrm{eqv}} \cdot K_{\omega 2}) & \text{if } \beta_{\mathrm{eqv}} < \beta_{\max} \\ \mathrm{sign}\,(\omega_{es}^{\mathrm{ref}} - \omega_{es}) \cdot (K_{\omega 1} - \beta_{\max} \cdot K_{\omega 2}) & \text{if } \beta_{\mathrm{eqv}} \geq \beta_{\max} \end{cases} \\ \omega_{es}^{\mathrm{ref}} = \Omega_{\mathrm{slp}} + \omega_{er}^{\mathrm{ref}} = \Omega_{\mathrm{slp}} + \mathrm{p} \cdot \omega_r^{\mathrm{ref}} \end{cases} \tag{8.90}$$

In case the limited hardware resources available do not allow the implementation of a supplementary division block (it consumes a significant amount of chip area), the division by $I_s$ can be replaced by a stepwise approximation. Therefore, the unique constant $K_I$ is replaced by a stepwise approximation that uses a set of different constants $K_{I1}$, $K_{I2}$, $K_{I3}$, . . . depending on the value of $I_s$. In this case, the parameters of the electrical drive dynamic response depend on the quality of the approximation, which in turn depends on the amount of available hardware resources. The functions (8.90) represent the enhanced version of the new sensorless speed control algorithm proposed in this book.

### 8.3.3 The complete control scheme

The complete sensorless induction motor control scheme includes a speed controller that operates according to (8.90), a current controller that implements the new method described in section 8.2, and a conversion block that interfaces the two controllers (Fig. 8.25), which transforms the quantities $\omega_{es}^{\text{ref}}$ and $I_s^{\text{ref}}$ into the reference current $\underline{i}_s$ for the current controller.

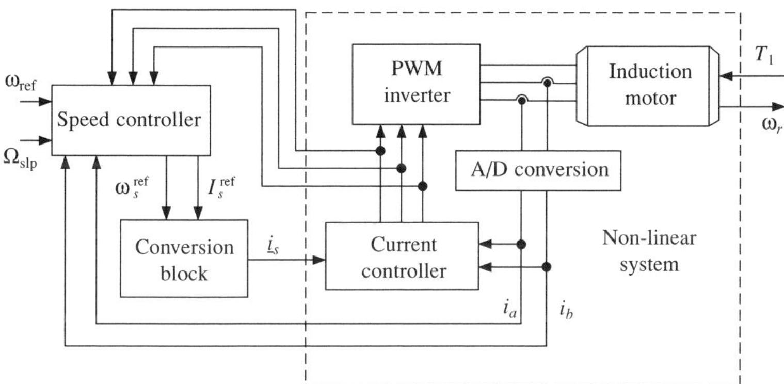


**Fig. 8.25** The block diagram of the sensorless control scheme

The current control principles formulated represent a generalisation of the method presented in [179) that leads to superior control performance. The new method requires a big computational effort that can only be performed with the aid of hardware implemented neural networks. The combined effect of the speed controller non-linearity and the slip estimation errors during the transients are very difficult to analyse mathematically but the overall system behaviour can be studied using computer simulations. MATLAB simulations prove that the drive system operates without significant speed oscillations, without stationary errors and with good dynamic performance.

## 8.4 Induction motor controller VHDL design

The design and hardware implementation of the motor controller is carried out using two main software resources: Workview Office 7.31 and Xilinx Foundation 1.4. Both of them are sophisticated tools for the design, simulation and testing of FPGA implemented circuits. Workview Office is a general software package produced by Viewlogic [2], [3] and it can be used in conjunction with FPGAs manufactured by a large variety of producers. It includes a flexible VHDL simulator that supports all the features described by the IEEE 1076-1993 standard definition of the language [10]. On the other hand, Xilinx Foundation is a software package specialised in developing applications using the FPGA families manufactured by Xilinx [5], [8], [14]. It is capable of optimising the hardware implementation according to speed and chip area requirements of particular applications, being more versatile than Workview Office from this point of view. However, Xilinx Foundation supports only a subset of the standard VHDL language, namely those statements and functions that can be synthesised and directly implemented into hardware. Furthermore, this software lacks a VHDL simulator. It performs simulations using the netlist files obtained after the synthesis stage, which limits its capabilities and slows down the design and test cycle. Consequently, the VHDL design and simulation have been performed using Workview Office while the implementation and timing verification have been carried out using Xilinx Foundation. The combined use of the two software packages improved the overall efficiency of the simulation, troubleshooting and synthesis stages in the design cycle. The VHDL description of the complete motor controller includes the current control strategy and the sensorless speed control algorithm. The model has been developed using a hierarchical approach and contains four tiers that consist of several specialised logic blocks (Fig. 8.26).

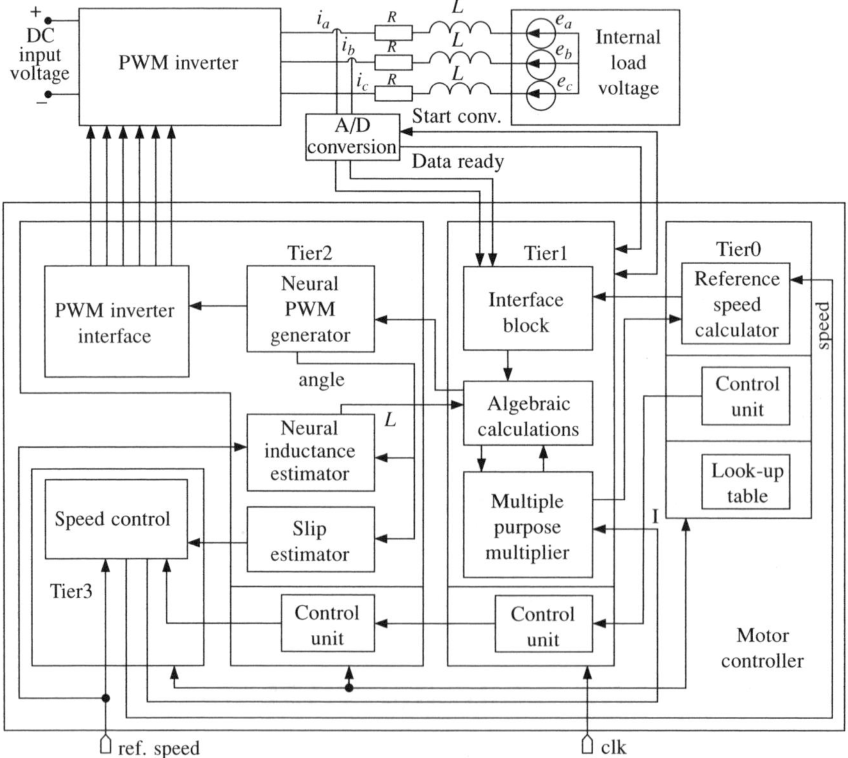


**Fig. 8.26** The functional blocks of the FPGA motor controller

The VHDL code related to these blocks utilises generic parameters to define the size of the registers, adders, subtracters, buses and other elements involved. This allows rescaling of the controller hardware structure according to the calculation precision imposed by the available types of FPGA circuits. The present version of the motor controller has been implemented into a Xilinx XC4010XL FPGA included on an XS40 test board, which contains a 12 MHz clock circuit. The control units included in the structure of tiers 0, 1 and 2 synchronise the operation of all the other logic blocks and control the information transfer between different tiers. They have been designed as clock-synchronised finite state machines (FSMs) using the specialised State Editor program included in the Xilinx Foundation package. Using this program, a FSM is graphically described as a state diagram, which can be automatically converted into a VHDL model. The other blocks in each tier have been directly described in VHDL using the register-transfer logic (RTL) manner. Therefore, they consist of registers interleaved with combinational logic structures: adders, subtracters, comparators, etc.

Each of the four tiers performs specific tasks:

- Tier0 generates a space vector of constant amplitude and variable angular speed. The angular speed corresponds to the stator current frequency. The vector is defined by its real and imaginary parts in the stator reference frame. Therefore, tier0 is a sine wave generator. It produces two variable frequency sine waves shifted with 90°.
- Tier1 carries out the algebraic calculations required by the control algorithm and controls the operation of the A/D converters that provide information on the motor currents. It determines the reference current space vector by multiplying the unit space vector generated by tier1 with the amplitude of the reference stator current. It also calculates the vectors $\underline{V}_{ni}$ and $V_{\Delta}$ involved in the current control algorithm and in the on-line inductance estimation process.
- Tier2 contains the neural network that generates the PWM switching pattern based on space vectors $\underline{i}_s$ and $\underline{V}_{ni}$. The angle calculation subnetwork, which is part of the complex neural network, is involved in the on-line inductance estimation algorithm. It also calculates the motor slip angle $\alpha_{eqv}$ that is used by tier3.
- Tier3 generates the reference stator angular frequency and the reference stator current amplitude using the external reference rotor speed and the motor slip angle as input information.

### 8.4.1 The reference speed calculator and sine wave generator (Tier 0)

Tier0 is described by a VHDL entity with three input ports (**clock, reset** and **speed_rate**) and three output ports: **cosx**, **cosy** and **start_tier1** as shown in the Code Fragment 8.3. The first two outputs are the projections on axes OX and OY (Fig. 8.27) of the unit vector rotating around the origin of the two-dimensional plane with the angular speed indicated by the input port **speed_rate**. The output port **start_tier1** informs tier1 when the calculations performed by tier0 have been completed and the data on the output ports **cosx** and **cosy** is ready.

```
-- Code Fragment 8.3
entity tier0 is
  port (
  clk: in STD_LOGIC;
  reset: in STD_LOGIC;
```

```
  speed_rate: in STD_LOGIC_VECTOR(16 downto 0);
  cosx: out STD_LOGIC_VECTOR (8 downto 0);
  cosy: out STD_LOGIC_VECTOR (8 downto 0);
  start_tier1: out STD_LOGIC
  );
end tier0;
```

The architecture of tier0 contains three components: a data processing unit (**data_tier0**), a control unit (**ctrl_tier0**) and a look-up table (**sin_rom**). The look-up table stores information about the waveform of the two sine waves that need to be generated. Traditionally, the look-up table contains the samples of the sine wave to be generated and the data processing unit reads the table in sequence and at the required speed. To reduce the memory size, only the information referring to a quarter of a sine wave period is stored in the memory. However, such a look-up table is still too large to be efficiently implemented in the same FPGA with the rest of the controller. To minimise the memory implementation size, the differential modulation technique has been used. This was made possible by the fact that the angle $\theta$ in Fig. 8.27 has a predictable variation in time due to its relation to the stator current angular frequency: $d\theta/dt = \omega_{es}$. It increases or decreases depending on the sign of $\omega_{es}$, but is not subject to sudden variations. Therefore, the values of **cosx** and **cosy** can be determined by adding or subtracting small increments to the values calculated during the previous calculation cycle. The increments take up fewer bits than the corresponding sine wave samples because they are small quantities. These small increments can be stored in a compact look-up table that requires much less hardware resources than the classical look-up table.

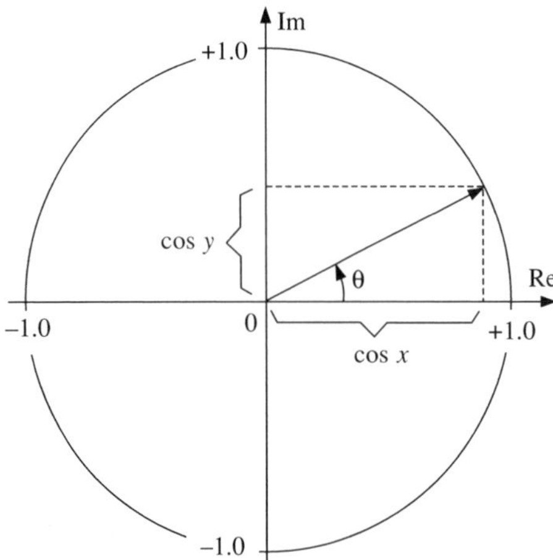


**Fig. 8.27** The real and imaginary components of the unit vector

The VHDL code describing the look-up table has been automatically generated by the specialised C++ program Sin_Rom.CPP presented in Appendix D. The program has two parameters defining the amplitude of the sine wave (**ampl**) and the number of entries in the look-up table (**n_steps**). Therefore, several versions of the look-up table can be generated by altering these two parameters. The optimal version depends on the required precision and on the available hardware resources. Different alternatives have been tested by simulation and physical implementation, and the solution given by **ampl** = 127 and **n_steps** = 64 has eventually been adopted. These two parameters define a sine wave with 265 samples per period and values between –127 and +127. The difference between two samples varies between 0 and 7, so 3 bits are sufficient for each memory location. In conclusion, the differential modulation technique reduces the size of the look-up table to 33 per cent because the initial 9-bit samples can be replaced by 3-bit sample differences. The VHDL model automatically generated by the C++ program is an entity with one input port (the address bus) and one output port (the data bus). The associated architecture contains a single process that produces the data corresponding to the address using the constant array of 64 std_logic_vector elements shown in Code Fragment 8.4

```
-- Code Fragment 8.4
LIBRARY IEEE;
USE IEEE.std_logic_1164.ALL;
USE IEEE.std_logic_unsigned.ALL;
ENTITY sin_rom IS
  PORT(
  A: IN std_logic_vector(5 DOWNTO 0);
  DO: OUT std_logic_vector(2 DOWNTO 0));
END sin_rom;

ARCHITECTURE sin_rom_arch OF sin_rom IS
  TYPE mem_data IS ARRAY (0 TO 63) OF std_logic_vector(2 downto 0);
  constant VD: mem_data :=
  ( ('0','0','0'),
  ('0','0','0'),
  ('0','0','1'),
  ('0','0','0'),
-- .................
  ('1','1','1'),
  ('1','1','0'),
  ('1','1','0'),
  ('1','1','0'),
  ('1','1','0'),
  ('1','1','1'),
  ('1','1','0'));
BEGIN
  PROCESS(A)
  begin
  DO<=VD(conv_integer(A));
  END PROCESS;
END sin_rom_arch;
```

The data processing unit inside the sine wave generator has a cyclical operation. During each cycle, it reads the look-up table in sequence and adds or subtracts the memory values to the current outputs in order to generate the required sine waves. The operation cycles are initiated by a clock divider modelled by the VHDL process **start_generator** inside the architecture of **data_tier0**. The corresponding VHDL code is:

```
-- Code Fragment 8.5
start_generator: process(reset,clk)
  variable counter_val: integer range 0 to 511;
  begin
  if reset='1' then
  counter_val:=2;
  start<='0';
  elsif clk='1' and clk'event then
  if counter_val=0 then
  counter_val:=UpperCount_tierf;
  start<='1';
  else
  counter_val:=counter_val-1;
  start<='0';
  end if;
  end if;
  end process;
end data_tierf_arch;
```

The clock division ratio specified by the constant **UpperCount_tierf** has been set to 79. Given the 12 MHz frequency of the clock signal, a number of 150 000 operation cycles are initiated every second. A complete sine wave period contains 256 samples, so each sample is repeatedly generated for a number of times that depends on the required sine wave frequency. To achieve this, a new value read from the ROM memory is added to the previous result only when the memory address changes. Thus, the speed of changing the memory address is proportional to the required frequency. The multiplication with the corresponding proportionality constant is performed by tier1, which transmits the result to tier0 on the input port **speed_rate**.

The value of **speed_rate** is added to the signal **adr_cosy** and the result is stored in the register **next_adr_cosy**. Based on the information stored in **adr_cosy** and **next_adr_cosy**, the correct memory address is generated, after which the value of **cosy** is updated. At the end of each operation cycle, **next_adr_cosy** is copied to **adr_cosy** so that a new value for **next_adr_cosy** can be calculated at the beginning of the next cycle. The addition is performed using the '+' operator defined in **std_logic_signed** package from IEEE library. The operators in this package have the advantage that the sign bit of the shorter operand is always extended on the empty positions as shown in Fig. 8.28. This simplifies the design process for complementary code adders and subtracters. They can be directly modelled by the corresponding algebraic equations.

Two signals generated by the control unit, **add_speed_rate** and **inc_adr**, indicate the moments when the two address values **adr_cosy** and **next_adr_cosy** are updated. They correspond to independent VHDL processes because the updating operations are carried out at separate moments in time:

```
-- Code Fragment 8.6
process(reset,add_speed_rate)
  begin
  if reset='1' then
  next_adr_cosy<=(others=>'0');
```

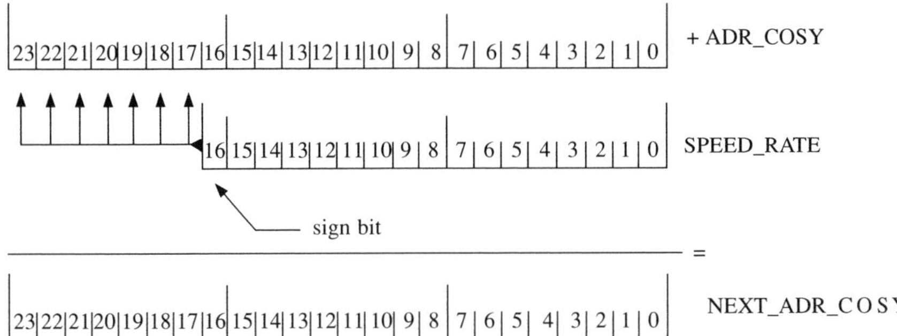


**Fig. 8.28** NEXT_ADR_COSY calculation

```
  speed_sign<='0';
  elsif add_speed_rate='1' and add_speed_rate'event then
  next_adr_cosy<=adr_cosy+speed_rate;
  speed_sign<=speed_rate(16);
  end if;
  end process;
process(inc_adr,reset) --It is the last process to be performed
  begin
  if reset='1' then
  adr_cosy<=(others=>'0');
  elsif inc_adr='1' and inc_adr'event then
  adr_cosy<=next_adr_cosy;
  end if;
  end process;
```

The signals **adr_cosy** and **next_adr_cosy** are 24-bit vectors whose most significant 8 bits correspond to the sample index relative to the beginning of the sine wave period (a number between 0 and 255). During each operation cycle, **next_adr_cosy** is compared to **adr_cosy**. If the most significant 8 bits in the two vectors are different, it means that the current sample index has changed after **next_adr_cosy** has been modified by adding **speed_rate**. Consequently, a new value is to be read from the memory and it has to be either added to or subtracted from **cosy**, depending on the slope of the sine wave around the current sample. Otherwise no operation is performed.

The look-up table has only 64 entries corresponding to a quarter of the complete sine wave period. As a result, the address required by the look-up table is made up of only 6 bits and varies between 0 and 63. The address needs to be calculated according to a certain algorithm, which locates the correct look-up table entry depending on the current sample index (between 0 and 255) and the sine wave frequency sign. The details of the address calculation algorithm are first presented for positive speeds and then it is extended to both positive and negative speeds. As the sample index increases from 0 to 255, the memory address varies according to Fig. 8.29. Thus, when the sample index is inside the interval [0; 63] (the first sine wave quarter), the bits 16 to 21 of **adr_cosy** are used as memory address:

$$\text{mem\_adr} = \text{adr\_cosy}(21 \text{ downto } 16) \tag{8.91}$$

For sample indices in the interval [64; 127] (the second sine wave quarter), the memory

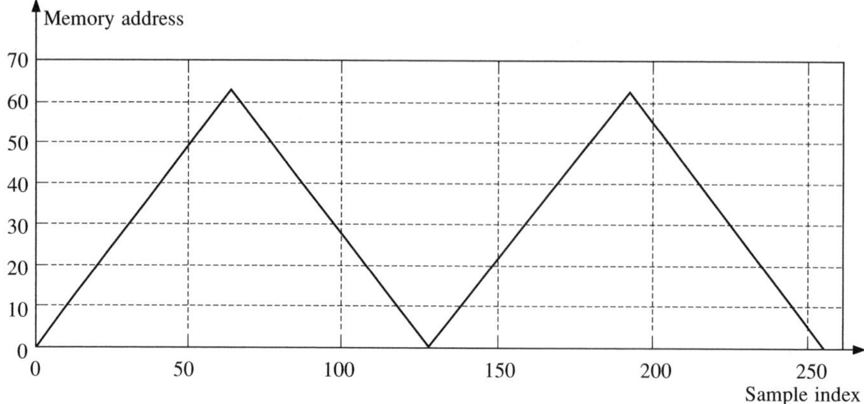


**Fig. 8.29** The correspondence between the sine wave sample index and the memory address

values need to be extracted in the reversed order. The address is in this case calculated using the formula:

$$mem\_adr = 127 - adr\_cosy \ (22 \text{ downto } 16) \tag{8.92}$$

The same addressing sequence is used for the second sine wave half because the two halves differ only by the most significant bit of **adr_cosy**, which is '0' during the first half and '1' during the second half. The bits 16 to 22 of **adr_cosy** undergo the same sequence of changes during the two sine wave halves (Fig. 8.30) and therefore the same calculations can be used to generate the entire waveform. Thus, formula (8.91) is applied for the third sine wave quarter, while formula (8.92) is used for the fourth quarter.

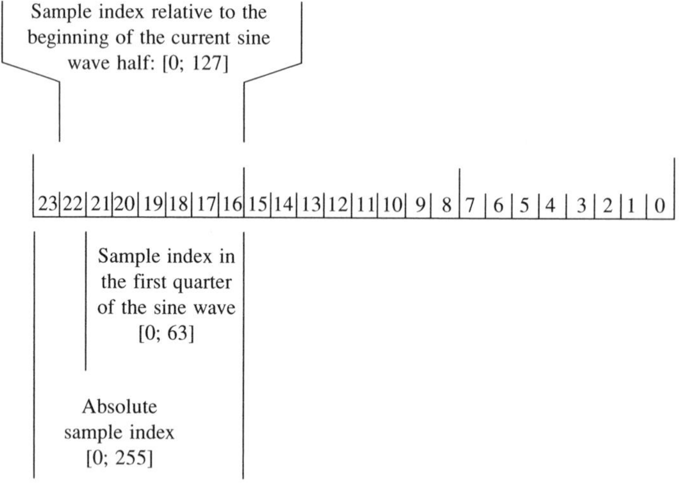


**Fig. 8.30** The bits of signal adr_cosy

The most significant bit of **adr_cosy** (the bit 23) is used to decide whether the new memory value has to be added to or subtracted from the current **cosy** value. These values are added during the first half and subtracted during the second as shown in Fig. 8.31.

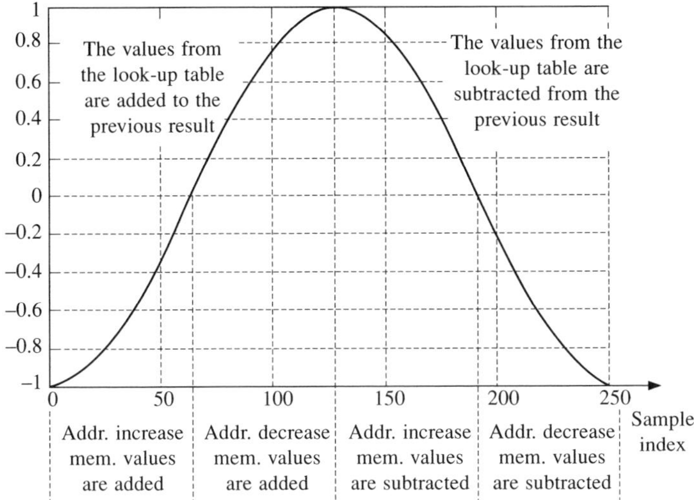


**Fig. 8.31** Sine wave generation algorithm (positive speeds)

The algorithm can be now extended for both positive and negative speeds. To do so, it must be noted that each location in the look-up table stores the difference between the next sine wave sample and the current sine wave sample for **positive sine wave slope** and **positive speed:**

TABLE [mem_adr] = | NextSample – CurrentSample | (8.93)

At negative speeds the sequence of samples is reversed so that the previous sample is calculated instead of the next sample:

TABLE [mem_adr–1] = | CurrentSample – PreviousSample | (8.94)

The vector **next_adr_cosy** is larger than **adr_cosy** at positive speed because **speed_rate** is a positive value. When the speed is negative, **speed_rate** is negative as well, and **next_adr_cosy** becomes smaller than **adr_cosy**. Thus, **adr_cosy** is used to generate **mem_adr** when the speed is positive, while **next_adr_cosy** is used to calculate **mem_adr-1** when the speed is negative.

The memory address used to update **cosx** is derived from signals **adr_cosx** and **next_adr_cosx** which are 8-bit long vectors. Their values are related to the values of **adr_cosy** and **next_adr_cosy** because the two sine waves are 90° shifted, which is translated into a sample index difference of 64. Consequently, **adr_cosx** is obtained adding 64 to the most significant 8 bits of **adr_cosy**, which can be reduced to adding '01' to the bits 22 and 23 of **adr_cosy** (Fig. 8.32). The vector **next_adr_cosx** is used only to generate the memory addresses at negative speeds because the transitions between two sine wave samples are already determined by the difference between **adr_cosy** and **next_adr_cosy**. Consequently, **next_adr_cosx** is calculated according to the simple equation

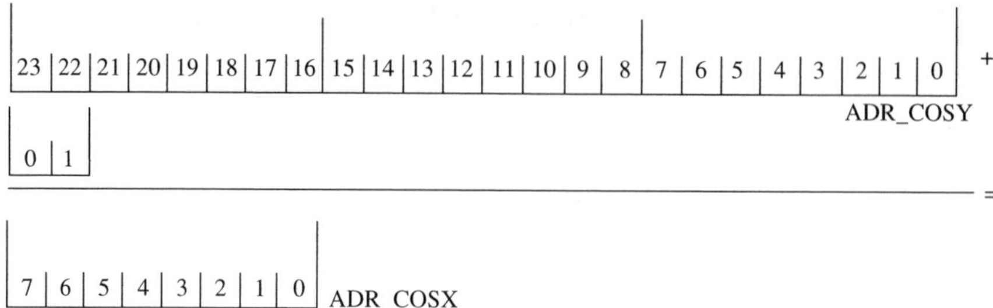


**Fig. 8.32** ADR_COSX calculation

$$\textbf{next\_adr\_cosx} = \textbf{adr\_cosx} - 1 \tag{8.95}$$

The sine wave generator calculates the address in two stages implemented as two VHDL processes. First, an internal memory address (**int_mem_adr**) is calculated based on the 7 bits which give the relative sample index to the beginning of the current sine wave half (Fig. 8.30). The internal memory address is compared to 63 during the second stage, and in case it surpasses this limit then equation (8.92) is used to calculate the equivalent address, which is confined between 0 and 63. When this upper limit is not surpassed, no calculation is performed. The final result **mem_adr** consists of the least significant 6 bits of the vector **x** generated at stage two.

Due to the large number of operation cycles performed every second compared to the number of sine wave samples, there are numerous cycles when the memory is not addressed because tier0 outputs do not need to be updated. During the cycles when the memory needs to be addressed, the operation is carried out twice: first time to update **cosy** and second time to update **cosx**. The signal controlling which of the two memory addresses is to be calculated at a certain moment (**adr_mux**) is generated by the control unit. This signal is '0' when the address corresponding to **cosy** is calculated, and it is '1' otherwise. Therefore, the calculation of **int_mem_adr** in the first VHDL process depends both on **adr_mux** and on the speed sign stored by the signal **speed_sign** as shown in the following code fragment.

```
-- Code Fragment 8.7
process(adr_cosy,adr_cosx,next_adr_cosy,next_adr_cosx,
  adr_mux,speed_sign)
  begin
  if speed_sign='0' and adr_mux='0' then
  int_mem_adr<=adr_cosy(22 downto 16);
  elsif speed_sign='0' and adr_mux='1' then
  int_mem_adr<=adr_cosx(6 downto 0);
  elsif speed_sign='1' and adr_mux='0' then
  int_mem_adr<=next_adr_cosy(22 downto 16);
  else
  int_mem_adr<=next_adr_cosx(6 downto 0);
  end if;
  end process;

  process(int_mem_adr)
  variable x: std_logic_vector(6 downto 0);
```

```
begin
x:=int_mem_adr;
if x(6)='1' then
x:="0111111"-('0' & x(5 downto 0));
end if;
mem_adr<=x(5 downto 0);
end process;

adr_cosx<=(adr_cosy(23 downto 22)+"01") & adr_cosy(21 downto 16);
next_adr_cosx<=adr_cosx-"01";
```

Signals **cosy** and **cosx** are updated inside two VHDL processes, which are activated by the signals **add_cosy** and **add_cosx** generated by the control unit. These processes decide whether to add or subtract the value read from the look-up table based on the speed sign and the most significant bit of **adr_cosy** and **adr_cosx**, respectively. This most significant bit indicates if the current sample is situated in the first or in the second half of the sine wave period. This information is correlated with the sign of the sine wave slope. If the slope is positive the new value has to be added to the output signal, otherwise it has to be subtracted. The **reset** signal is part of the sensitivity list of the two processes so it initialises the outputs at the beginning of the circuit operation. The two outputs **cosx** and **cosy** are also periodically reinitialised to the correct values whenever **adr_cosy** is zero (Code Fragment 8.8). This mechanism improves the system robustness by avoiding the accumulation of errors caused by possible electromagnetic interference generated by the power transistors in the PWM inverter.

```
-- Code Fragment 8.8
process(adr_cosy)
  -This reset ensures that errors are periodically eliminated
  begin
  if adr_cosy(23 downto 16)="00000000" then
  periodical_reset<='1';
  else
  periodical_reset<='0';
  end if;
  end process;

  process(add_cosy,reset,adr_cosy,periodical_reset)
  begin
  if (reset='1') or (periodical_reset='1') then
  int_cosy<=(8=>'1',0=>'1',others=>'0');
  elsif add_cosy='1' and add_cosy'event then
  if (adr_cosy(23) xor speed_sign)='0' then
  int_cosy<=int_cosy+('0' & data);
  else
  int_cosy<=int_cosy-('0' & data);
  end if;
  end if;
  end process;
--The value of 'cosx' is reset whenever the memory address is 0 and
--mux_adr=0. When mux_adr=1 it means cosx will be increased.
--Therefore it mustn't be reset any longer.
```

```
process(add_cosx,reset,adr_cosy,adr_mux,periodical_reset)
begin
if reset='1' or (periodical_reset='1' and adr_mux='0') then
int_cosx<=(others=>'0');
elsif add_cosx='1' and add_cosx'event then
if (adr_cosx(7) xor speed_sign)='0' then
int_cosx<=int_cosx+('0' & data);
else
int_cosx<=int_cosx-('0' & data);
end if;
end if;
end process;
cosy<=int_cosy;
cosx<=int_cosx;
```

As previously mentioned, the control unit has been designed as a finite state machine using the State Editor included in the Xilinx Foundation software package. Thus, the operation of ctrl_tier0 was initially described by a state diagram, which has subsequently been converted into a VHDL model. The elements of a typical diagram are states, transitions, transition conditions, actions, the reset signal, the clock signal, input ports and output ports.

- The transitions between states are triggered by the clock signal. The state machine can be defined as either active on the rising clock edge or active on the falling clock edge.
- A condition assigned to a transition inhibits the state change until the condition is fulfilled. All conditions need to comply with the VHDL syntax because they are included as they were written in the automatically generated VHDL model of the FSM.
- The actions performed by the state machine are changes of the output ports. There are three different types of actions entry actions, state actions and exit actions. The changes occur at different moments in time: at the transition from the previous state to the current state (entry action), during the current state (state action) or at the transition between current state and next state (exit actions).
- The reset signal is a special input port that brings the state machine in its original state. It can be defined as synchronous or asynchronous. The synchronous reset brings the FSM in the initial state only when the correct clock edge occurs, while the effect of an asynchronous reset signal is instantaneous.
- There are two types of output ports: registered and combinational. The registered outputs are modelled as registers and therefore the effect of any action is valid until the next action modifies the port. The combinational outputs have no memory. The effect of any action lasts as long as the FSM is in the state linked to the respective action, after which the output returns value specified for the original state (the state forced by the **reset** signal).

The model of ctrl_tier0 has been defined as a state machine with six states that is active on the falling clock edge and uses an asynchronous reset signal. The control unit has five output ports: **adr_mux**, **add_cosx**, **add_cosy**, **add_speed_rate** and **inc_adr**. The port **adr_mux** is registered while the others are combinational (Fig. 8.33). Each operation cycle of ctrl_tier0 begins when the **start** signal is activated by the process in

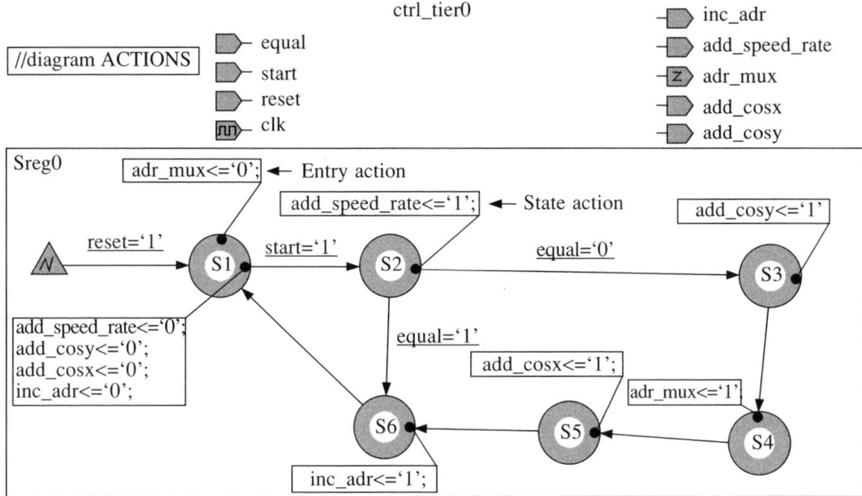


**Fig. 8.33** The state diagram of ctrl_tier0

Code Fragment 8.5 contained in the data processing unit. The first action carried out during the control unit operation cycle is to trigger the calculation of **next_adr_cosy** by adding **speed_rate** to its previous value. The vectors **adr_cosy** and **next_adr_cosy** are compared by the data processing unit and the signal **equal** is set accordingly. This requires the inclusion of a comparator in the structure of the data processing unit. The corresponding VHDL model is described by Code Fragment 8.9.

```
-- Code Fragment 8.9
process(adr_cosy,next_adr_cosy)
  begin
  if adr_cosy(22 downto 16) /= next_adr_cosy(22 downto 16) then
  equal<='0';
  else
  equal<='1';
  end if;
  end process;
```

If the most significant 8 bits of **next_adr_cosy** and **adr_cosy** are different, then the values of **cosx** and **cosy** need to be updated. During states S1, S2 and S3 the variable **adr_mux** is set to '0' so that **cosy** can be updated when **add_cosy** is activated. During states S4, S5 and S6 **adr_mux** is set to '1' to calculate the memory address corresponding to **cosx**. The output vector **cosx** is updated during the state S5. During state S6, the signal **inc_adr** is activated and, as shown by Code Fragment 8.6, the vector **adr_cosy** is updated. If the vectors **adr_cosy** and **next_adr_cosy** are equal then the operation cycle comprises only the final action linked to the state S6. Consequently, there are short operation cycles and long operation cycles depending on the value of the signal **equal** generated by the data processing unit. These two cycle types are illustrated by the simulation results in Fig. 8.34.

The complete model of the sine wave generator has been simulated using Workview

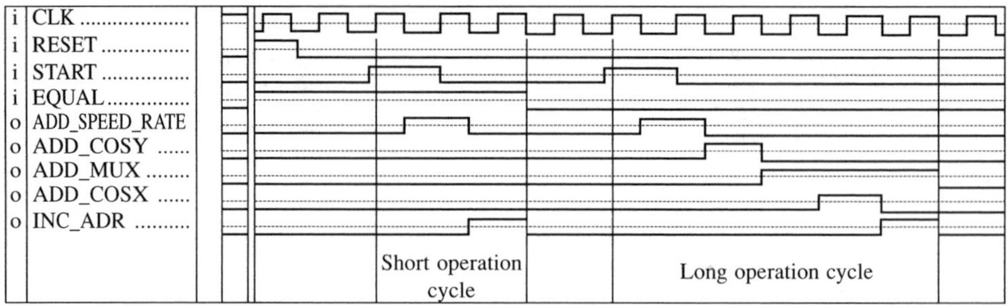


**Fig. 8.34** Control unit simulation results

Office and the results have been exported in MATLAB to draw the graphs. Figure 8.35 illustrates the operation corresponding to four different frequencies and proves the correct generation of the two sine waves, **cosx** and **cosy**. It also demonstrates the correct transition from the waveforms corresponding to one frequency to the waveforms corresponding to another frequency.

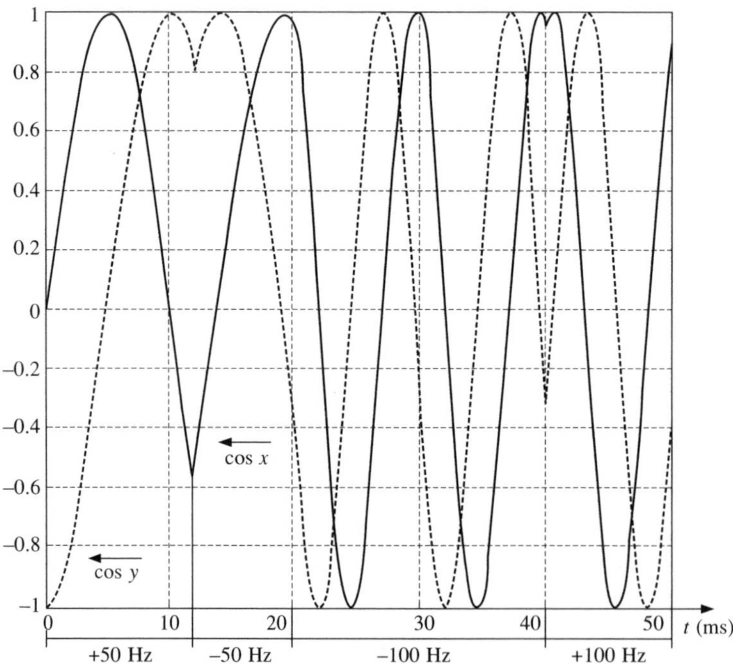


**Fig. 8.35** The operation of tier0

### 8.4.2 The multiplication and algebraic calculations (Tier 1)

Tier1 is a complex processing module composed of a control unit (ctrl_tier1) and a data processing unit (data_tier1) that performs several calculations and control tasks required by both tier0 and tier2:

- Controls the operation of the two A/D converters connected to Hall sensors that measure the stator currents of the motor.
- Uses the digital output from the A/D converters to calculate the stator current space vector.
- Determines the derivative of the current space vector and uses this information to calculate the non-inductive space vector $\underline{V}_{ni}$ .
- Calculates the space vector $\underline{V}_{\Delta} = \underline{u}(k) - \underline{V}_{ni}(k-1)$, which is used by tier2 to estimate the inductance in the equivalent *R–L–e* circuit.
- Calculates the vector **speed_rate** used by tier0.
- Calculates the rectangular coordinates of the reference stator current by multiplying the unit vector produced by tier0 with the amplitude calculated by tier3.
- Performs adjustments of the numerical values supplied to the neural network located in tier2, so that the network operation speed is maximised.

To achieve all these tasks, tier1 needs to perform several multiplications. There are numerous multiplier hardware architectures reported in the literature [233], [234], [151], [91]. They differ in hardware complexity and operating speed. The fastest multipliers use combinational architectures but unfortunately they have very large hardware complexity. Sequential architectures are more compact but at the same time they are slower. To optimise both the speed of the motor controller and its complexity, a single fast multiplier is used by tier1 to perform all the multiplications. Therefore, all the signals involved in multiplications have been multiplexed to the inputs of this multiplier.

The VHDL multiplier model developed implements the $2^{N_{sl}}$ – radix multiplication algorithm which is flexible as it allows good control over the trade-off between speed and circuit complexity. The multiplication is carried out in several stages using groups of $N_{sl}$ bits at a time, where the number $N_{sl}$ is the multiplier's step length. If $N_{sl}$ is 1 the classical Booth architecture is obtained. This generates a very compact but slow multiplier. If the value of $N_{sl}$ is larger than 1 then faster multipliers are obtained but they occupy more space on the chip. The fastest architecture is obtained when $N_{sl}$ equals the length of the second operand. In this case, the corresponding hardware multiplier has a purely combinational structure, which makes it space inefficient but very fast.

The VHDL multiplier model uses three generic parameters that define the length of the two operands and the step length, as shown in Code Fragment 8.10. These parameters allow the adaptation of the multiplier to any application requirements referring to speed, operand size and hardware complexity.

```
-- Code Fragment 8.10
entity smultiplier is
  generic(n,m,step_length: integer);
  port (
  a: in STD_LOGIC_VECTOR (n-1 downto 0);
-- Can be only positive
  b: in STD_LOGIC_VECTOR (m-1 downto 0);
-- Can be both positive and negative
  prod: out STD_LOGIC_VECTOR (n+m-1 downto 0);
  clk,start: in STD_LOGIC;
  ready: out STD_LOGIC
  );
end smultiplier;
```

The fact that all the multiplications required by the motor control algorithm involve a signed operand and an unsigned operand has been exploited to optimise the structure of the controller. Thus, a specialised multiplier has been created, which has an unsigned input bus (**b**) and a signed one (**a**). The multiplication process is composed of a series of simple operation cycles. Each cycle consists in multiplying the operand **a** with the least significant $N_{sl}$ bits of **b** and adding the result into a shift register. Both the result and the operand **b** are then shifted with $N_{sl}$ positions to the right. The architecture comprises two shift registers, a control unit and a reduced size combinational multiplier with input buses of width equal to the size of **a** and $N_{sl}$, as illustrated in Fig. 8.36. The multiplication process is triggered by the **start** input signal. When this signal is active (is '1') both the control unit and the multiplication result register are reset. When **start** returns to '0', the control unit initiates the multiplier operation by activating the signal **first_step** which causes the operand **b** to be loaded into the corresponding shift register. After **b** has been loaded, the series of calculation cycles commences. During each cycle, the **load_step** signal is activated first and then **shift_step** is activated. After the shift, the most significant $N_{sl}$ bits of the two registers in Fig. 8.36 are padded with zeroes.

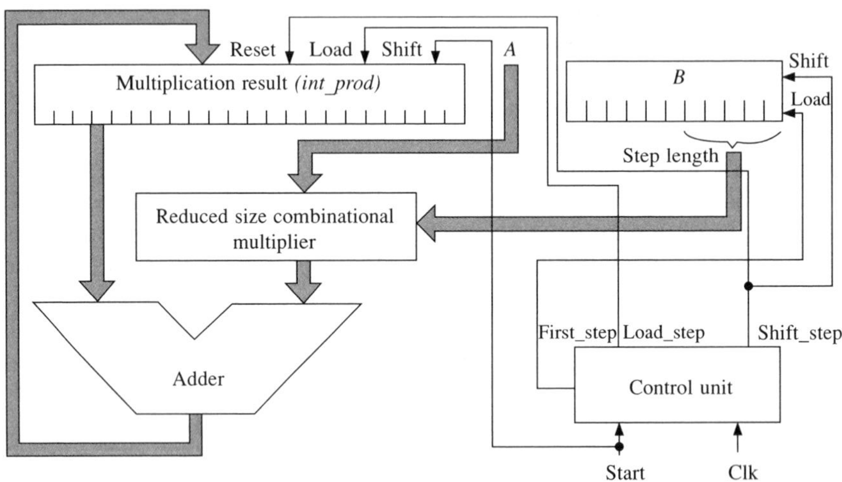


**Fig. 8.36** Flexible multiplier structure

The number of necessary calculation cycles is given by the VHDL constant **nsteps** that depends on the length of **b** (parameter **m**) and on the step length $N_{sl}$, according to equation

$$\text{nsteps} = \text{ceiling}\ (\text{m}/N_{sl}) \tag{8.96}$$

where the function 'ceiling' generates the smallest integer that is larger than its argument. Signal **count** is loaded with value **nsteps** when the input signal **start** is activated, and it is decreased at each calculation cycle simultaneously with adding the partial multiplication result to the result register. If **count** is larger than 1 then the two registers are shifted and a new cycle is initiated, otherwise the calculations are stopped and the **ready** signal is activated.

The VHDL model of the control unit, shown in Code Fragment 8.11, operates with five different states (s0, s1, s2, s3, s4), each of them activating one of the control signals previously discussed.

```
-- Code Fragment 8.11
type state is (s0,s1,s2,s3,s4);
  signal s: state;
constant nsteps: integer := (m+step_length-1)/step_length;
process(clk,start)
  begin
  if start='1' then
  s<=s0;
  count<=nsteps;
  elsif clk='1' and clk'event then
  case s is
  when s0 => s<=s1;
  when s1 => s<=s2;
  when s2 => if count>1 then
  s<=s3;
  count<=count-1;
  else
  s<=s4;
  end if;
  when s3 => s<=s2;
  when s4 => null;
  when others => null;
  end case;
  end if;
  end process;
  first_step <= '1' when s=s1 else '0';
  load_step <= '1' when s=s2 else '0';
  shift_step <= '1' when s=s3 else '0';
  ready_int <= '1' when s=s4 else '0';
  ready<=ready_int;
```

This algorithm is applicable only to positive values. Therefore, if the operand **b** is positive then the multiplication result is the value stored in the result register after the calculation sequence has finished. Otherwise, this result has to be transformed into a valid two's complement codification of the negative multiplication result. This transformation is based on the next considerations:

- If **b** is a negative number then '$2^m$ – **b**' which is its two's complement is a positive number.
- If **b** is replaced by '$2^m$ – **b**' in the multiplication process, the result is '($2^m$ – **b**) · **a**' that has the same module as the correct result but it has the opposite sign.
- The correct multiplication result is obtained by reversing the sign of the previous expression. Therefore, the calculation formula is: **a** · **b**–$2^m$ · **a**.

The VHDL implementation of this principle is described by the process in Code Fragment 8.12, where the information is transferred from the internal register **int_prod** to the output port **prod** in two manners, depending on the most significant bit of **b**. If **b**(**m** – 1) = '0' then the operand **b** is positive and the internal information is copied to the output port, while if **b**(**m** – 1) = '1' then the previous calculation formula is used.

```
-- Code Fragment 8.12
process(a,b,int_prod)
  begin
  if b(m-1)='0' then
  prod<=int_prod(n+m-1 downto 0);
  else
  prod<=int_prod(n+m-1 downto 0) - (a & zeroes(m));
  end if;
  end process;
```

Figure 8.37 presents the pipelined architecture of data_tier1 that includes the multiplier previously discussed. The first operation performed is the calculation of the rectangular components $i_x$ and $i_y$ of the stator current space vector. The calculation is carried out using a modified form of the classical conversion equations. Thus, the two components are replaced by equivalent values that are rescaled to limit their maximal values and to limit the number of bits necessary to be allocated for each of them. The rescaling technique allows a good control over the number of bits used by each internal signal, but on the other hand decreases the calculation precision of data_tier1. Furthermore, the rescaling factors need to be taken into account by the subsequent calculations that involve the equivalent quantities. The simulation and the synthesis results showed that rescaling with 0.5 offers the best trade-off between precision and hardware complexity. Therefore, the calculations are performed according to:

$$\begin{cases} i_x = \dfrac{3}{2} \cdot i_a & \Rightarrow i_x^{\text{eqv}} = \dfrac{i_x}{2} = \dfrac{3}{4} \cdot i_a \\[2ex] i_y = \dfrac{\sqrt{3}}{2} \cdot (2i_b + i_a) & \Rightarrow i_y^{\text{eqv}} = \dfrac{i_y}{2} = \dfrac{\sqrt{3}}{2} \cdot \left(i_b + \dfrac{i_a}{2}\right) = 0.866 \cdot \left(i_b + \dfrac{i_a}{2}\right) \end{cases} \tag{8.97}$$

The multiplication with 0.866 required in (8.97) is performed by the multiplier integrated in data_tier1. Once the two rectangular components have been determined, the current error vector $\Delta\underline{i}_{\text{ref}}(k) = \underline{i}_{\text{ref}}(k+1) - \underline{i}(k)$ and the current variation $\Delta\underline{i}(k) = \underline{i}(k) - \underline{i}(k-1)$ are calculated by simple subtracters. The components of $\Delta\underline{i}(k)$ are multiplied with the estimated inductance $L$ and subtracted from the corresponding voltage components to determine the vector $\underline{V}_{ni}$, which is used by both the neural network generating the PWM signal and by data`tier1 to calculate the vector $\underline{V}_{\Delta}$. The adapter blocks included in Fig. 8.37 enhance the operational precision of the angle subnetwork inside the neural network contained by tier2.

The angle subnetwork calculates the argument of a space vector based on its rectangular coordinates. The number of input bits of the angle subnetwork is smaller than the total number of bits of the two coordinates. Therefore, it uses only the most significant **n** bits of these coordinates. If the two values are small numbers, the most significant bits are all '0' or all '1' (depending on the sign of the numbers) and an incorrect result is generated.

The adapter simultaneously shifts the two coordinate values to the left until their leftmost $n$ positions contain significant bits. Shifting a binary number to the left is equivalent to a multiplication by a power of two. As the two coordinates are simultaneously multiplied with the same power of two, the adapter amplifies the module of the vector but leaves the vector argument unchanged.

The VHDL model of the adapter has two main input ports **inbusa** and **inbusb** (the initial two coordinates) and two main outputs **outbusa** and **outbusb**, as shown in Code Fragment 8.13. The generic parameter '*n*' defines the width of the input and the output buses. It has to be set in accordance with the width of the VHDL signals to which it is connected. All the signals in this tier have correlated widths that are primarily determined by a generic parameter '*ni*' which determines the width of the buses $i_a$ and $i_b$ presented in Fig. 8.37.

```
-- Code Fragment 8.13
entity adapter is
  generic(n: integer);
  port (
  inbusa: in STD_LOGIC_VECTOR (n-1 downto 0);
  inbusb: in STD_LOGIC_VECTOR (n-1 downto 0);
  outbusa: out STD_LOGIC_VECTOR (n-1 downto 0);
  outbusb: out STD_LOGIC_VECTOR (n-1 downto 0);
  clk: in STD_LOGIC;
  ld: in STD_LOGIC;
  ready: out STD_LOGIC
  );
end adapter;
```

The additional input **ld** triggers the shifting process while **ready** signals the moment when the process is finished. Each step of the process is synchronised by the clock signal **clk**. The method to determine the end of the process is to test the most significant two bits in each of the two partial results. If any of the two pairs contains different bits then the process must stop to avoid an overflow that would change the sign of at least one of the coordinates.

The process must also be stopped if all the bits are zero in the same time. This happens when both input coordinates are simultaneously zero, which would cause an infinite shifting process. The architecture of the adapter contains two VHDL processes: the first shifts the two input values in a synchronised manner, while the second verifies the existence of non-zero bits and communicates the result through the internal signal **not_all_zero**:

```
-- Code Fragment 8.14
architecture adapter_arch of adapter is
  signal int_busa,int_busb: STD_LOGIC_VECTOR(n-1 downto 0);
  signal not_all_zero: std_logic;
begin
  process(ld,clk,inbusa,inbusb)
  begin
  if clk='1' and clk'event then
  if ld='1' then
  int_busa<=inbusa;
  int_busb<=inbusb;
  ready<='0';
  elsif (int_busa(n-1)=int_busa(n-2)) and
  (int_busb(n-1)=int_busb(n-2))
  and (not_all_zero='1') then
```

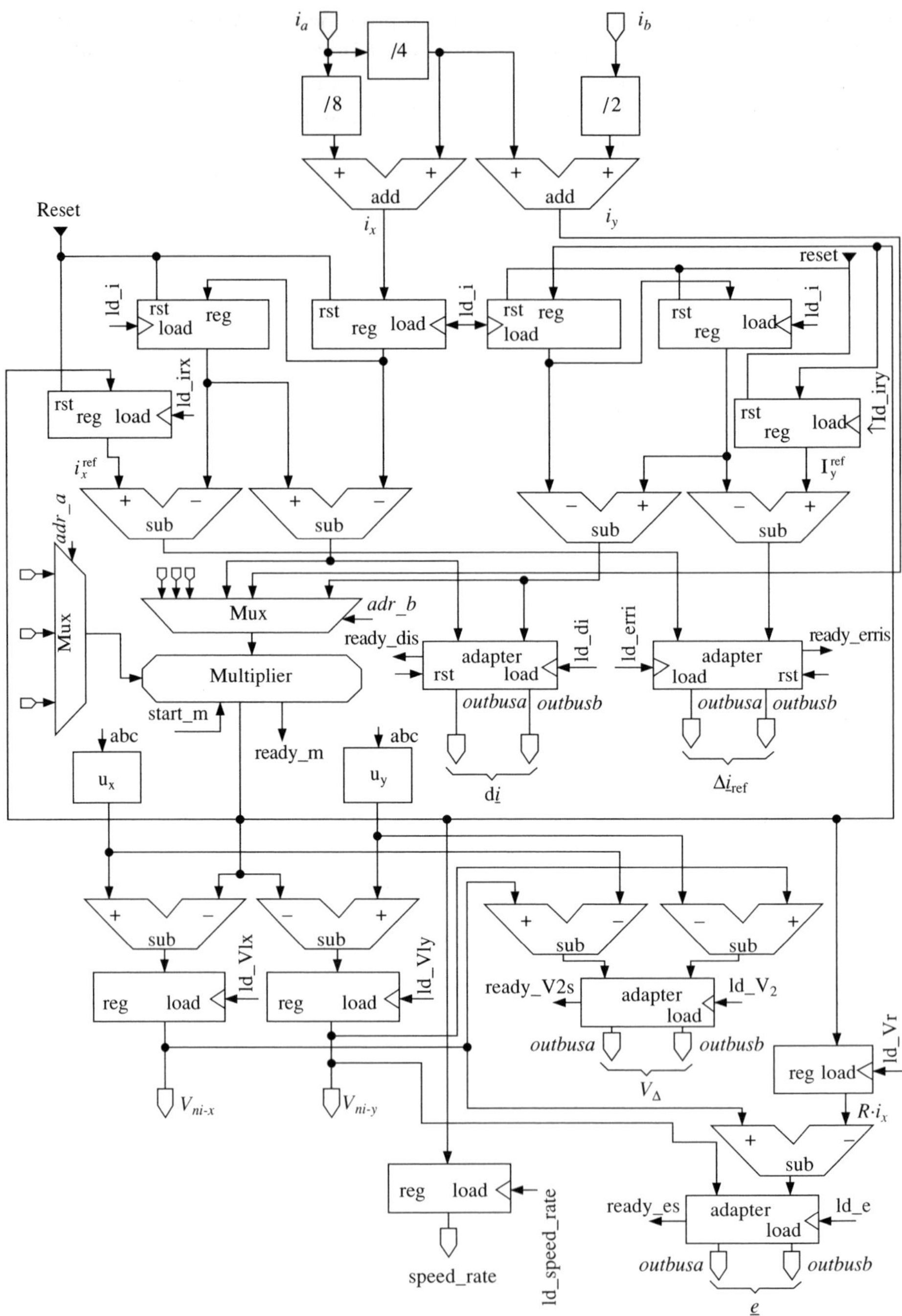


**Fig. 8.37** The structure of data_tier1

```
int_busa<=int_busa(n-1) & int_busa(n-3 downto 0) & '0';
int_busb<=int_busb(n-1) & int_busb(n-3 downto 0) & '0';
else
```

```
  ready<='1';
  end if;
  end if;
  end process;
process(int_busa,int_busb)
  variable x: std_logic;
  begin
  x:='0';
  for i in 0 to n-1 loop
  x:=x or int_busa(i);
  x:=x or int_busb(i);
  end loop;
  not_all_zero<=x;
  end process;
  outbusa<=int_busa;
  outbusb<=int_busb;
end adapter_arch;
```

Figure 8.38 presents the state diagram that describes the operating sequence of tier1, which is controlled by ctrl_tier1 modelled as a pair of correlated state machines. The first dictates the sequence of mathematical operations performed by tier1 while the other controls the interface with the A/D converters. The A/D circuits TLC1550 produced by Texas Instruments [1] have been used, as they offer the advantage of integrating the sample-and-hold circuit, an internal clock oscillator and the digital converter itself in the same chip. Moreover, this type of chip can be easily interfaced with other digital circuits due to the 3-state parallel port, and it can be addressed as an external memory device. Thus, it has an **RD** input pin that triggers the conversion and an active low **EOC** output pin that signals the end of the conversion. The interface state machine activates **RD** in state S28 and then waits for the conversion to finish. After each conversion, it loads the information in the specialised registers, as shown in Fig. 8.38 correlated with Fig. 8.37.

The operation of the first state machine in Fig. 8.38 is described by a linear sequence of states which controls the multiplier, the associated multiplexers and loads each partial result in the specially allocated register. The activity of the two state machines is correlated by means of the internal signal **RdNow** that is used to indicate the moments when the analogue to digital conversion is finished. Each operation cycle of ctrl_tier1 is triggered by the **start_tier1** signal, which has a frequency of 150 kHz and it is generated by ctrl_tier0. As a result, the A/D converters are activated with the same frequency and therefore the sampling frequency of the motor controller is 150 kHz as well.

### 8.4.3 The PWM generation and on-line inductance estimation (Tier 2)

The data processing unit of tier2 named data_tier2 (Fig. 8.39) contains the PWM signal generator, which includes the new neural network architecture presented at the beginning of this example, the on-line inductance estimator and the motor slip calculator. All these three digital structures use information provided by the angle subnetwork. Thus, the inductance estimation is performed by comparing the arguments of vectors $\Delta \underline{i}(k)$, $\Delta \underline{i}(k-1)$, $\underline{V}_{\Delta}(k)$, the motor slip is calculated as the difference between $\arg\{\underline{e}\}$ and $\arg\{\underline{i}_s\}$,

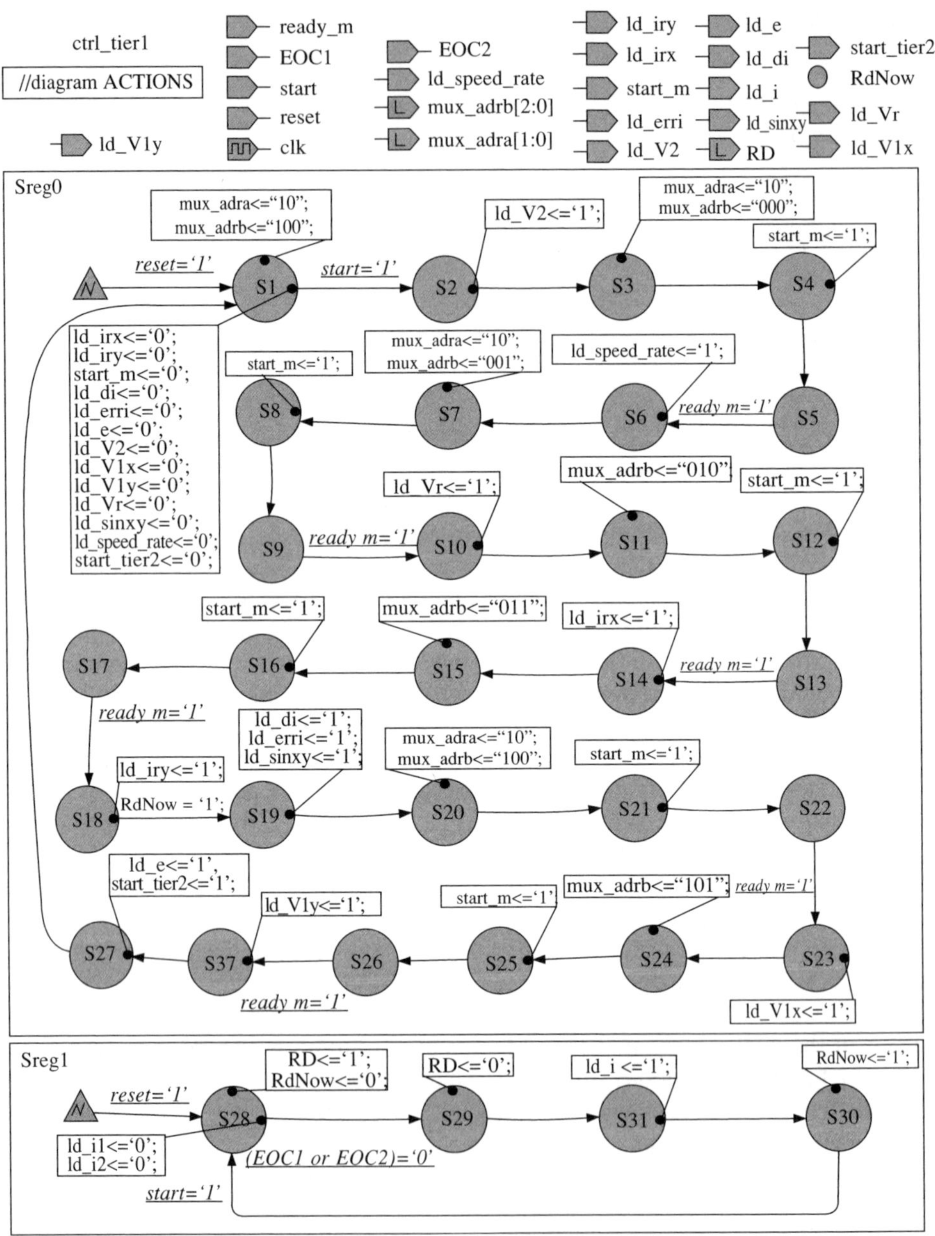


**Fig. 8.38** The state diagram of ctrl_tier1

while the PWM generation requires the calculation of $\arg\{\Delta \underline{i}_{\text{ref}}\}$. This implementation solution reduces the hardware complexity of the motor controller because the same neural structure is used for three different purposes.

The structure of data_tier2, shown in Fig. 8.39, includes three registers that are connected to the angle subnetwork and store the output codes associated with the vectors $\Delta\underline{i}(k)$, $\Delta\underline{i}(k-1)$, $\underline{V}_{\Delta}(k)$ calculated by tier1. They provide this information to an analysis module that controls the inductance updating block, which increments or decrements the value of the inductance, depending on the relative position between the three vectors. The induction estimation result is loaded into a register whose 'load' input is validated by comparing the reference rotor frequency with the upper limit of 10 Hz. It was shown in an earlier section that the induction estimation errors increase with increasing stator frequency. It is also demonstrated that stator frequency increases linearly with the rotor steady-state angular speed. Thus, the inductance estimation errors can be maintained low if the estimation process is performed only at low rotor reference speed. The 'frequency check' block in Fig. 8.39 compares the reference speed with the upper limit and validates the **estimate** signal generated by ctrl_tier2 only if the reference speed is situated below this limit. Otherwise, the estimated inductance $\hat{L}$ is unchanged.

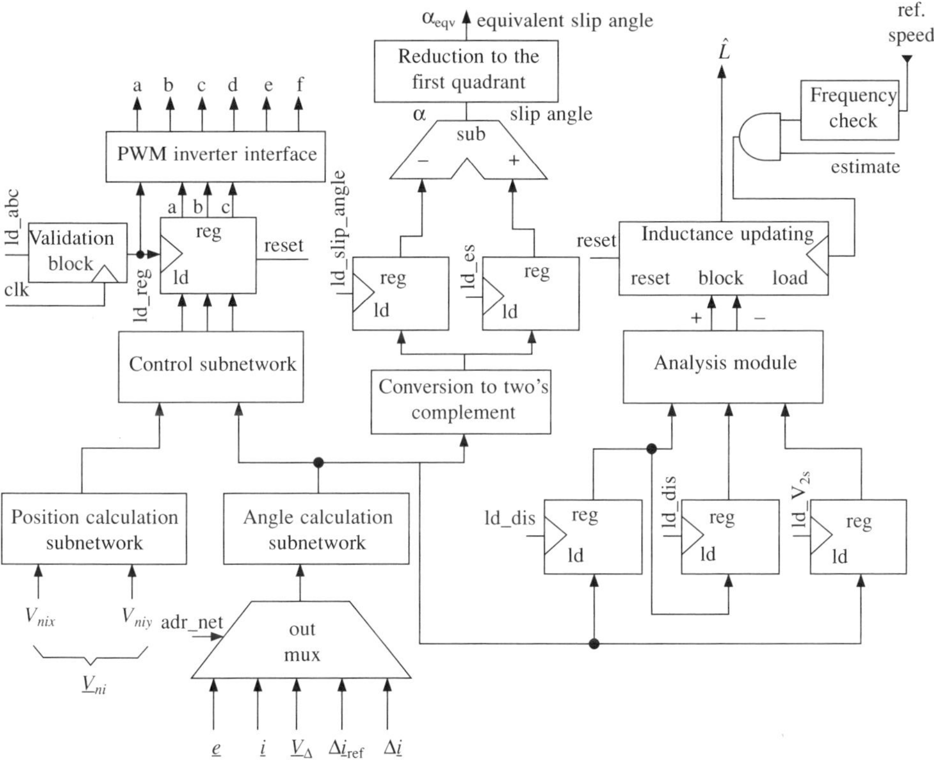


**Fig. 8.39** The RTL description of data_tier2

According to the induction estimation algorithm discussed in Chapter 4, the analysis module has to determine which vector ($\Delta\underline{i}(k)$, $\Delta\underline{i}(k-1)$ or $\underline{V}_{\Delta}(k)$) is situated between the other two. As demonstrated earlier, the angle subnetwork divides the 360° interval into a number of equal sectors. All the output codes associated with space vectors that belong to a given group of consecutive sectors correspond to binary codes that share a certain number of identical bits. One space vector is situated between the other two if it is included in the group of sectors delimited by the two vectors. Therefore, the relative position of the space vectors can be determined by analysing the codes associated with the angle subnetwork.

The method is illustrated by an example in Table 8.3 that involves three vectors $\underline{v}_2$, $\underline{v}_4$, $\underline{v}_6$ situated in the sectors 2, 4 and 6. It can be calculated that $\underline{v}_4$ lies between $\underline{v}_2$ and $\underline{v}_6$ because the code associated with $\underline{v}_4$ shares the bits $b_5$ and $b_0$ with the codes corresponding to $\underline{v}_2$ and $\underline{v}_6$. On the other hand, the vectors $\underline{v}_2$ and $\underline{v}_4$ share the bits $b_5$, $b_2$, $b_1$, $b_0$. Therefore, $\underline{v}_6$ is not situated between $\underline{v}_2$ and $\underline{v}_4$ because the code of $\underline{v}_6$ does not share the bits $b_2$ and $b_1$ with the other two codes.

**Table 8.3** The output codes of an angle subnetwork with $n = 6$ neurones

| Sector index | Angle interval | Code | | | | | |
|---|---|---|---|---|---|---|---|
| | | $b_5$ | $b_4$ | $b_3$ | $b_2$ | $b_1$ | $b_0$ |
| 1 | [−15°; 15°) | 0 | 0 | 0 | 0 | 0 | 0 |
| 2 | [15°; 45°) | 1 | 0 | 0 | *0* | *0* | 0 |
| 3 | [45°; 75°) | 1 | 1 | 0 | 0 | 0 | 0 |
| 4 | [75°; 105°) | 1 | 1 | 1 | *0* | *0* | 0 |
| 5 | [105°; 135°) | 1 | 1 | 1 | 1 | 0 | 0 |
| 6 | [135°; 165°) | 1 | 1 | 1 | *1* | *1* | 0 |
| 7 | [165°; 195°) | 1 | 1 | 1 | 1 | 1 | 1 |
| 8 | [195°; 225°) | 0 | 1 | 1 | 1 | 1 | 1 |
| 9 | [225°; 255°) | 0 | 0 | 1 | 1 | 1 | 1 |
| 10 | [255°; 285°) | 0 | 0 | 0 | 1 | 1 | 1 |
| 11 | [285°; 315°) | 0 | 0 | 0 | 0 | 1 | 1 |
| 12 | [315°; 345°) | 0 | 0 | 0 | 0 | 0 | 1 |

The PWM inverter interface transforms the 3 bits 'abc' defining the desired inverter output voltage into a vector with six control signals 'abcdef' that are transmitted to the power transistors. The edges of the signals controlling the transistors in the same inverter leg (a–d, b–e, and c–f) do not simultaneously occur so that short-circuits are avoided.

Thus, a transistor is turned on at 2.5 μs after its counterpart in the inverter leg has been turned off. This is achieved by using the hardware structure in Fig. 8.40 where the control signals are generated by AND gates whose outputs depend both on the current bits and on the previous bits generated by the neural network.

The previous bits are stored into a 6-bit register that is loaded at 2.5 μs after the neural network new output has been transmitted to the interface block. If one of the bits was previously '1' while the current value is '0', then the corresponding AND gate output changes from '1' to '0' immediately. If the previous value was '0' and the current value is '1' the AND gate output transition cannot occur immediately because the gate inputs are different for a period of 2.5 μs. The 2.5 μs delay is generated by a down-counter that is reset by the same signal **ld_reg** which loads the 'abc' register (Fig. 8.39).

The signal **ld_reg** is generated by the validation block, which is a modulo 9 counter. Therefore, the frequency of **ld_reg** is ten times smaller than the frequency of **ld_abc**. The signal **ld_abc** is activated once during each operation cycle of the motor controller. The operation cycles are initiated by tier0 with a frequency of 150 kHz, which entails that the signal **ld_reg** has a frequency of 15 kHz. Therefore, in this configuration the frequency of the PWM signal generated by the motor controller is 15 kHz.

The control unit of **tier2** (Fig. 8.41) synchronises the operation of the multiplexer connected to the angle subnetwork in Fig. 8.39 with the activity of the registers and the

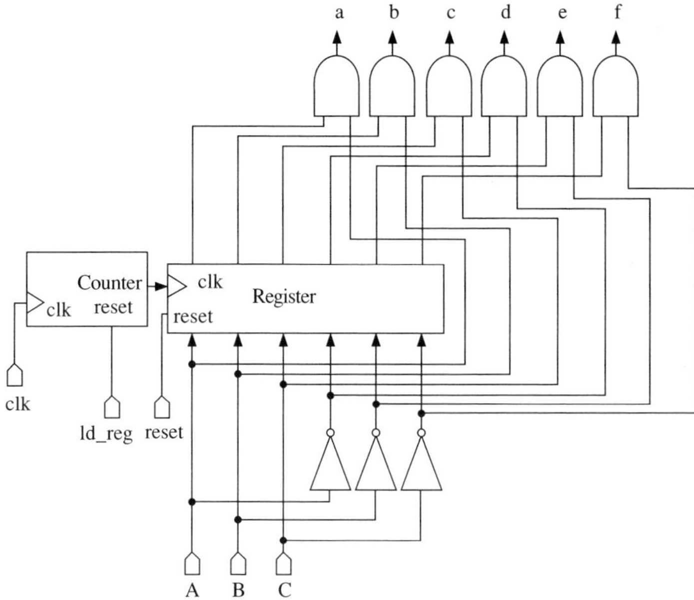


**Fig. 8.40** The PWM inverter interface

operation of the inductance updating block. The operation cycle of the control unit is described by Fig. 8.41 and it consists of the following steps:

1. The stator current vector is supplied to the angle subnetwork and after the calculations the 3-bit vector abc is loaded in the corresponding register.
2. The shifted components of vector $\underline{e}$ are analysed and the corresponding value is stored in the specialised register by activating signal **ld_es**.
3. The angle of the current space vector $\underline{i}_s$ is determined and the quantity proportional with the slip angular frequency is determined.
4. The code corresponding to the argument of the vector $\underline{V}_\Delta$ is calculated and stored in a register.
5. The argument of $\Delta\underline{i}(k)$ is calculated and it replaces the value of $\Delta\underline{i}(k-1)$ which is transferred into a second register (Fig. 8.39).
6. Once the angle codes associated with the three vectors $\Delta\underline{i}(k)$, $\Delta\underline{i}(k-1)$, $\underline{V}_\Delta(k)$ are known, their relative position is analysed by the corresponding combinational module and then the estimated value of $L$ is updated. It can be increased, decreased or left unchanged.

Each operation cycle starts when the necessary input information, calculated by tier1, is available on the input ports of tier2. The appropriate moment to start the operation of tier2 is indicated by the signal **start_tier2** generated by tier1. On the other hand, ctrl_tier2 waits for the four adapters included inside tier1 to finish their shifting tasks. As shown in Fig. 8.41, the signals **ready_erris**, **ready_es, ready_V2s** and **ready_dis** generated by the adapters in Fig. 8.37 are used to test the validity of the input before it is processed by the angle subnetwork.

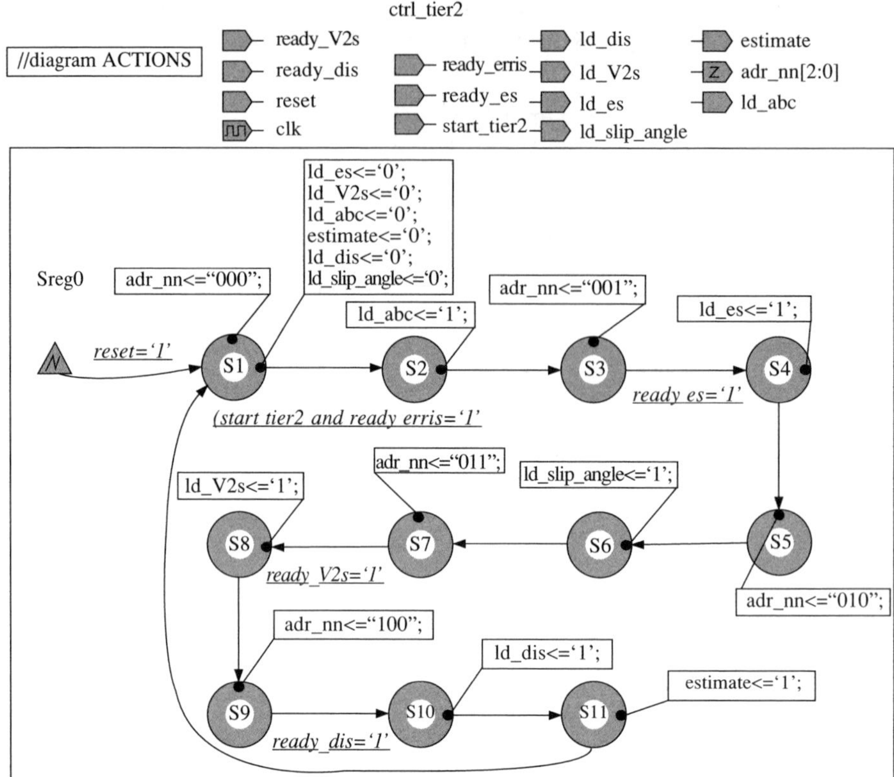


**Fig. 8.41** The state diagram of ctrl_tier2

## 8.4.4 The implementation of the speed control strategy (Tier 3)

Tier3 calculates the stator frequency and the stator current amplitude using the slip angle and the rotor reference speed. The calculations are performed according to the control principles discussed. The simplified function $F_I$ using a single proportionality constant $K_I$ has been used to minimise the hardware structure of this tier. To simplify the VHDL model even further, the multiplication with $K_I$ has been modelled as a set of shifts and additions as follows:

```
-- Code Fragment 8.15
process(beta) -- Multiplication by 0.101B=0.625
  variable x: std_logic_vector(7 downto 0);
begin
  x:=beta & "0000";
  x:=beta+beta(3) & beta(3) & beta & "00";
Ki_times_beta<=x;
end process;
```

The approach is advantageous when $K_I$ contains a large number of '0' bits but a small number of '1' bits. The current increments calculated using $F_I$ are accumulated in a specialised register. The accumulation process can be inhibited if $I_s$ decreases under the limit $I_{s\text{-min}}$ or increases over $I_{s\text{-max}}$, thereby maintaining the current amplitude within the acceptable range of values (Fig. 8.42). The speed derivative function $F_\omega$ has been

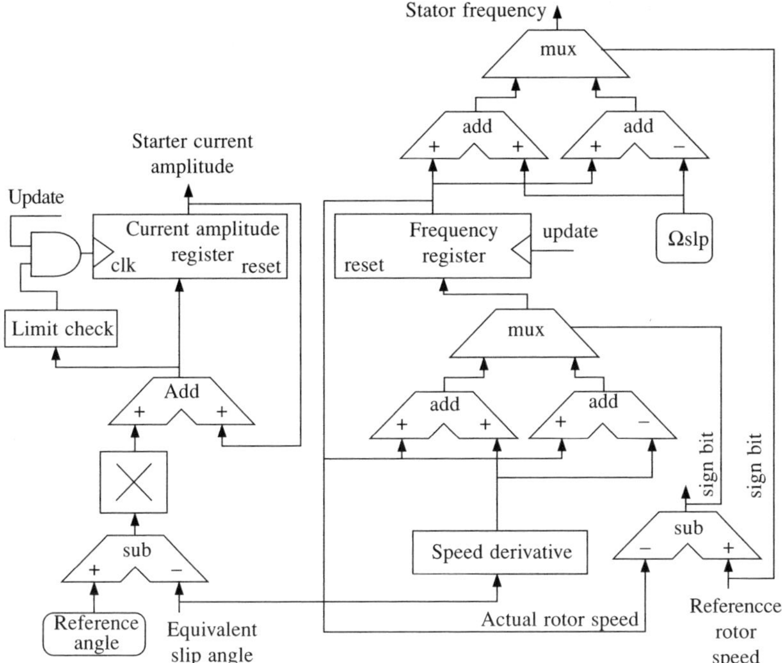


**Fig. 8.42** The structure of tier3

implemented as a look-up table. The angle neural network divides the 360° interval into 36 sectors, which means that only nine sectors are allocated to each quadrant. Therefore, the look-up table is small and can be modelled by a **case** VHDL statement:

```
-- Code Fragment 8.16
process(alpha)
begin
  case alpha is
  when "0000"=> F_omega<="00011";
  when "0001"=> F_omega<="00011";
  when "0010"=> F_omega<="00011";
  when "0011"=> F_omega<="00100";
  when "0100"=> F_omega<="00110";
  when "0101"=> F_omega<="01000";
  when "0110"=> F_omega<="01010";
  when "0111"=> F_omega<="01100";
  when "1000"=> F_omega<="01110";
  when others=> F_omega<="01111"
  end case;
end process;
```

The approach has the advantage that non-linear versions of $F_\omega$ can be implemented with the same hardware resources as the piecewise linear versions. The values generated by $F_{\omega}$ are always positive and are added or subtracted depending on the difference between the reference speed and the calculated speed of the rotor. Furthermore, the constant quantity $\Omega_{slp}$ needs to be added to or subtracted from the previously obtained result depending on the sign of the reference speed. All these situations are analysed by the simple interconnection of adders, subtracters and multiplexers shown in Fig. 8.42.

The operation of tier3 does not require a control unit. All the results are updated when the input signal **update** is activated by tier2. This input signal is connected to the signal **estimate** generated by ctrl_tier2. The signal **estimate** is activated at the end of the operation cycle of tier2 after the motor slip angle has been calculated. Therefore, it indicates an appropriate moment for tier3 to perform its calculations as the slip angle $\alpha_{eqv}$ is one of the two input data used by this tier.

### 8.4.5 The complete motor controller simulations

A VHDL test-bench can be developed by combining the model of the complete controller with the VHDL model of the three-phase induction motor presented earlier. Several simulations have been performed using a Workview Office software package with different values of the generic parameters involved in the controller description, in order to test its correct operation. The simulations demonstrate the controller capability to generate correct PWM signals (Fig. 8.43), to accurately control the stator current and to determine the motor equivalent inductance.

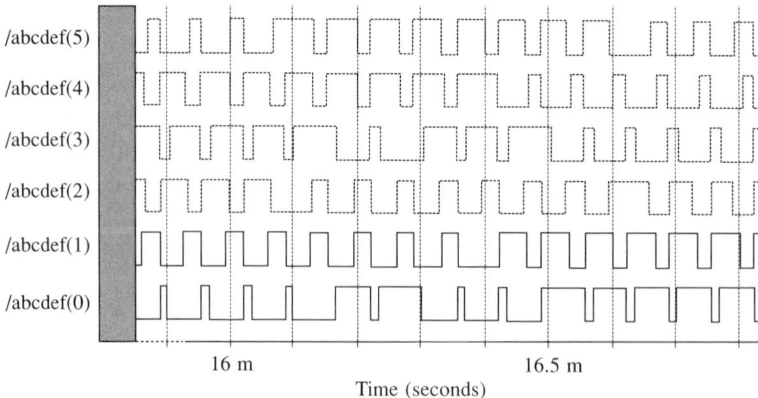


**Fig. 8.43** Motor controller simulation results

The precision of the neural network generating the PWM signal is restricted by the limited number of Voronoi cells. However, given an adequately high number of cells, the operation imprecision is sufficiently low to have a negligible effect on the overall operation of the drive system. The on-line induction estimation process is affected by the limited precision of the angle subnetwork involved (±5° error), and as result, the final estimated inductance is smaller than the correct value. Figure 8.44 presents a comparison between the simulation results performed in ideal conditions (perfectly accurate angle calculations), and the simulation results obtained with the controller model that performs angle calculations using the hardware implemented angle subnetwork. The inductance estimation inaccuracy causes errors in the calculation of vectors $\underline{V}_{ni}$ and $\underline{e}$ but this does not affect their average value over several operation cycles of the controller. Thus, the effect of the inductance estimation error over the motor speed control is minimal due to the high inertia of the rotor that filters out the ripples of the control signals, originated by the induction estimation errors. The accuracy of the inductance estimation can be increased by increasing the number of neurones included in the angle subnetwork. However, the simulations performed proved that adapting this approach brings only a minimal improvement to the global behaviour of the drive system which is not justified due to the increase of the total hardware complexity of the motor controller. The computer simulations demonstrated the adequate operation of the motor controller including the neural network, and a satisfactory capability to control the operation of a three-phase induction motor. After the successful implementation of the motor controller into a Xilinx XC4010XL FPGA, it has been tested in conjunction with a small three-phase induction motor (less than 0.5 kW).

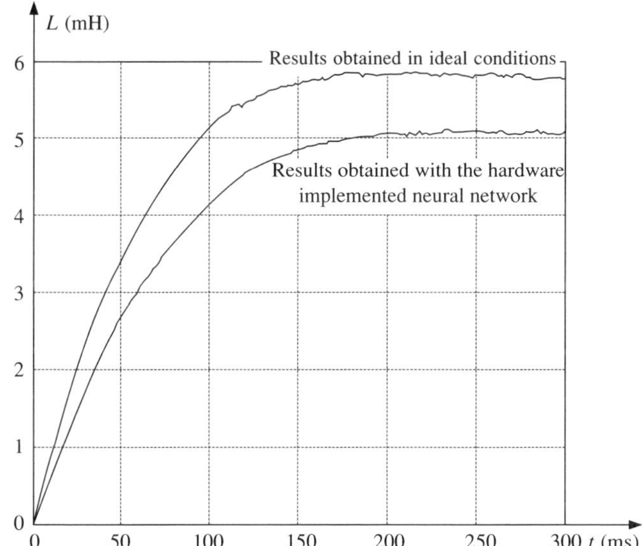


**Fig. 8.44** Comparison of inductance estimation simulation results

## 8.5 FPGA controller experimental results

This section presents experimental results relating to the performance of a complete three-phase induction motor drive system controlled by the new neural FPGA controller.

### 8.5.1 The drive system

Practical tests have been performed using the FH2 MkIV test-bench produced by TecQuipment [4]. The test-bench offers the facility to mount up to two electrical machines, d.c. and a.c. on the same shaft (Fig. 8.45) and includes speed and torque sensors for testing the motor operation.

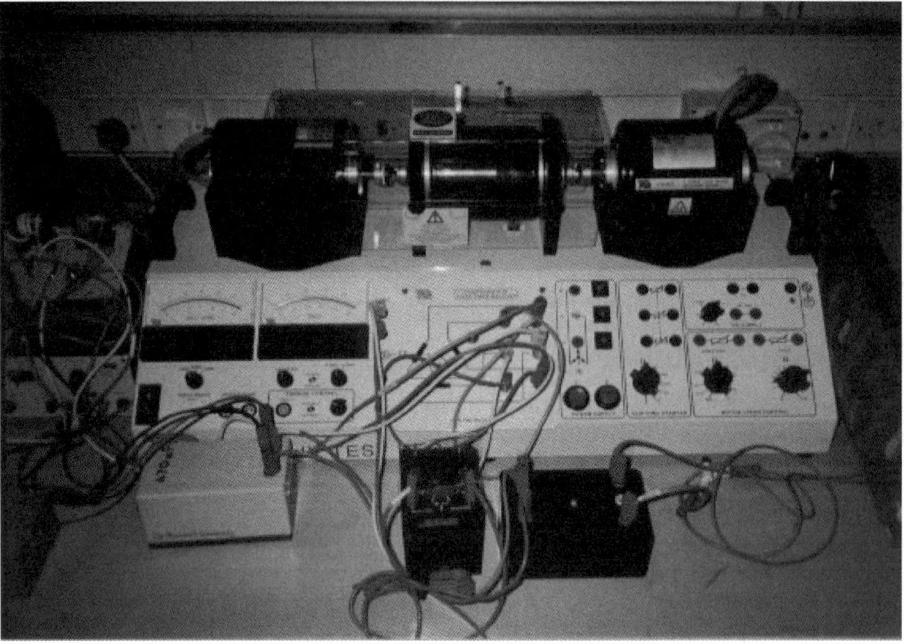

**Fig. 8.45** The FH2 MkIV test-bench

The laboratory test-bench configuration available includes the FH90 four-pole three-phase cage induction motor and the FH50 d.c. motor (not used in the experiments). The stator windings are reconfigurable, both Δ-connection and Y-connection being possible. Nevertheless, TecQuipment recommends that the Δ-connection is used. The rated line voltage is 220 V in this configuration, while the line currents have values of up to 1 A, depending on the load torque. The practical experiments proved that the speed control principles discussed in section 8.3.2 are valid for both Y-connection and Δ-connection. This experimental conclusion is supported by the theoretical possibility of transforming any Δ-connected load supplied by a sinusoidal voltage system into an equivalent Y-connected load. The high frequency harmonics contained by the PWM voltage are filtered by the motor inductances and therefore the corresponding current harmonics are negligible. As a result, only the fundamental harmonics of the voltage and current need to be taken into account and the Y-connected equivalent *R–L–e* circuit is applicable to Δ-connected motors as well.

The experimental setup that includes the FH2 MkIV test-bench and the FH90 induction motor is presented as a block diagram in Fig. 8.46. The motor is supplied by a PWM inverter bridge SKM40GD132D produced by Semikron [7] which contains 1200 V IGBT transistors. The bridge is supplied with d.c. voltage by a diode rectifier via a low-pass filter. The input voltage of the rectifier can be adjusted using an autotransformer, which allows the smooth control of the d.c.-link voltage. The IGBT transistors in the inverter are controlled by the XC4010XL FPGA controller on the XS40 test board. This FPGA is a low voltage device that associates the voltage level of 3.3 V with logic '1' [8].

The voltage level of the control signals is adapted in two stages to the electrical characteristics of the power transistors. First, the CMOS interface in Fig. 8.46 amplifies

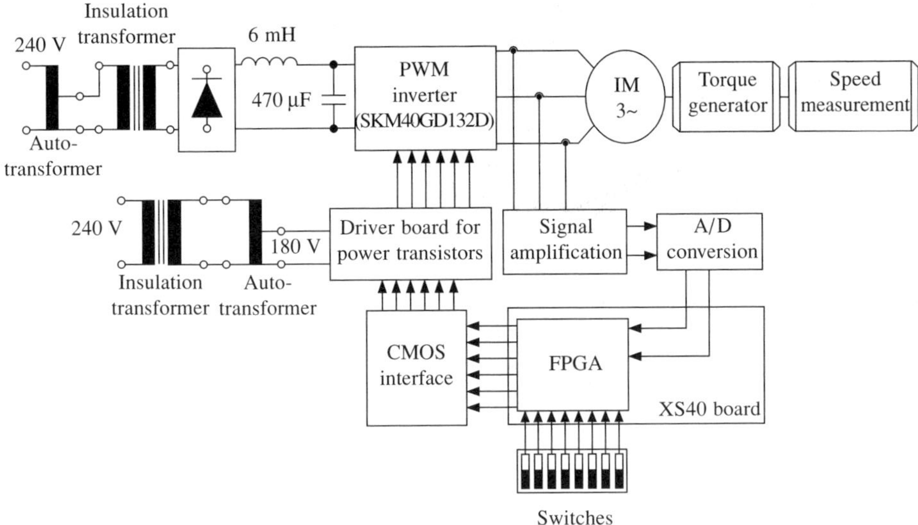


**Fig. 8.46** The schematic of the experimental setup

the output signals of the FPGA to 5 V and supplies them to the transistor driver board. At the same time, the CMOS interface protects the control circuits against the damaging effects of any failure that may occur in the power circuits. In the second stage, the driver board amplifies the control signals to 15 V, which is the control voltage level, recommended by the IGBT manufacturer.

Two of the motor currents are measured using Hall effect transducers that generate a voltage proportional to the measured current, which are then amplified using simple operational amplifiers and transmitted to the TLC1550 10-bit A/D converters produced by Texas Instruments [1]. The binary codes produced by the A/D converters are transmitted to the FPGA controller. As mentioned in Chapter 6, the VHDL controller model contains a series of generic parameters defining the size of several internal modules. Many of these parameters are correlated and depend on the width '*ni*' of the two current input buses '$i_a$' and '$i_b$'. To implement the entire motor controller in a single XC4010XL FPGA, the generic parameter '*ni*' was limited to eight. Thus, only the eight most significant bits are used by the FPGA controller in this configuration. The reference speed of the motor is set in two's complement code using a set of eight switches. Consequently, the rotor electrical angular frequency $\omega_{er}$ can be set at values between –128 and +127 Hz, corresponding to mechanical speeds between –3840 rev/s and +3810 rev/s. The CMOS interface and the operational amplifiers have been implemented on a single interface board, illustrated in Fig. 8.47 together with the XS40 board.

The parameters of the speed control algorithm implemented by tier3 (see Chapter 6) have been determined based on practical experiments with the motor. The equivalent parameters of the FH90 motor were initially determined. The stator resistance was directly measured, the result being 95 Ω, which is a large value for a three-phase induction motor. The equivalent inductance $L$ has been approximated by measuring the voltages across the motor and currents at 50 Hz stator frequency, with the rotor locked (zero speed). In these conditions, the internal voltage $\underline{e}$ has a small value and the total impedance of the motor is mainly due to the resistance $R$ and the equivalent inductance.

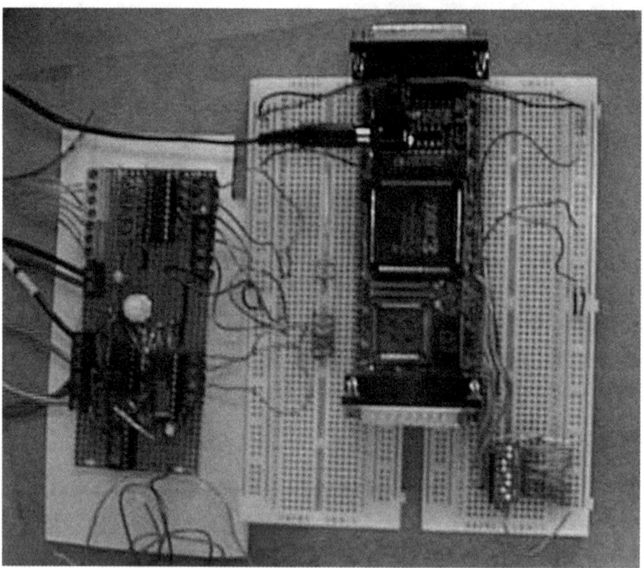

**Fig. 8.47** The XS40 board and the interface board

The measurements and the calculations showed that $L$ is approximately 220 mH. The critical slip angular frequency was determined while keeping the stator current constant. This was achieved by connecting the rotor windings to variable resistors as shown in Fig. 8.48. The obtained result was $\omega_{\text{slp}}^{k} = 95$ rad/s corresponding to a speed of 1050 rev/s (the rated speed is 1500 rev/s). The equivalent slip angle $\alpha_{\text{eqv}}^{\text{ref}}$ has been arbitrarily set at 65° (according to the considerations in Chapter 4, it can have any value between 45° and 90°). The equation

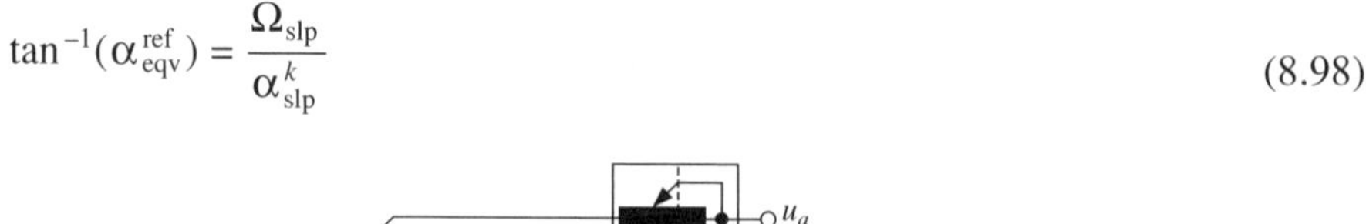

$$\tan^{-1}(\alpha_{\text{eqv}}^{\text{ref}}) = \frac{\Omega_{\text{slp}}}{\alpha_{\text{slp}}^{k}} \tag{8.98}$$

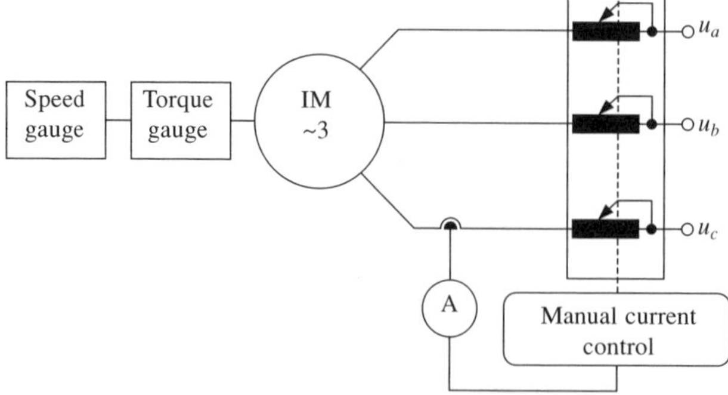


**Fig. 8.48** The measurement of the critical slip at constant current

demonstrated in Chapter 4, yields the value $\Omega_{\text{slp}} = 44.3$ rad/s, corresponding to a slip frequency of 7 Hz. The other parameters that define the speed control algorithm ($K_{\omega 1}$, $K_{\omega 2}$, $\beta_{\max}$, $K_I$) were determined by tuning in practical experiments.

In conclusion, the main characteristics of the adopted VHDL motor controller are:

- The sampling frequency (tier1) $f_s = 150$ kHz.
- The PWM frequency (tier2) $f_{PWM} = 15$ kHz.
- The maximal rotor speed allowing inductance estimation 600 rev/min (10 Hz).
- The width of the input buses $i_a$ and $i_b$ (tier1) $ni = 8$.
- The step length of the multiplier $N_{sl} = 4$.
- The stator resistance (defined by a constant in the model of tier1): $R_s = 95\ \Omega$.
- The equivalent slip angle $\alpha_{eqv} = 65°$.
- The reference slip angular frequency $\Omega_{slp} = 44.3$ rad/s.
- The parameters defining the speed control dynamics: $K_{\omega 1} = 650\ s^{-1}$, $K_{\omega 2} = 200\ s^{-1}$, $\beta_{max} = 0.7$ rad, $K_I = 1$ A/s.
- The switching delay between the two transistors in the same inverter leg: 2.5 μs.
- The input clock frequency $f_{clk} = 12$ MHz.
- The number of samples used to generate the reference sine wave (tier0): 256.
- The number of bits for each entry of the look-up table (tier0): 3.
- The number of bits used for the reference motor speed: 8.
- The number of inverter output voltages used: 6 out of 7 (the null voltage is not used).
- The number of triangular Voronoi cells of the position neural subnetwork: 54.
- The number of sectors of the angle subnetwork: 36.
- The number of bits used to code the input signals of the position subnetwork: 5.
- The number of bits used to code the input signals of the angle subnetwork: 5.

In this configuration, the implementation of the controller took up 98 per cent of the hardware resources available on the XC4010XL FPGA. The values of all the parameters can be easily modified by altering a series of constants and generic parameters in the VHDL code describing the model of the motor controller. Consequently, given the appropriate FPGA, the controller can be adapted in terms of hardware complexity and operation accuracy to the requirements of a large range of particular applications.

Two sets of experiments were carried out. The first set verified the PWM voltage generation and the current control accuracy, while the second set referred to the speed control performance of the drive system.

### 8.5.2 Current and voltage control tests

A special version of the controller VHDL model was created for the first set of experiments. This version lacks tier3 so the frequency of the stator current is identical to the frequency specified by means of the eight switches and the stator current amplitude is constant. This approach simplifies the testing procedure because it checks the operation of tiers 0, 1 and 2 without the feedback signals calculated by tier3 and therefore any operational error can be easily located.

Figure 8.49 presents four of the PWM control signals generated by the FPGA motor controller. They were monitored using a four-channel Hewlett Packard digital oscilloscope. The waveforms demonstrate the correlation between two of the signals that control the transistor on the same inverter leg (**ctrl(5)** and **ctrl(2)**). Thus, the two signals have complementary values: when one of them is '0' the other is '1'. The 2.5 μs delay generated by the interface block contained in tier2 is not visible in Fig. 8.49 due to the inappropriate time scale (50 μs/div), but it can be observed in Fig. 8.50 where the

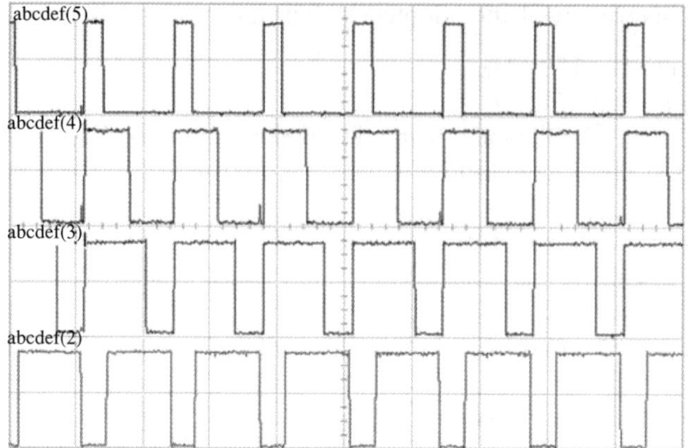


**Fig. 8.49** Four of the FPGA output signals controlling the transistors in the PWM inverter

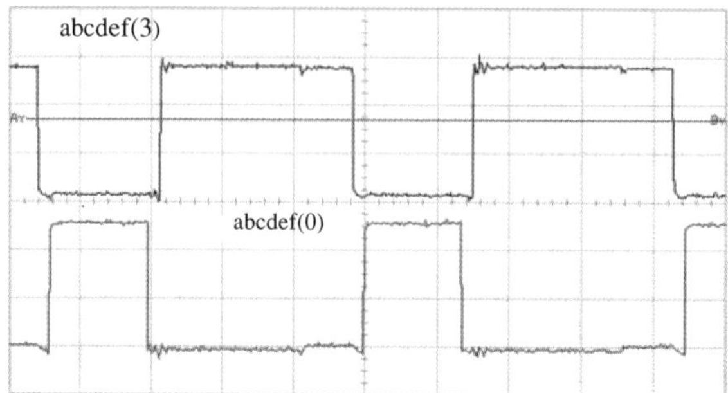


**Fig. 8.50** Switching delays produced by tier2 to avoid the short-circuits in the PWM inverter

time scale is 15 μs/div. The overall PWM voltage across the motor can be seen in Fig. 8.51.

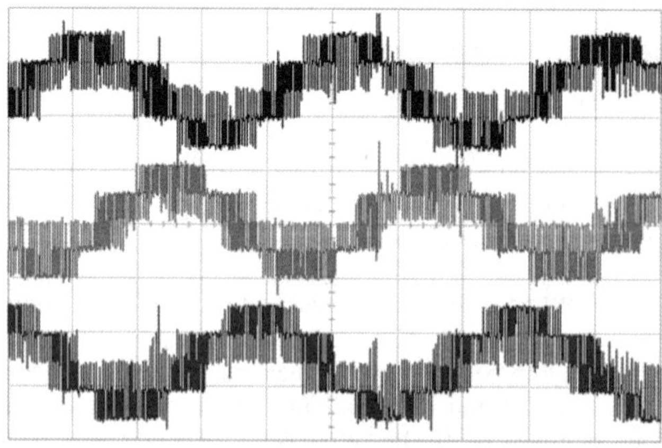

**Fig. 8.51** The PWM voltage generated by the inverter for a reference stator frequency of 50 Hz

Figure 8.52 illustrates one of the motor currents corresponding to the operation mode generated by the voltages in Fig. 8.51. The voltage signal in Fig. 8.52 is acquired from one of the Hall transducers and its amplitude of 200 mV corresponds to current amplitude of 0.4 A. Figure 8.53 presents the d.c.-link voltage in the same operation conditions and demonstrates that the gamma filter composed of the 6 mH inductor and 470 μF capacitor illustrated in Fig. 8.46 is capable of maintaining the d.c. voltage level within acceptable limits.

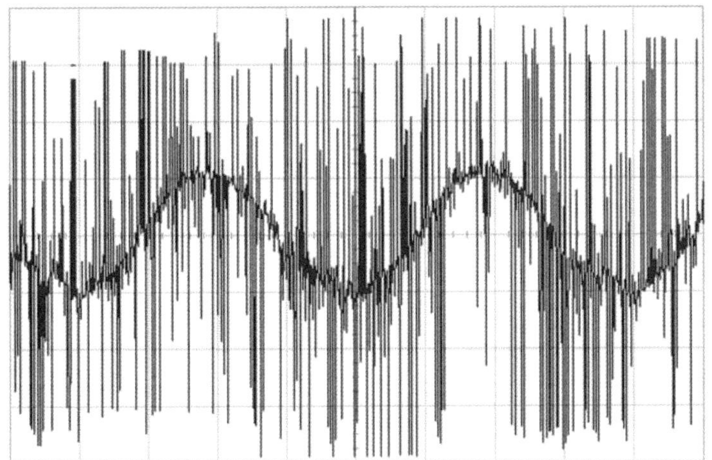

**Fig. 8.52** One of the stator currents (the stator is Y-connected) measured using a Hall transducer

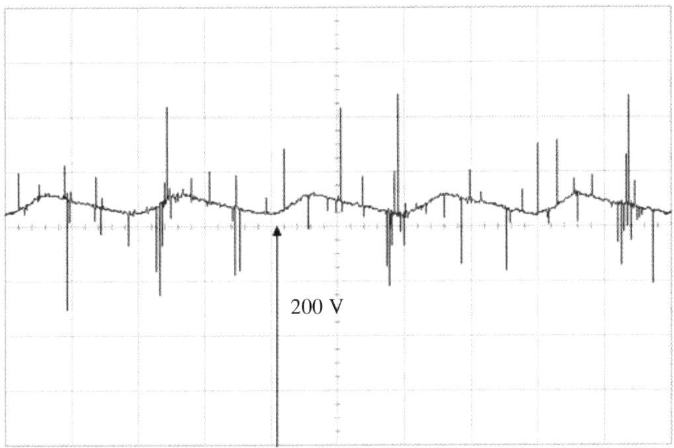


**Fig. 8.53** The voltage ripple on the d.c. link

### 8.5.3 Speed control tests

The second set of experiments, demonstrating speed control, were carried out using the entire VHDL model of the controller, as presented in the previous chapter. The tests were performed with the stator Δ-connected, but similar results are obtained when the stator winding is Y-connected. The drive system has been tested both in steady-state and in transient operation.

Figure 8.54 compares the steady-state operation of the induction motor with or without the new FPGA controller connected. The reference speed of the controller is 1500 rev/s (the rated speed of the FH90 motor). The graphs demonstrate that the controller is capable of maintaining the rotor speed almost constant despite large variations of the load torque. The improvement brought aboard by the new controller is seen in the improved speed–torque characteristic. The improvement can be quantified as the speed increase produced by the controller for each value of the load torque.

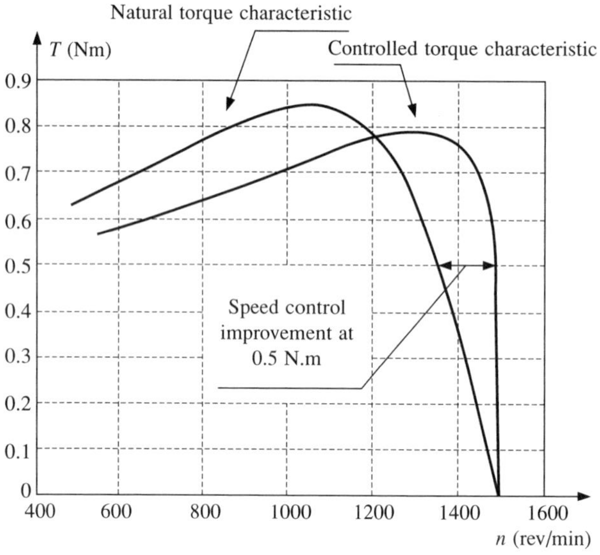


**Fig. 8.54** The static mechanical characteristic of the motor with and without digital controller

Figure 8.55 presents the motor transient response to a fast variation of the torque. A perfect step variation of the load torque could not be achieved due to the limitations of the test-bench. However, Fig. 8.55 shows the capability of the FPGA controller to maintain the speed almost constant while the load torque undergoes significant variations relative to the motor rated power. Thus, the increase of the torque causes a slight slow down of the rotor and this leads to an increase of the motor slip above the imposed value. Therefore, the slip compensation mechanism included in the controller increases the motor current and boosts the active torque reducing the slip frequency to the initial value. The transient process takes approximately 0.8 s.

Figure 8.56 illustrates the motor start-up when the load torque is null. The transient response lasts for approximately 0.65 s, after which the motor speed is constant at 1500 rev/min.

The load torque during the transient operation illustrated in Figs. 8.57 and 8.58 is proportional to the rotor speed and therefore motor acceleration is more difficult and the transient response slower. The result in Fig. 8.58 can be compared with the motor starting characteristic when no speed control system is used (Fig. 8.57). Clearly, the FPGA speed controller improves the dynamic response of the drive system.

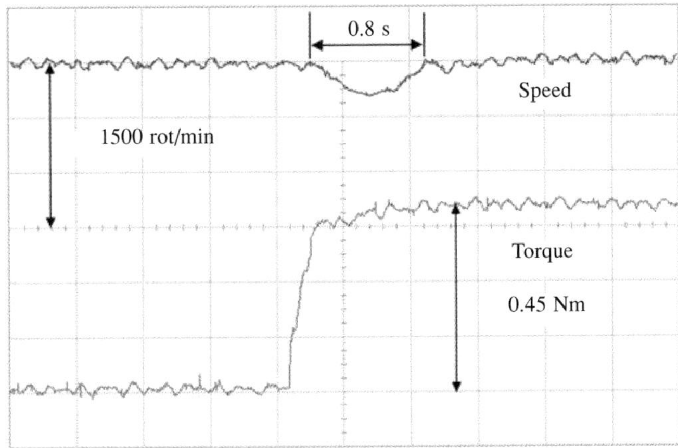


**Fig. 8.55** The drive system response to a step change of the load torque

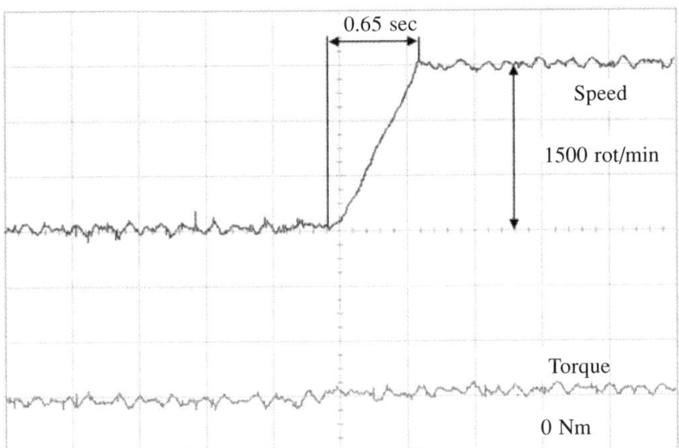


**Fig. 8.56** The motor start without external load torque

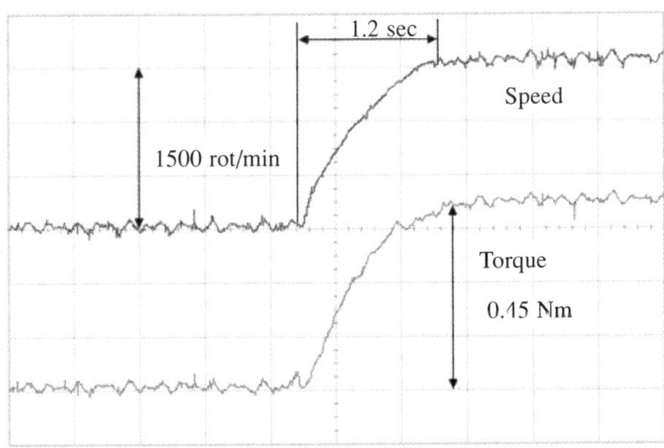


**Fig. 8.57** The natural motor start under load without the FPGA controller

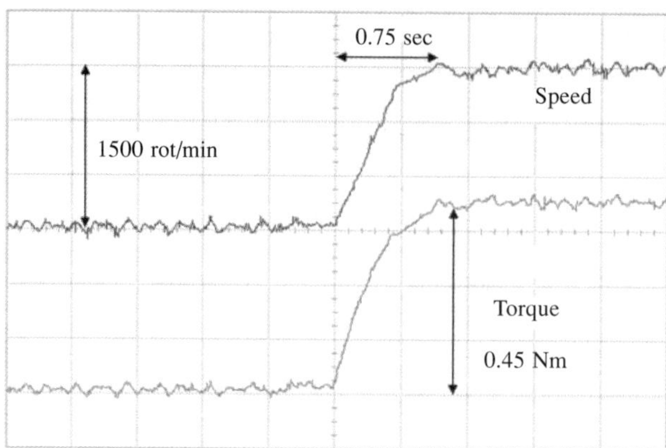


**Fig. 8.58** The motor start under load when controlled by the FPGA device

The improvement is limited by two factors:

- The motor parameters.
- The available hardware resources.

The parameters of the motor FH90 are not suitable for sensorless control with high response speed because the stator resistance is large (95 Ω) and therefore the resistive voltage component is large. Furthermore, the motor power is small (approximately 0.5 kW) so the amplitude of the internal voltage $\underline{e}$ is smaller than the resistive voltage component. As a result, the calculated equivalent slip angle $\alpha_{eqv}$ is inaccurate and affected by large fluctuations during motor operation. These fluctuations tend to cause system instability, which can be counteracted only by increasing the time constants of the system, equivalent to limiting its overall dynamic performance. Nonetheless, the limitations imposed by the motor parameters are not particular to the new control method developed, they affect most of the sensorless speed control algorithms.

# 9

# Fuzzy logic control of a synchronous generator set

## 9.1 System representation

Before commencing an in-depth discussion about the control system, it is appropriate to discuss the plant, i.e. the elements to be controlled. In this chapter mathematical models of some of the control plant elements are described. The plant comprises a power electronic system, a permanent magnet synchronous generator and a diesel engine. Two control units are presented in this example, a fuzzy logic controller (FLC) for the engine and a PWM controller for the output inverter. The analysis of the system is divided into two parts as shown in Fig. 9.1. Ensuing sections in this chapter discuss the control plant elements relating to the fuzzy logic controller, namely the rectifier, the generator and the diesel engine. The analysis of the inverter and the PWM controller is presented as a separate section.

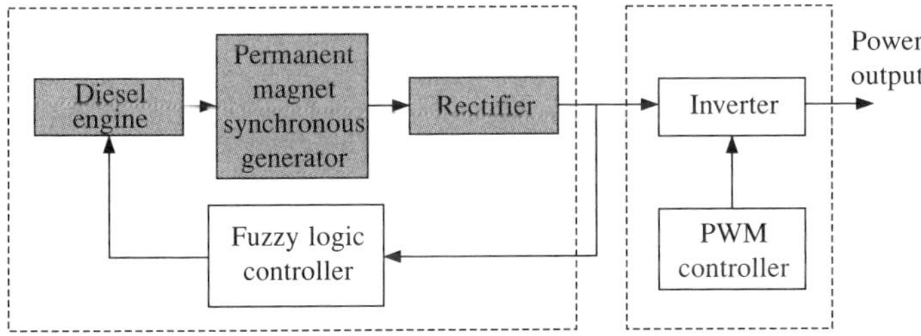


**Fig. 9.1** Block diagram of complete system

### 9.1.1 Rectifier circuit

A circuit diagram of a three-phase uncontrolled (diode) rectifier attached to a Y-connected sinusoidal voltage source is shown in Fig. 9.2. This circuit is used to convert the a.c. output of the generator terminals into d.c. power. The phases are marked *a*, *b* and *c*. The branches of the rectifier bridge are connected to the generator terminals, therefore the line to line voltage $v_{l-l}$ of the rectifier is equal to the generator terminal voltage $v_t$.

#### *9.1.1.1 Voltage analysis*

Figure 9.3 shows the waveforms of various voltages taken from the circuit in Fig. 9.2. Waveforms $v_{a-n}$, $v_{b-n}$ and $v_{c-n}$ are the line–neutral voltages of each phase. $v_{dc}$ is the

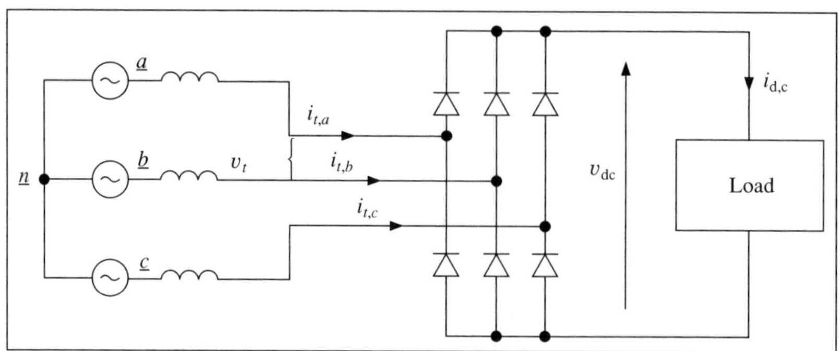


**Fig. 9.2** Three-phase uncontrolled rectifier

instantaneous rectifier output voltage while $V_t$ is the r.m.s. value of the line–line voltage, $v_t$. The mean output voltage $V_{dc}$ can be obtained by integrating $v_{dc}$ over a period of $\pi/3$ radians as follows:

$$V_{dc} = \frac{1}{\left(\frac{\pi}{3}\right)} \int_{-\frac{\pi}{6}}^{\frac{\pi}{6}} \sqrt{2}\, V_t \cos \omega t \cdot d\omega t = \frac{3\sqrt{2}}{\pi} V_t$$

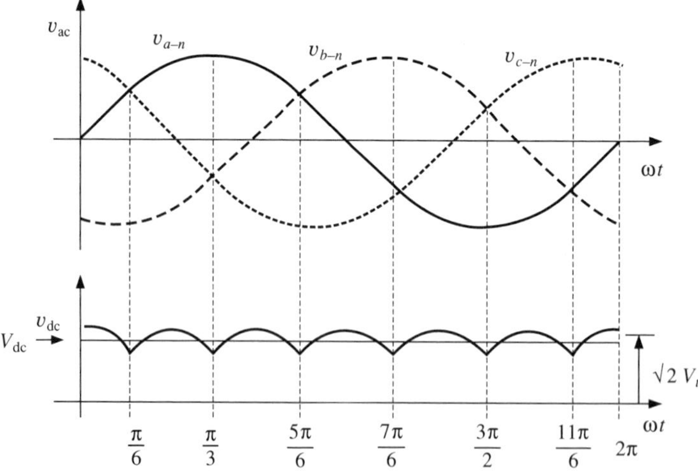


**Fig. 9.3** Voltage waveforms of rectifier circuit

### *9.1.1.2 Current analysis*

The current waveforms of the rectifier circuit are shown in Fig. 9.4. The line current waveform $i_{t,a}$ is not sinusoidal, therefore the r.m.s. value cannot be obtained simply with a division by $\sqrt{2}$. From the definition of *root mean square*, the r.m.s. of $i_{t,a}$ is given by

$$\text{r.m.s.} : I_t = \sqrt{\frac{i_1^2 + i_2^2 + i_3^2 \ldots + i_n^2}{n}}$$

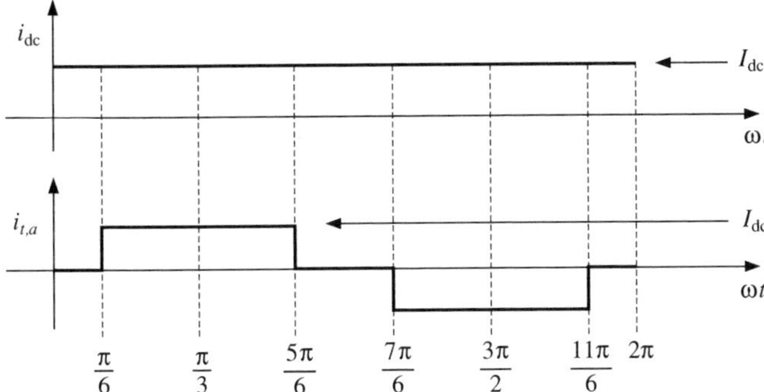


**Fig. 9.4** Current waveform of rectifier circuit

Taking samples at every $\pi/6$ interval,

$$I_t = \sqrt{\frac{[4 \times (I_{dc})^2] + [4 \times (-I_{dc})^2] + [4 \times 0^2]}{12}}; \quad I_t = \sqrt{\frac{2}{3}} \cdot I_{dc}$$

## 9.1.2 Synchronous generator

### 9.1.2.1 Voltage analysis

The operation of the synchronous generator is based on Faraday's law of electromagnetic induction which states that: *The total electromagnetic force (e.m.f.) generated in a closed circuit is equal to the negative time rate of change of the flux linkages linking the circuit.*

This law was derived from Faraday's observation that an e.m.f. is induced in a conductor or circuit placed in a magnetic field when there is a change of flux linkages linking the circuit. The flux changes can be the result of a relative motion between the circuit and the flux or the effect of a varying magnetic flux. Faraday's law can be represented by the following equation:

$$e(t) = -\frac{d}{dt}\lambda$$

where $e(t)$ is the generated e.m.f. and $\lambda$ is the flux linkage. In a synchronous machine, the magnetic flux is provided either by a permanent magnet or an electromagnet, placed inside a set of windings as shown in Fig. 9.5. The flux linkages linking the winding are varied by physically rotating the magnet. The rotor can be a permanent magnet, or an electromagnet with a field winding. Electromagnet generators are easier to control and therefore more common. The stator houses the armature windings.

The armature windings are the windings in which the e.m.f. is induced. Figure 9.5 shows a three-phase synchronous machine, with armature windings marked $aa'$, $bb'$, and $cc'$. If the angular velocity of the rotating field in radians per second is $\omega$, then:

$$\varphi = \omega t \tag{9.1}$$

Considering only phase $a$, if the rotor moves by an angle $\varphi$, the flux linking $aa'$ is given by,

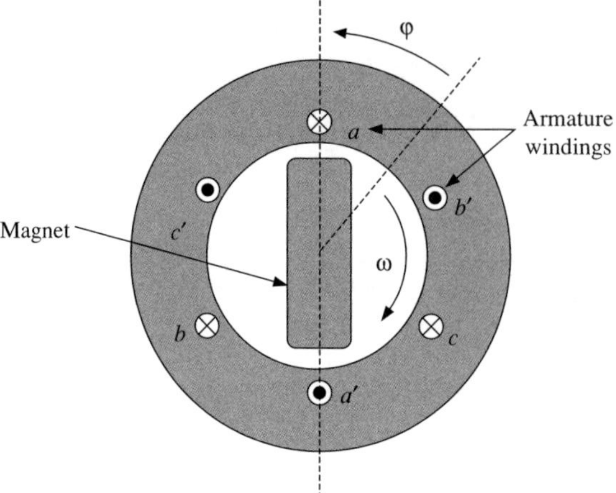


**Fig. 9.5** Three-phase synchronous machine

$$\lambda = N\phi \cos \varphi \tag{9.2}$$

where $N$ is the number of turns in $aa'$ and $\phi$ is the airgap flux. Therefore, the e.m.f. induced per phase is given by

$$e_a = -\frac{d\lambda}{dt} = -\frac{d\lambda}{d\varphi} \cdot \frac{d\varphi}{dt} = (N \cdot \phi \cdot \sin \varphi) \cdot \frac{d\varphi}{dt} \tag{9.3}$$

Substituting $\varphi$ for $\omega t$ and $d\varphi/dt$ for $2\pi f$ yields

$$e_a = 2\pi f\, N\, \phi \sin \omega t \tag{9.4}$$

In practice, the constructions of the machine, such as the pitch and distribution of the coils, assert influence on the waveshape of the induced e.m.f. A reduction factor known as the *winding factor* $K_W$ has to be applied to account for this. Therefore, the induced e.m.f. per phase becomes

$$e_a = 2\pi f N \phi K_w \sin \omega t \tag{9.5}$$

which can also be written as

$$e_a = \hat{E}_i \sin 2\pi f t \tag{9.6}$$

where $\hat{E}_i$ is the peak value of the induced e.m.f. and $f$ is its frequency.

Using phase $a$ as reference and because phases $a$, $b$ and $c$ are displaced by 120°, the voltages for each field are therefore given by

$$e_a = \hat{E}_i \sin 2\pi f t \tag{9.7}$$

$$e_b = \hat{E}_i \sin(2\pi f t - 120°) \tag{9.8}$$

$$e_c = \hat{E}_i \sin(2\pi f t + 120°) \tag{9.9}$$

In machines with more than a pair of poles, the generated electrical frequency $f$ is given by

$$f = p \,.\, n_{\text{mech}} \tag{9.10}$$

where $n_{mech}$ is the mechanical synchronous speed (revolutions per second) and $p$ is the number of machine pole-pairs.

### 9.1.3 Equivalent circuit model

Dynamic models of the synchronous machine can be derived from *the generalised theory of electrical machines* and are widely used in the design of classical control systems. The design of the controllers is derived from the plant model. However, because the control system presented in this example is not developed from a model of the plant, the accuracy of the model does not directly influence the performance of the design. Instead, the model is used to study the response of the design and to verify its functionality through simulations. Although a dynamic model would provide a more comprehensive analysis, the steady-state model is sufficient for the task.

An equivalent circuit model of a synchronous generator can be derived to study the voltage and current characteristics of the machine. The flow of current in the field winding $I_f$ induces a flux $\phi_i$ in the airgap which links with the armature winding. Similarly, the flow of current in the armature winding also produces a flux, $\phi_a$. The armature flux $\phi_a$ comprises two components, the *armature reaction flux* $\phi_{ar}$, and the *leakage flux* $\phi_{al}$. The larger component, $\phi_{ar}$, is established in the airgap and links with the field winding, while $\phi_{al}$ does not exist in the airgap and links only with the armature winding. Therefore, the resultant flux in the airgap $\phi_{ag}$ is made up of $\phi_i$ and $\phi_{ar}$.

$$\phi_{ag} = \phi_i + \phi_{ar} \tag{9.11}$$

$\phi_{ag}$ induces a voltage in the stator winding, $E_{ag}$ which lags $\phi_{ag}$ by 90°. According to the superposition theorem, $E_{ag}$ comprises two components, induced by the corresponding flux component. Hence,

$$E_{ag} = E_i + E_{ar} \tag{9.12}$$

$E_i$ is the induced e.m.f. also known as the internal generated voltage of the machine. A diagram of the vectors concerned is shown in Fig. 9.6(a). It shows the armature reaction voltage $E_{ar}$, lagging $\phi_{ar}$ and $I_a$ by 90°. In other words, $-E_{ar}$ leads $I_a$ by 90°. A voltage leading a current by 90° can be represented as the voltage drop across a reactance. Hence $-E_{ar}$ can be represented as the voltage drop across a reactance, say $X_{ar}$, where $X_{ar}$ is known as the *magnetising reactance*. Therefore, by substituting $(-E_{ar})$ with $(I_a jX_{ar})$, (9.12) can be rewritten as

$$E_i = I_a jX_{ar} + E_{ag} \tag{9.13}$$

Figure 9.1(b) shows an equivalent circuit of the model described by (9.13). To include the terminal voltage per phase $V_\phi$ (instead of just the airgap voltage, $E_{ag}$), the leakage reactance $X_{al}$ and the effective armature winding resistance $R_a$ have to be taken into account. The effective resistance of $R_a$ is approximately 1.6 times its d.c. resistance as a result of skin effect and operating temperature effects. Figure 9.6(c) shows the complete equivalent circuit of a synchronous machine.

$X_{ar}$ and $X_{al}$ can be lumped together as a single reactance denoted by $X_s$, which is known as the *synchronous reactance*.

$$X_s = X_{ar} + X_{al} \tag{9.14}$$

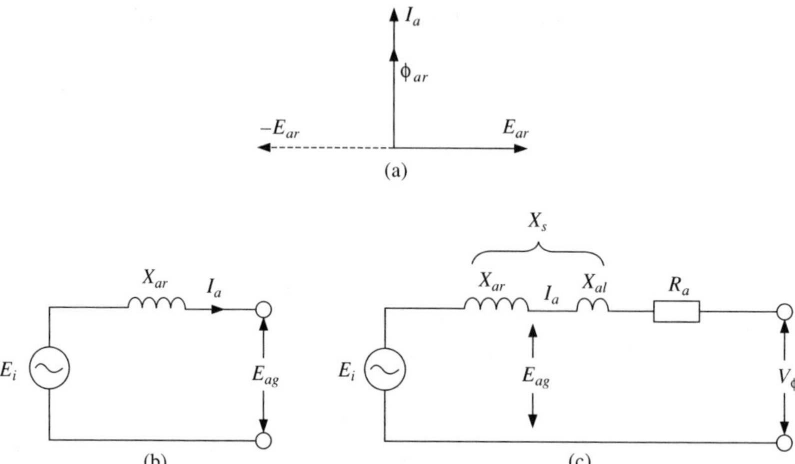


**Fig. 9.6** Equivalent circuit model of a synchronous generator

The synchronous impedance is given by

$$Z_s = R_a + jX_s \tag{9.15}$$

The steady-state mathematical model of the synchronous machine can therefore be described by the equation

$$E_i = I_a(R_a + jX_s) + V_\phi \tag{9.16}$$

The equivalent circuit model is a steady-state model and does not provide any information about the dynamics of the machine. Transient response to load changes, faults and other disturbances cannot be properly analysed using this model. Dynamic models of the synchronous machine can be derived from *the generalised theory of electrical machines* and are widely used in the design of classical control systems. The design of the controllers is derived from the plant model.

### 9.1.4 The generator–rectifier model

From (9.5) and (9.10), the peak value of the internal generated voltage is found to be

$$\hat{E}_i = 2\pi N\phi K_w p n_{\text{mech}} \tag{9.17}$$

Because the generated voltage is sinusoidal, its r.m.s. value is given by

$$E_i = \sqrt{2}\pi N\phi K_w p n_{\text{mech}} \tag{9.18}$$

For an electromagnet generator the flux in the machine is controlled by its field current. The relationship between the flux and the field current can be observed by performing an *open circuit test* on the machine and plotting the results on an '$E_i$ vs. $I_f$' graph which is sometimes called the open circuit characteristic (OCC) plot. Details of the test can be found in any text book on electrical machines. Figure 9.7 shows a typical OCC plot of a synchronous generator. The curve is observed to be linear until some saturation starts to occur at high field current. This is due to iron saturation in the synchronous generator.

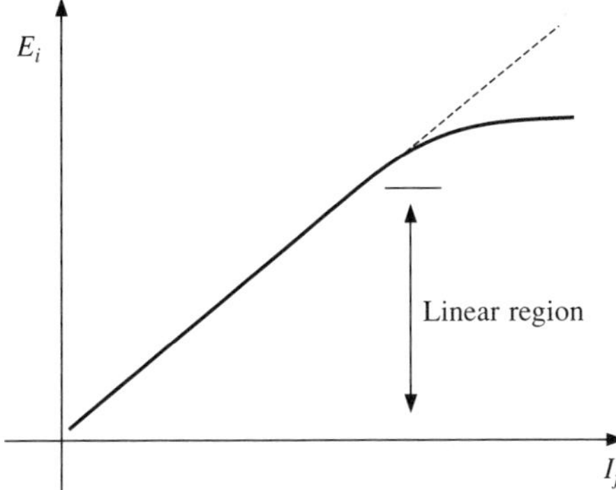


**Fig. 9.7** A typical open circuit plot for a synchronous generator

From (9.18) and the OCC plot, it can be assumed that flux $\phi$ is directly proportional to the field current $I_f$ in the linear region. Therefore, $\phi \propto I_f$

$$\phi = G \cdot I_f \tag{9.19}$$

where $G$ is the proportionality constant.

From (9.16), (9.18) and (9.19), the synchronous generator can be described by the steady-state model,

$$E_i = K \cdot G \cdot I_f \cdot n_{\text{mech}} \tag{9.20}$$

$$V_\phi = E_i - I_a\,(jX_s + R_a) \tag{9.21}$$

where $K = \sqrt{2}\,\pi N K_w p$.

Equation (9.20) gives the phase voltage of the generator output. The generator's terminal voltage may or may not be the same as its phase voltage, depending on the configuration of its windings. The three phases of a synchronous generator can either be connected in a star (Y) configuration or delta ($\Delta$) configuration as shown by Fig. 9.8.

For a Y-connected generator,

$$V_t = \sqrt{3} V_\phi \text{ and } I_t = I_a \tag{9.22}$$

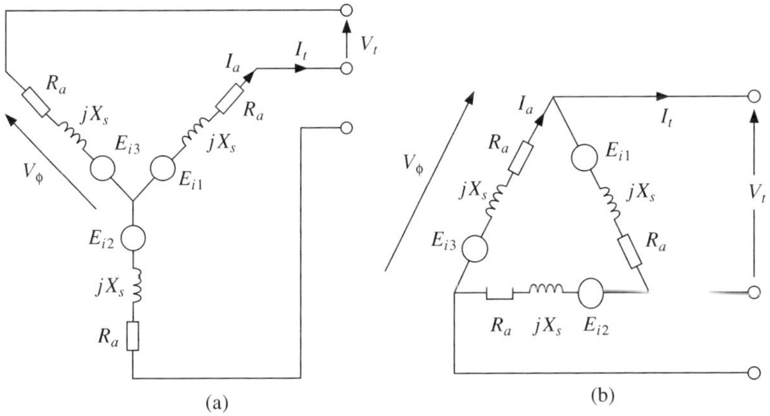

**Fig. 9.8** Configuration of generator windings: (a) star-connection; (b) delta connection

For a Δ-connected generator,

$$V_t = V_\phi \text{ and } I_t = \sqrt{3}I_a \tag{9.23}$$

where:

$V_t$ is the generator's terminal voltage
$I_t$ is the generator's terminal current
$V_\phi$ is the generator's phase voltage
$I_a$ is the armature current or the generator's phase current.

Figure 9.9 shows a block diagram of the synchronous generator model based on the equivalent circuit model. $V_t$ is the complex r.m.s. voltage of the generator terminal. For

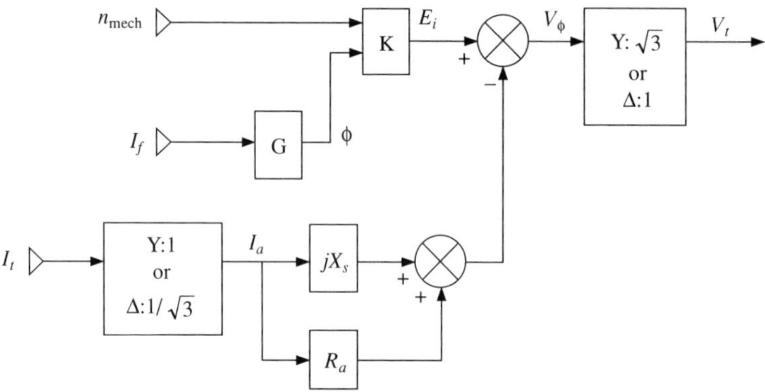


**Fig. 9.9** Block diagram based on equivalent circuit model

the present application, this model has to be further developed to include a three-phase rectifier and also to consider the torque characteristics as well as the power factor. A phasor diagram of the generator voltages and currents is shown in Fig. 9.10. The angle between the phase voltage $V_\phi$ and the internal generated voltage $E_i$ is known as the power angle δ. For generator operation (as opposed to motor operation), the power angle is always positive, as shown in Fig. 9.10. The diagram also shows a lagging power factor condition whereby the current $I_a$ lags $V_\phi$ by an angle θ. The power factor is given by cos θ.

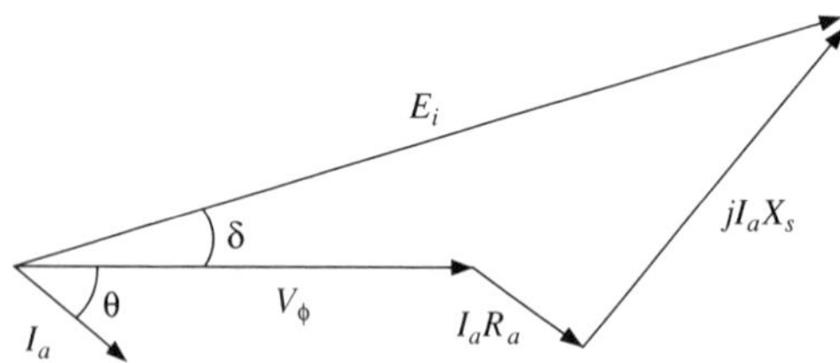


**Fig. 9.10** Phasor diagram of a synchronous generator

Using $V_\phi$ as reference, the equivalent circuit model can be expressed as

$$V_\phi \angle 0° = E_a \angle \delta - I_a \angle \theta \, (R_a + jX_s)$$

$$V_\phi = \{E_i \cos\delta - I_a R_a \cos\theta + I_a X_s \sin\theta\} + j\,\{E_i \sin\delta - I_a X_s \cos\theta - I_a R_a \sin\theta\}$$

Because $V_\phi$ is the reference, its imaginary component is zero, therefore,

$$V_\phi = E_i \cos\delta - I_a R_a \cos\theta + I_a X_s \sin\theta \tag{9.24}$$

and $E_i \sin\delta - I_a X_s \cos\theta - I_a R_a \sin\theta = 0$

$$\delta = \sin^{-1}\left[\frac{I_a}{E_i}(X_s \cos\theta + R_a \sin\theta\right] \tag{9.25}$$

For a three-phase synchronous generator, power is given by $P_{\text{real}} = 3V_\phi I_a \cos\theta$.

Power is the product of torque and speed, hence the induced torque is

$$\tau_{\text{ind}} = \frac{3V_\phi I_a}{\omega_{\text{mech}}}\cos\theta$$

or

$$\tau_{\text{ind}} = \frac{6\pi V_\phi I_a}{n_{\text{mech}}}\cos\theta \tag{9.26}$$

where $\omega_{\text{mech}}$ is the mechanical speed in radians per second and $n_{\text{mech}}$ is the mechanical speed in revolutions per second.

Figure 9.11 shows the model that is developed to represent the synchronous generator and rectifier. The inputs of this model are the rotational speed $n_{\text{mech}}$, field current $I_f$, power factor angle $\theta$ and the current at the d.c. link $I_{\text{dc}}$, used as a measure of the electrical load. The outputs are the d.c. link voltage $V_{\text{dc}}$ and the induced torque $\tau_{\text{ind}}$ which loads the engine. To account for the generator winding configuration (Y or $\Delta$), conversions between the phase values and terminal values of voltage and current are based on (9.22) and (9.23). The generator phase voltage is obtained from (9.24) and (9.25) while the induced torque is from (9.26).

### 9.1.5 The diesel engine

The second model to be developed is one for the diesel engine. The operation of diesel engines involves a large number of complex processes, with many delays, lags and non-linearities. It is therefore difficult to develop a comprehensive mathematical model to study and analyse the diesel engine. Even manufacturers of control systems and speed governors for diesel engines are known to focus on empirical methods rather than theoretical modelling for analysis. In this book, a model of a diesel engine is required to provide a complete representation of the control plant. The aim of this exercise is to develop a 'good enough' model for the functional verification of the control system. Since detailed studies are not necessary at this stage, an approach which favours reduced complexity and incorporates a certain degree of approximation is preferred. For the purpose of speed control studies, the diesel engine can be represented as a combustion system and inertia as shown by the block diagram in Fig. 9.12. The combustion system produces an engine torque $\tau_E$ as a function of the fuel rack position $y$. The engine torque is used to oppose the load torque $\tau_L$ and the difference $\Delta\tau$ is converted into acceleration/deceleration $\Delta n_{\text{mech}}$ in the lumped inertia (engine inertia + load incria).

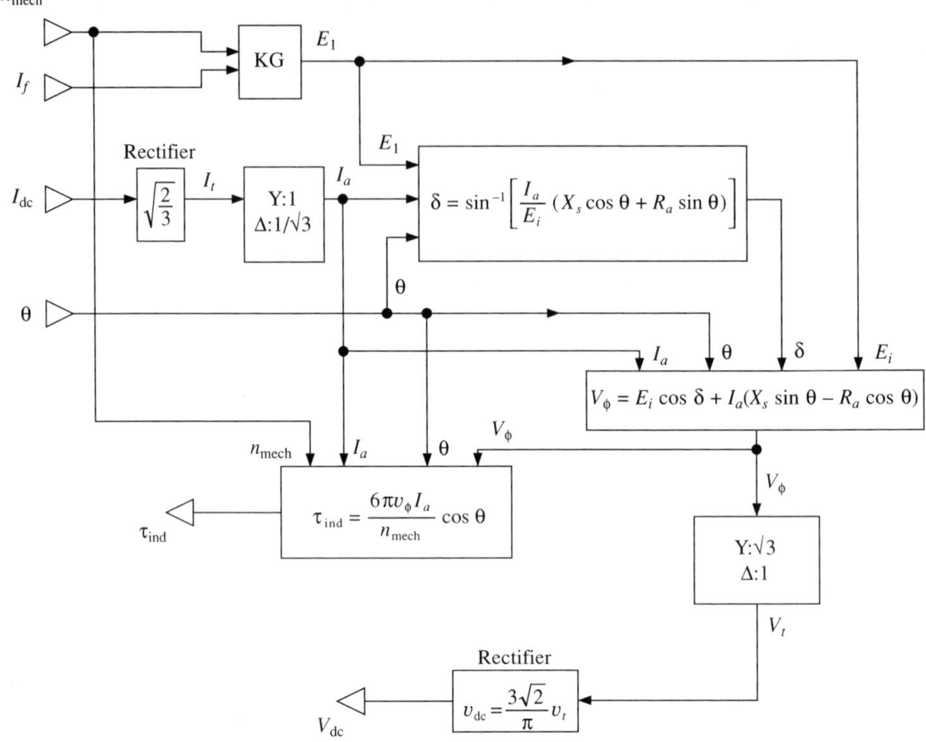


**Fig. 9.11** Model of generator–rectifier system

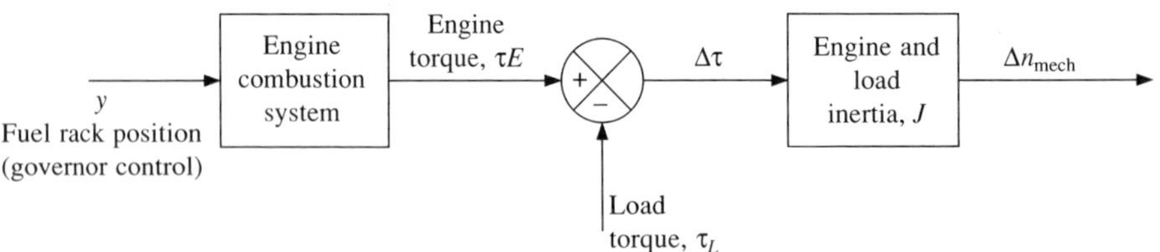


**Fig. 9.12** Diesel engine and load representation

When the speed of the engine changes, it moves along the engine's torque curve as shown in Fig. 9.13. Each engine's torque curve is specific to a fuelling rate $q$. For a given speed, the developed engine torque $\tau_E$ can be increased by increasing $q$. An equilibrium point is reached when the engine's torque curve intersects the load's torque curve.

If the engine torque is written as

$$\tau_E = f(n_{mech}, q) \tag{9.27}$$

and the equilibrium point for a given operating condition is when

$$n_{mech} = \tilde{n}_{mech}; \quad q = \tilde{q}$$

then, expanding (9.27) into a Taylor series about the equilibrium point yields

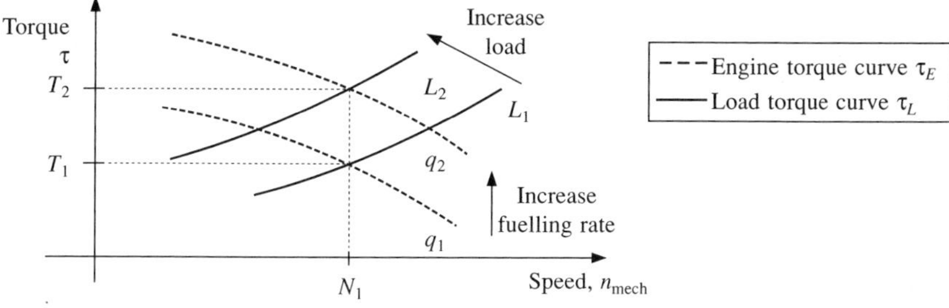


**Fig. 9.13** Torque–speed characteristics

$$\tau_E = f(\tilde{n}_{mech}, \tilde{q}) + \left[ \frac{\partial f}{\partial n_{mech}}(n_{mech} - \tilde{n}_{mech}) + \frac{\partial f}{\partial q}(q - \tilde{q}) \right]$$

$$+ \frac{1}{2!}\left[ \frac{\partial^2 f}{\partial n^2_{mech}}(n_{mech} - \tilde{n}_{mech})^2 + \frac{\partial^2 f}{\partial n_{mech}\partial q}(n_{mech} - \tilde{n}_{mech})(q - \tilde{q}) + \frac{\partial^2 f}{\partial q^2}(q - \tilde{q})^2 \right] + \ldots$$

If the operating condition is within the vicinity of the equilibrium point, it can be assumed that the higher order terms are negligible, therefore,

$$\tau_E - f(\tilde{n}_{mech}, \tilde{q}) = \left( \frac{\partial f}{\partial n_{mech}} \right)(n_{mech} - \tilde{n}_{mech}) + \left( \frac{\partial f}{\partial q} \right)(q - \tilde{q})$$

$$\tau_E - \tilde{\tau}_E = \left( \frac{\partial f}{\partial n_{mech}} \right)(n_{mech} - \tilde{n}_{mech}) + \left( \frac{\partial f}{\partial q} \right)(q - \tilde{q}) \tag{9.28}$$

Using 'δ' to denote small changes about the equilibrium point, this can be written as

$$\delta\tau_E = \left( \frac{\partial \tau_E}{\partial n_{mech}} \right)_q \delta n_{mech} + \left( \frac{\partial \tau_E}{\partial q} \right)_{n_{mech}} \delta q \tag{9.29}$$

At this juncture, a bold step is taken to further simplify the situation by extending the use of the equation above over the entire range of analysis. It is acknowledged that (9.29) is a linearised approximation which is only valid within the proximity of the operational point from which the equation is derived. However, this linear model is assumed to be capable of providing at least a qualitative 'feel' of the engine's torque–speed characteristics, even if the quantitative results are only rough approximations. It has to be pointed out that these models are not developed to be used as a starting point for the design of the control system. The controllers presented are model-free designs, therefore there is no danger of the limitations of the models being translated into the control system. Writing

$$\left( \frac{\partial \tau_E}{\partial n_{mech}} \right)_q = A \text{ and } \left( \frac{\partial \tau_E}{\partial q} \right)_{n_{mech}} \delta q = B\delta y$$

where $\delta y$ is the change in fuel rack position $y$, (9.29) can be written as

$$\delta\tau_E = A \cdot \delta n_{mech} + B \cdot \delta y$$

By extending the linear properties over the entire range of analysis, it can be assumed that $A$ and $B$ are constants. Therefore, from the *superposition principle*, the engine's torque curve can be represented by the linear equation:

$$\tau_E = A \cdot n_{\text{mech}} + B \cdot y \tag{9.30}$$

The net accelerating torque is given by the difference between the engine's torque and the load torque:

$$\Delta\tau = \tau_E - \tau_L$$

Since $\Delta\tau = J\dfrac{d\omega_m}{dt}$

where $J$ is the combined moment of inertia of the engine and generator and $\omega_m$ is the mechanical speed in radians per second. It can be shown that the discrete equivalent can be written as

$$\frac{C(n_{\text{mech}} - n_{\text{mech}} \cdot z^{-1})}{T} = \frac{\Delta\tau}{J}$$

$$\Delta n_{\text{mech}} = \frac{\Delta\tau}{J} D \tag{9.31}$$

where:
$C$ is a constant
$T$ is the sampling period
$D = (T/C)$
$n_{\text{mech}}$ is the mechanical speed in revolutions per second.

A block diagram of a discrete engine model based on (9.30) and (9.31) is shown in Fig. 9.14.

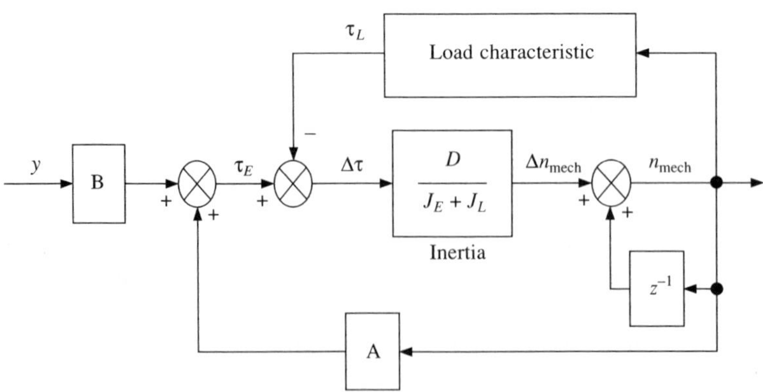


**Fig. 9.14** Linearised discrete model of the diesel engine

## 9.2 VHDL modelling

Although VHDL is a hardware description language and is used primarily for circuit design, it has the basic properties of any software programming language. It is therefore capable of implementing mathematical models. Two VHDL components are designed to represent the models of the generator–rectifier and the diesel engine developed in this chapter. The complete source codes of the models can be found in Appendix E. Since the models are initially designed for the purpose of computer simulations and not for hardware implementation, it is possible to declare the signals and variables as the type REAL. This gives the design a greater freedom in numerical analysis. It is also possible to incorporate complex mathematical functions such as trigonometry in the design.

The VHDL design of the generator and rectifier model is configured with the entity name **Generect**. It is designed as a synchronous circuit with five input signals (including **CLK**). Three output signals provide simulated information on the generator's phase voltage **Vph** and induced torque **TG** as well as the d.c. voltage at the rectifier's output **Vdc**. As already touched upon earlier in this chapter, the voltages at the generator's terminals are largely determined by the nature of its winding configuration. Two constants, **YD–CURR** and **VD–VOLT**, which have values defined in the declaration section of the design architecture sets the configuration of the model. In this, they are defined with values that set the model as a Y-connected generator.

The VHDL model for the diesel engine is configured with the entity name **Engine**. The input signals of the model are load torque **TL**, control signal to the fuel actuator **Y**, the clock signal **CLK** and the sampling period **PRD** while the engine's torque **TE** and speed **NE** make up the output signals. It implements the mathematical equations of the engine and, like **Genrect**, the section of the code that does this is self-explanatory. A slight addition to the model, to prevent the calculated output values from 'running away' in the event of simulation errors, is that the output signals are designed to operate within a fixed boundary of values. The maximum and minimum limits of these values are set within the code.

Figure 9.15 (a) and (b) show the block diagram of **Genrect** and **Engine** respectively. The two components have their respective input and output ports and can be connected

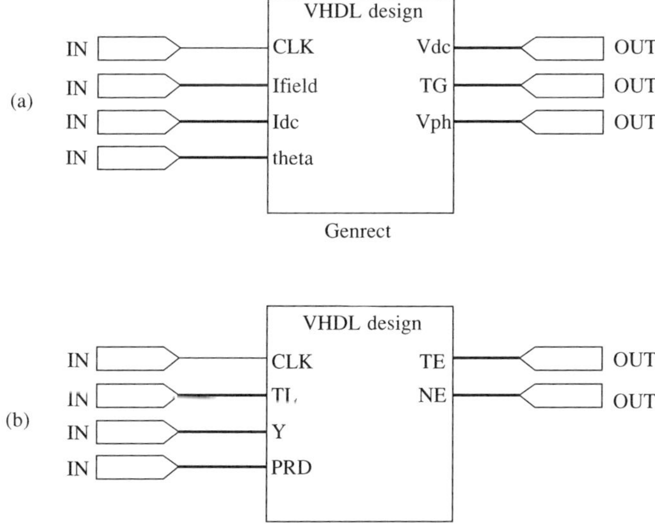


**Fig. 9.15** Block diagram of VHDL models: (a) Genrect; (b) Engine

directly to the VHDL designs of the control system for simulation purposes. The use of the plant models in conjunction with the control system in a functional simulation will be discussed later. The PWM inverter is presented in the following chapter. It includes some functional and performance simulations as well as a description of the inverter control system design.

### 9.2.1 PWM inverter design and analysis

While the previous chapter discussed the mathematical representation of the diesel engine, synchronous generator and rectifier, this chapter presents an analysis of the control plant, namely the power inverter system shown as shaded blocks in Fig. 9.16. A common application of power inverters is in a.c.–d.c.–a.c. systems, such as variable speed control of a.c. motors, uninterruptible power supplies and variable speed wind energy conversion systems. In this example, the inverter is used to convert the d.c. voltage at the d.c. link into a constant 50 Hz sinusoidal output voltage.

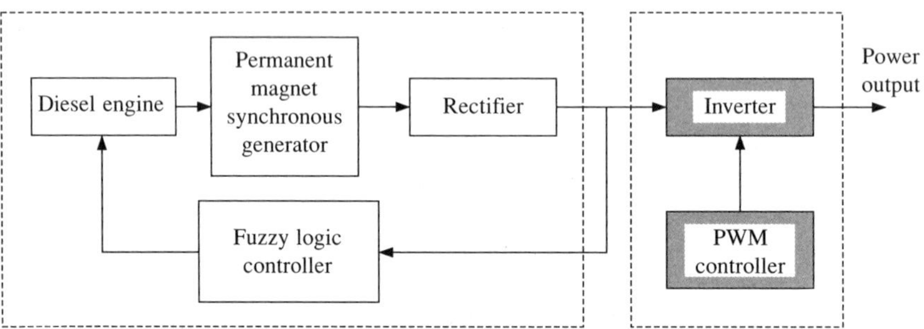


**Fig. 9.16** Block diagram of the complete system

The diagrams in Fig. 9.17 show the configurations of single-phase and three-phase switch-mode inverters. They are made up of a number of power switches (four for single phase and six for three phase) arranged in a bridge configuration. At present, IGBTs are the most widely used type of power switches for inverter applications due to their controllability and relatively high voltage ratings. Various control strategies can be used to control the power switches such that a sinusoidal waveform can be obtained at the output. However, the raw output voltage usually contains other high frequency harmonic

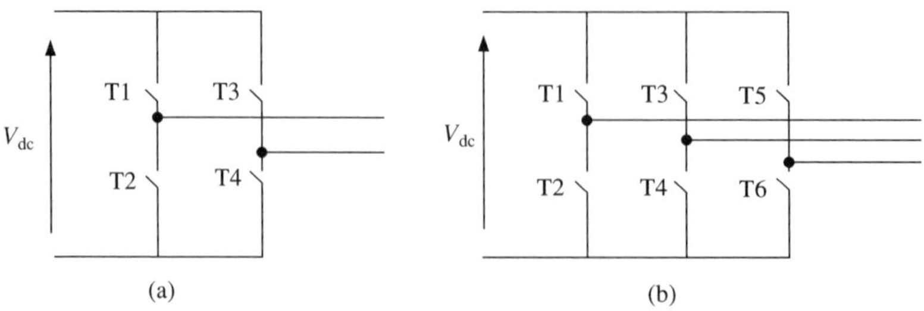


**Fig. 9.17** Switch-mode inverter: (a) single phase; (b) three phase

components, which have to be eliminated before the desired voltage waveform can be recovered. This is normally achieved using a low-pass filter. The amount of harmonic distortion at the output depends largely on the type of switching pattern. Examples of commonly used switching patterns are square wave, quasi square wave and pulse width modulation (PWM). The switching strategy used is pulse width modulation, a technique that produces waveforms, which tend to contain less of the low order harmonics than other switching arrangements.

### 9.2.2 Fuzzy synchronous generator control

There are a number of reasons for using fuzzy logic in this application, the primary advantage being the flexibility offered by fuzzy logic. The backbone of any FLC is embodied in a set of fuzzy rules, with two implications:

- The fact that the control strategy is represented by a set of rules and not an elaborated set of equations. This allows the designer to change the basic characteristics of the controller with minimal fuss, simply by redefining the rules. The structure of other components in the FLC remains intact and hence the effort to redesign the hardware configuration is significantly reduced. This is a great advantage in research applications where some further studies into alternative control strategies are expected.
- The *fuzzy* aspect of the rules, which is dealing with the imprecise definition of the system. This allows vagueness in the design of the control system to be tolerated to a certain degree and eliminates the need for a well-defined mathematical model of the plant. The control plant in this project is not a trivial system to model with significant accuracy, since it comprises a number of different non-linear components such as the diesel engine and synchronous generator. Models with reduced complexity can be employed for analysing the functional characteristics of the FLC.

The following sections describe the work involved in developing the desired controller for the generator set based on the concepts introduced at the beginning of this chapter.

#### *9.2.2.1 Control block*

Figure 9.18 shows a block diagram of the FLC and a stand-alone generator set. The components of the FLC are represented by the shaded blocks. There are two inputs to the FLC. Its final functionality is determined by the choice of inputs and the definition of the *fuzzy rule base*. This allows the system to be studied under different control strategies and specifications using the same hardware. Only three sections have to be modified: the two interfacing blocks and the rule base. In this experiment, the d.c. voltage at the output of the rectifier is used as an input to the FLC. The plan is to test the system with a basic control strategy: maintaining the d.c. voltage within a small range, over varying load currents. The control output $u$ is used to control the rate of flow of the fuel to the diesel engine.

The tasks involved in designing a typical FLC can be loosely summarised as follows:

- Decide on an overall strategy based on the design criteria.
- Identify an implementation technology.
- Identify the I/O variables (knowledge base).
- Define the membership functions for the variables (knowledge base).

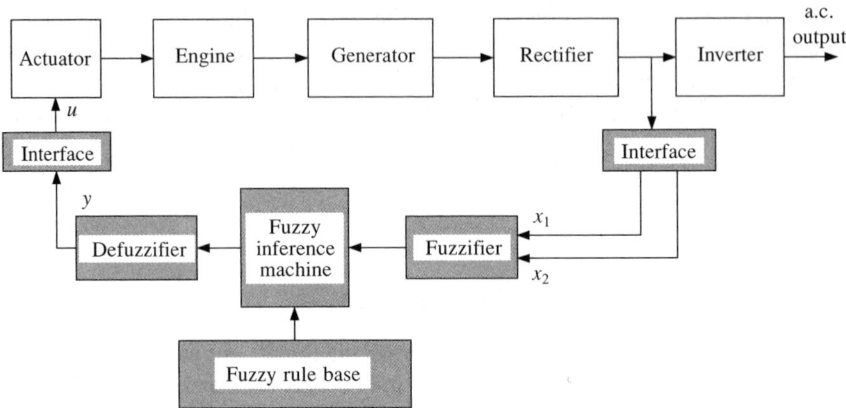


**Fig. 9.18** Block diagram of fuzzy logic controller and control plant

- Formulate a fuzzy rule base.
- Choose a method of inference.
- Choose a defuzzification technique.

This design procedure forms a general guide and may well vary from one design to another, depending on the individual aims and requirements.

#### 9.2.2.2 Overall strategy

The design has to be able to accommodate different control strategies and yet not be too inherently complex for hardware implementation. Two input variables ($x_1$ and $x_2$) are used. This is enough to give the system a certain amount of possible capabilities. Too many inputs add complications to the design as the number of fuzzy rules expands exponentially with the number of input variables. There are methods to handle large amounts of input variables but these are not discussed in this book. A generic structure of the FLC is first designed and the specific functionality which is determined by the *knowledge base* and the formulation of the *rule base* is designed at a later stage.

#### 9.2.2.3 Implementation technology

The most common method of implementing fuzzy controllers at the moment is with microprocessors. In this project, a different implementation technology has been chosen. One of the objectives of this book is to investigate the implementation of FLCs into FPGAs. This is a major consideration, as the design would ultimately have to be able to fit into the chosen FPGA(s). The required 'area space' or gate count becomes a significant design criteria.

#### 9.2.2.4 Membership functions

The present design utilises three types of functions – Γ-function, *L*-function and Λ-function – all of which have already been presented. These functions have been proven to produce good results for control applications and can be easily implemented into hardware. The universe of discourse of the input variables is partitioned into five fuzzy sets or *linguistic values* ($B^1$ to $B^5$), while the output variable can take any of the seven linguistic values ($D_1$ to $D_7$). Graphical representations of the membership functions are shown in Figs. 9.19 and 9.20.

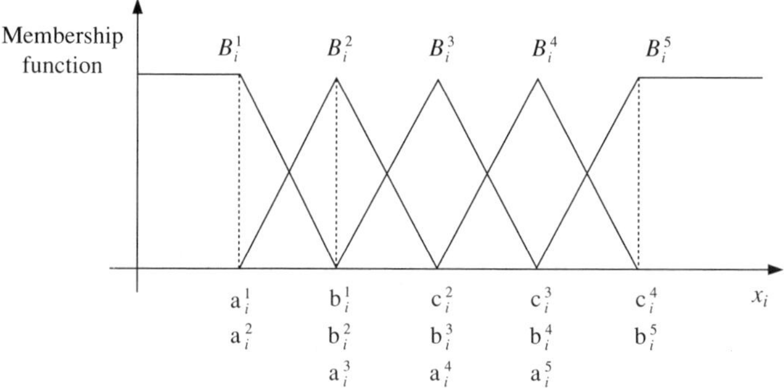


**Fig. 9.19** Membership function of input variables

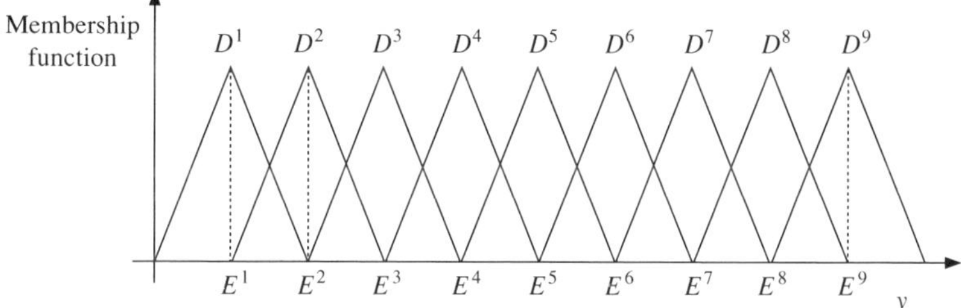


**Fig. 9.20** Membership function of output variable

The following equations define the membership functions for the linguistic values associated with the input variables.

$$B_i^j = \begin{cases} 1 & x_i < a_i^j \\ \dfrac{(x_i - b_i^j)}{(a_i^j - b_i^j)} & a_i^j \leq x_i \leq b_i^j \quad \text{for } j = 1 \\ 0 & x_i > b_i^j \end{cases} \tag{9.32}$$

$$B_i^j = \begin{cases} 0 & x_i < a_i^j \\ \dfrac{(x_i - a_i^j)}{(b_i^j - a_i^j)} & a_i^j \leq x_i \leq b_i^j \quad \text{for } j = 5 \\ 1 & x_i > b_i^j \end{cases} \tag{9.33}$$

$$B_i^j = \begin{cases} 0 & x_i < a_i^j \\ \dfrac{(x_i - a_i^j)}{(b_i^j - a_i^j)} & a_i^j \le x_i \le b_i^j \\ \dfrac{(x_i - c_i^j)}{(b_i^j - c_i^j)} & b_i^j < x_i \le c_i^j \\ 0 & x_i > c_i^j \end{cases} \quad \text{for } j = 2, 3, 4 \tag{9.34}$$

where $i = 1, 2$ and $j = 1, \ldots, 5$.

The crisp values of the input variables are mapped onto the fuzzy plane using the equations above. It gives each input variable a membership function relating to the fuzzy sets, $B_i^1$ to $B_i^5$. It has to be pointed out that in these equations, $B_i^j$ is used to denote the *membership function*. The reader should be aware that in this book, the symbol $B_i^j$ is used for both, the linguistic value as well as the membership function. Strictly speaking, the linguistic value should be denoted using $B_i^j$, while the membership function using $\mu_{Bi,j}(x_i)$.

In this example, '$\mu$' and '$(x_i)$' are dropped from the denotation of membership function, thus,

$$\mu_{Bi,j}(x_i) = B_i^j = 0.5$$

is used to indicate that $x_i$ belongs to the linguistic value $B_i^j$ by a membership function of the value 0.5.

The universe of discourse of the output variable is divided into seven linguistic values. The membership functions of the output values are intentionally made to be symmetrical, as this will simplify the defuzzification computation. $E^1$ to $E^7$ are the mean of each function and act as the weightings to the weighted average method of defuzzification.

#### 9.2.2.5 Fuzzy rule base

Each input variable can take any of the five linguistic values, therefore 25 (= 5 × 5) rules are formulated. The rules have the typical fuzzy rule structure, using linguistic *variables* in both the antecedent and consequent, and are expressed in IF-THEN manner. They map the input states onto 25 output conditions ($C_1$ to $C_{25}$). The fuzzy rules have the general form,

$$R^k: \text{IF } x_1 \text{ is } A_1^k \text{ AND } x_2 \text{ is } A_2^k, \text{ THEN } y \text{ is } C^k \tag{9.35}$$

where:
$R^k$ ($k = 1, 2, \ldots, 25$) is the $k$th rule of the fuzzy system
$x_1, x_2$ are the input linguistic *variables*
$y$ is the output linguistic *variable*
$A_i^k$ ($i = 1, 2$; $k = 1, 2, \ldots, 25$) is the $k$th fuzzy set defined in the $i$th input space and $A_i^k$ can take any linguistic *value* associated with $x_i$,
$C^k$ is the output condition inferred by the $k$th rule.

If the denotation is such that the linguistic variables:

$x_i$ for $i = 1, 2$

have the following linguistic values:

$$B_i^j \quad \text{for } i = 1, 2;\ j = 1 \text{ to } 5$$

then the rule base can be represented by a fuzzy associative memory (FAM) table (Table 9.1).

**Table 9.1** Fuzzy associative memory table

| $x_1$ \ $x_2$ | $B_2^1$ | $B_2^2$ | $B_2^3$ | $B_2^4$ | $B_2^5$ |
|---|---|---|---|---|---|
| $B_1^1$ | $C^1$ | $C^2$ | $C^3$ | $C^4$ | $C^5$ |
| $B_1^2$ | $C^6$ | $C^7$ | $C^8$ | $C^9$ | $C^{10}$ |
| $B_1^3$ | $C^{11}$ | $C^{12}$ | $C^{13}$ | $C^{14}$ | $C^{15}$ |
| $B_1^4$ | $C^{16}$ | $C^{17}$ | $C^{18}$ | $C^{19}$ | $C^{20}$ |
| $B_1^5$ | $C^{21}$ | $C^{22}$ | $C^{23}$ | $C^{24}$ | $C^{25}$ |

### 9.2.2.6 *Inference engine*

The FLC design in this project incorporates Mamdani's implication method of inference, which is one of the most popular methods in fuzzy control applications. In essence, Mamdani's implication for the fuzzy rule of (9.35) is given by

$$\mu_C(y) = \max_k [\min[\mu_{A_1^k}(x_1), \mu_{A_2^k}(x_2)]] \quad k = 1, 2, \ldots, 25 \tag{9.36}$$

The implication has a simple min–max structure which makes it easy to incorporate into hardware. The block diagram in Fig. 9.21 provides an overview of the controller's internal structure. Two input variables are fuzzified, producing the corresponding linguistic values and membership functions ($B_i^j$). The first phase of Mamdani's implication involves *min*-operation since the antecedent pairs in the rule structure are connected by a logical 'AND'. All the rules are then aggregated using a *max*-operation.

### 9.2.2.7 *Defuzzification technique*

Different defuzzification techniques have different levels of complexities. There have been several studies into the methodology to guide the selection of defuzzification techniques based on the criteria of the design [207]. The dominant criteria in this design lies in the implementation stage. In order to implement the system into an FPGA, it would be advantageous for the defuzzification technique to be fairly straightforward and not to involve a large number of complex calculations. The weighted average method is viewed to be an appropriate technique for systems involving hardware implementation. Due to the fact that the output membership functions are symmetrical in nature, the mean of the fuzzy sets can be used as weightings for the defuzzification process. This technique requires several multiply-by-a-constant operations and only one division process. The fact that the multipliers have constant values further reduces the complexity of the hardware structure.

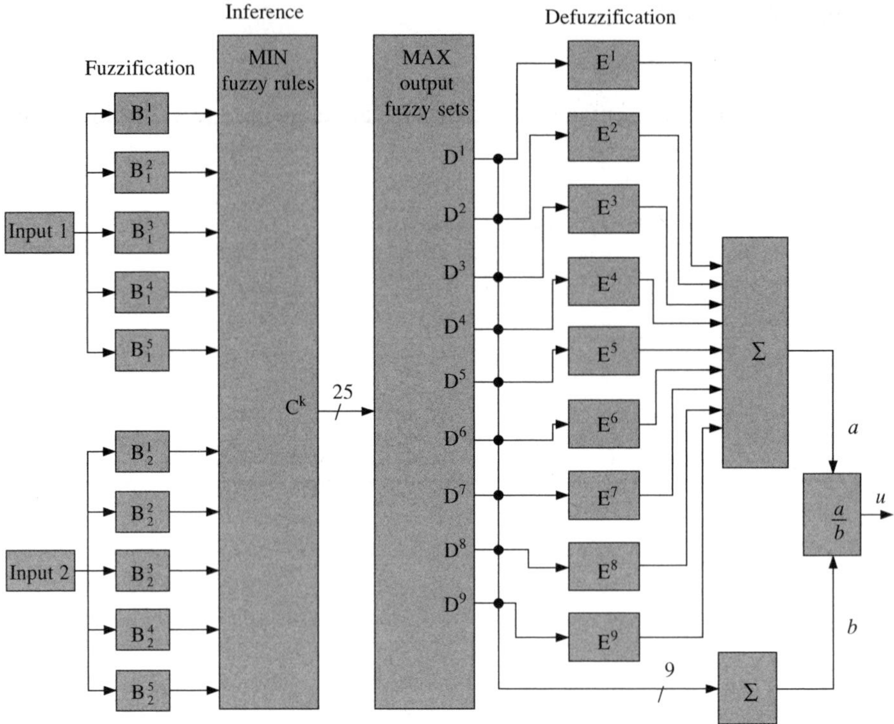


**Fig. 9.21** Block diagram of the operations in a fuzzy logic controller

### 9.2.3 Rule base design

It has been mentioned that the rule base moulds the functionality of an FLC. The rules are most likely to be formulated based on some level of human understanding of the plant, and although a fuzzy logic control system supports heuristics, the process of constructing the rules is still the subject of considerable studies. The most basic way of constructing the rule base is by trial and error, but this usually involves heavy computation and has a low efficiency. Numerous other techniques have been suggested, some of which involve automatic generation of the rule base using computational methods such as genetic algorithms [141], [116] and evolutionary programming [128]. The design of FLCs can also be based on conventional control structures such as PID and sliding mode control. Description and examples of such structures in fuzzy control can be found in [177], [157]. The design of the FLC in the present work is based on PI controllers because it is highly suitable for the governing (of the d.c. voltage) system required. The rule base is constructed from the control law of a PI system.

#### *9.2.3.1 PI control*

The proportional–integral (PI) controller is a well-known system in control engineering. It is, in essence, a lag compensator characterised by the transfer function

$$G(s) = K\left(1 + \frac{1}{T \cdot s}\right) \tag{9.37}$$

where:
$G(s)$ is the gain
$K$ is the control parameter
$T$ is the time constant.

The control law is given by the equation

$$u_{PI} = K_p \cdot e + K_I \cdot \frac{1}{T} \int_0^t e \cdot dt \tag{9.38}$$

where:
$u$ is the control signal
$e$ is the error, given by $e$ = (input value) – (reference value).

Differentiating (9.38) gives

$$\frac{du}{dt} = K_P \cdot \frac{de}{dt} + K_I \cdot e \tag{9.39}$$

In discrete-time systems, (9.39) can be written as

$$u(kT) - u(kT - T) = K_P \cdot \{e(kT) - e(kT - T)\} + K_I \cdot e(kT)$$

$$\Delta u = K_P \cdot \Delta e + K_I \cdot e \tag{9.40}$$

where:
$\Delta u$ is the change in $u$ over one sampling period
$\Delta e$ is the change in $e$ over one sampling period.

The characteristic of a PI controller can be represented by the phase plane diagram shown in Fig. 9.22. A diagonal line where $\Delta u = 0$ divides the area where $\Delta u$ is positive and $\Delta u$ is negative.

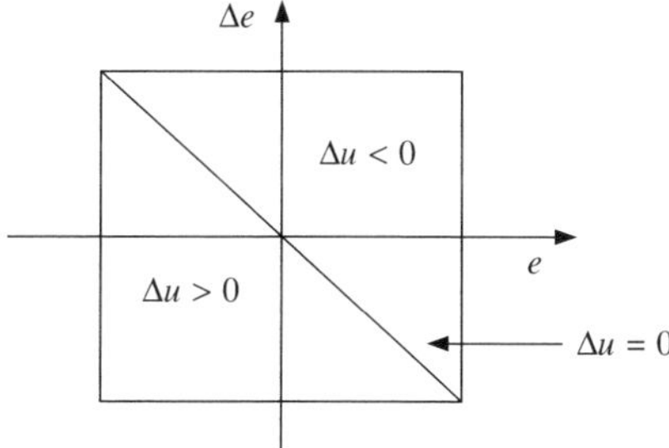


**Fig. 9.22** Characteristic of PI controller

### *9.2.3.2 PI-like fuzzy control*

At this stage, the control law in (9.40) is not in fuzzy terms. In order to design a fuzzy controller based on the PI control structure, the following definitions are made:

Let $E$ be the linguistic variable for the error $e$
$\Delta E$ be the linguistic variable for the change of error $\Delta e$
$U$ be the linguistic variable for the control output $u$.

Based on the conventions in Section 6.1.4, the following sets are defined:

$\underline{LE}$ = {Negative Big, Negative, Zero, Positive, Positive Big}
$\underline{L\Delta E}$ = {Negative Big, Negative, Zero, Positive, Positive Big}
$\underline{LU}$ = {Negative Big, Negative, Negative Small, Zero, Positive Small, Positive, Positive Big}

The corresponding PI control law in IF-THEN rules has the form:

$$R^k : \text{IF } E \text{ is } A_1^k \text{ and } \Delta E \text{ is } A_2^k, \text{ THEN } U \text{ is } C^k \tag{9.41}$$

where:
$A_1^k$ can take any linguistic value in the set $\underline{LE}$
$A_2^k$ can take any linguistic value in the set $\underline{L\Delta E}$
$C^k$ can take any linguistic value in the set $\underline{LU}$.

To implement this design into the FLC, let:

- $x_1 = E$
- $x_2 = \Delta E$
- $\{B_i^1, B_i^2, B_i^3, B_i^4, B_i^5\}$ = {Negative Big, Negative, Zero, Positive, Positive Big} for $i$ =1, 2
- $\{D^1, D^2, D^3, D^4, D^5, D^6, D^7\}$ = {Negative Big, Negative, Negative Small, Zero, Positive Small, Positive, Positive Big}.

Table 9.2 shows the FAM table of the design.

**Table 9.2** FAM table for FLC design

| E \ $\Delta E$ | NB | N | Z | P | PB |
|---|---|---|---|---|---|
| NB | $R^1$ $u$ = PVB | $R^2$ $u$ = PB | $R^3$ $u$ = P | $R^4$ $u$ = PS | $R^5$ $u$ = Z |
| N | $R^6$ $u$ = PB | $R^7$ $u$ = P | $R^8$ $u$ = PS | $R^9$ $u$ = Z | $R^{10}$ $u$ = NS |
| Z | $R^{11}$ $u$ = P | $R^{12}$ $u$ = PS | $R^{13}$ $u$ = Z | $R^{14}$ $u$ = NS | $R^{15}$ $u$ = N |
| P | $R^{16}$ $u$ = PS | $R^{17}$ $u$ = Z | $R^{18}$ $u$ = NS | $R^{19}$ $u$ = N | $R^{20}$ $u$ = NB |
| PB | $R^{21}$ $u$ = Z | $R^{22}$ $u$ = NS | $R^{23}$ $u$ = N | $R^{24}$ $u$ = NB | $R^{25}$ $u$ = NVB |

| | |
|---|---|
| NVB | Negative Very Big |
| NB | Negative Big |
| N | Negative |
| NS | Negative Small |
| Z | Zero |
| PS | Positive Small |
| P | Positive |
| PB | Positive Big |
| PVB | Positive Very Big |

### 9.2.3.3 *Interfacing blocks*

Figure 9.23 shows a block diagram demonstrating the implementation of the FLC in a stand-alone generator system. The control plant in the diagram represents the engine, generator and rectifier system. The notation '$z^{-1}$' is used to mark a delay in the signal by one sampling period (the subject of Z-transform can be found in [184]).

In this application, the input interface converts the d.c. voltage at the output of the plant into *error* and *change of error* which are used as the two inputs to the FLC.

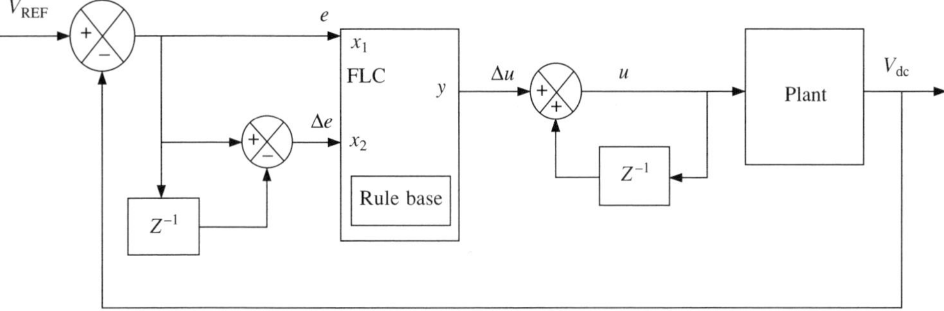


**Fig. 9.23** Block diagram of the control system

Another interface converts the output into the required value for the plant. The characteristics of the interfacing blocks can be described by the following equations:

*Input interface:*

$$e = V_{REF} - V_{dc}$$

$$x_1 = e$$

$$x_2 = x_1 - x_1 \cdot z^{-1}$$

*Output interface:*

$$\Delta u = y$$

$$u = \Delta u + u \cdot z^{-1}$$

Once the design issues of the FLC have been resolved, the next step is to consider the implementation scheme. In this project, the design is achieved using VHDL.

### 9.2.4 VHDL description

The FLC is designed and modelled in an EDA environment using VHDL. The controller is broken down into 'components', with each component performing a specific function such as fuzzification, inference, etc. This means that the design of each component can be modified or simply combined with one another to form a complete system. Two levels of design descriptions are presented in this example. A behavioural level description is used to test the functional validity of the design. At this level, the design is relatively generic and not restricted by any particular technological constraints. Once the functionality of the design is verified, it is subsequently modified for implementation into the chosen technology. This involves two main tasks: converting the code into structural level and optimising the design. The amount of optimisation required will depend on the target technology. In a general sense, the smaller the targeted device the greater the amount of effort required. The target device in this case is the Xilinx XC4010XL FPGA, which has an equivalent logic gate count of 10 000. This is comparatively small as FPGAs with millions of gates are nowadays available on the market.

The FLC is divided into five VHDL components. Figure 9.24 shows a diagram of the

FLC architecture. Each component is depicted with its VHDL code file '<filename>.vhd'. The functionality of components **Interface1**, **Interface2** and part of the component **Infer** will determine the characteristics of the FLC. Input and output variables are designed with a resolution of 8 bits.

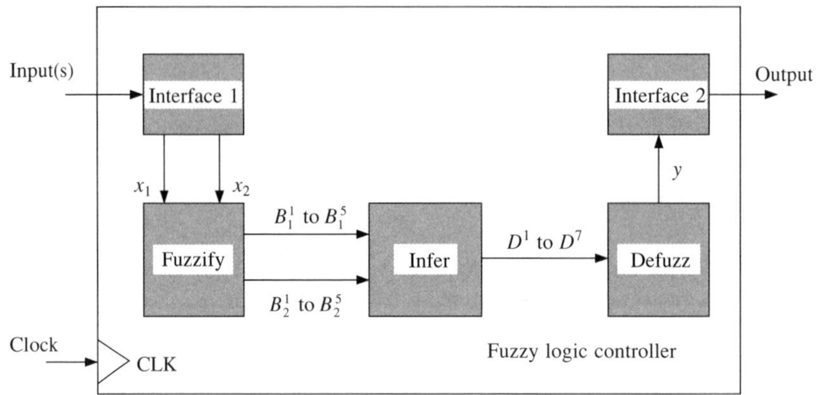


**Fig. 9.24** Components of the fuzzy logic controller

The ports and signals can be declared either as type **integer** or **std_logic_vector**. An example of the two ways of declaration mentioned here are:

```
port(x1 :in integer;
     x2 :in integer;
     y  :out integer);
```

and

```
port(x1 :in std_logic_vector(7 downto 0);
     x2 :in std_logic_vector(7 downto 0);
     y  :out std_logic_vector(7 downto 0));
```

The choice of type will depend on whether it is easier to write the VHDL code using **integer** or **`std_logic_vector`**. At *synthesis* level there is little difference between the two types. However, when developing the design for hardware implementation, the *range* of the **integer** must be specified during declaration. In this example both types are used, depending on the conditions. Conversion functions exist in VHDL to convert between the types [10]. They are:

```
conv_integer( ); and conv_std_logic_vector( );
```

#### *9.2.4.1 Interfacing components*

The interfacing circuits are vital parts of the FLC. They form the link between the core of the FLC and the control plant and, together with the rule base, shape the characteristics of the complete control system. The FLC is used as a PI-like controller in the stand-alone generator system, using the d.c. voltage as an input.

Components **Interface1** and **Interface2** are used to implement the input and output interfacing functions. Part of the code in the VHDL file **Interface1.vhd** is shown below (the complete contents of the VHDL files can be found in Appendix E):

```
process(CLK)
...
  begin
...
PAST_VAR := NOW_VAR;
 NOW_VAR := error;
  ...
DIFF := NOW_VAR - PAST_VAR;
-- Normalised input values:
x1 <= NOW_VAR ;
x2 <= DIFF * GAIN ;
...
end process;
```

In the final version of the design, **Interface2** is embedded within the code for **Defuzz**. The following code section implements the output interface function:

```
--Output interface
...
U_var := Uz + Y ;
Uz := U_var ;
...
U <= U_var ;
...
```

#### *9.2.4.2 Fuzzify*

The function of the component **Fuzzify** is to convert the crisp input variables into fuzzy values. The membership functions of the input variables are defined in the 'constant-declaration' part of **Fuzzify** as shown below:

```
architecture Fuzzify_arc of Fuzzify is
constant a1:integer := -60;
constant b1:integer := -30;
constant a2:integer := -10;
constant b2:integer := -30;
constant c2:integer := 0;
...
```

The membership values of the input variables to the linguistic values are then assigned based on these constant values. Thus, the membership functions can be slightly modified by redefining the constants. However, the shape of the membership functions would remain more or less the same, i.e. *L*–, Γ- and Λ-functions as shown in Fig. 9.25.

#### *9.2.4.3 Fuzzy inference*

The component **Infer** performs Mamdani's *min–max* inference to obtain the resulting fuzzy output, as a consequence of the rule base. A model, as shown in Fig. 9.26, is

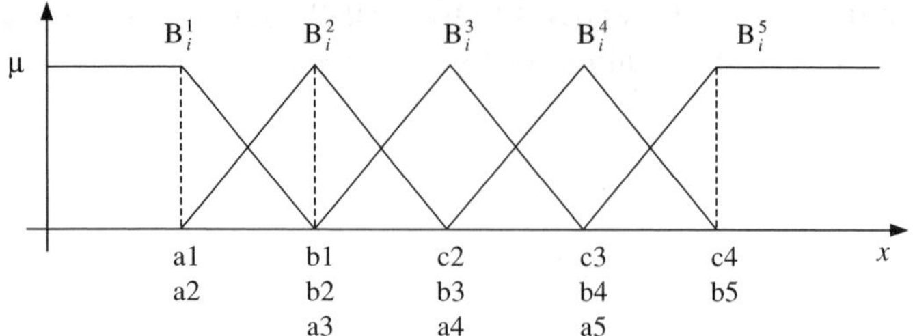


**Fig. 9.25** Membership function of input variables

developed to represent the inference engine in such a way that it can be easily adopted into the hardware description. The left-hand side of the diagram shows the implication of 25 output conditions by the fuzzy rule antecedent using *min*-operators. Each output condition models the consequence of a single rule. Therefore, the rule

$$R^1 : \text{IF } x_1 \text{ is } B_1^1 \text{ and } x_2 \text{ is } B_2^1\text{, THEN } U \text{ is } C^1$$

becomes

$$C^1 = \min[B_1^1, B_2^1].$$

The *max*-operator is used to take into account the combined effect of all the rules. The 25 output conditions are aggregated into seven linguistic values ($D^1$ to $D^7$) based on the conditions set by the rules. This operation is depicted by the right-hand side blocks in Fig. 9.26.

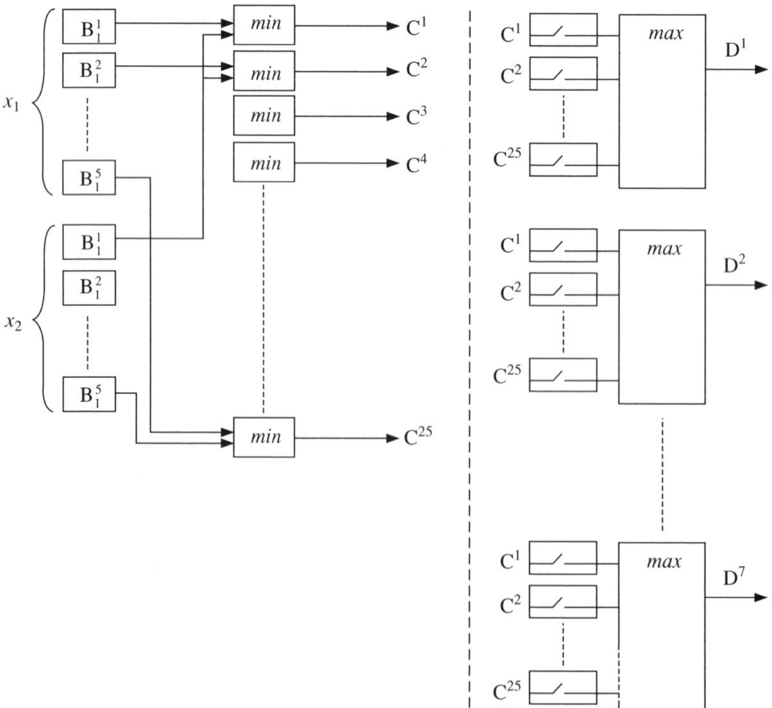


**Fig. 9.26** Model of the inference engine

For the purpose of illustration, switches are used to model the association of the output conditions to the linguistic values. A switch is set 'ON' if an output condition is associated with the linguistic value for the block of *max*-operation in question. This means that the collective effect of the rule base can be modelled by defining the status of the switches. For example, in the design of the FVG (refer to FAM table in Table 9.2), the membership function of the linguistic value $D^1$ is the aggregate of conditions $C^{20}$, $C^{24}$ and $C^{25}$. This is modelled by 'connecting' the three output conditions to the *max*-operation of $D^1$ and 'disconnecting' the other conditions, as illustrated by the block in Fig. 9.27. The operation is described by the function:

$$D^1 = \max[C^{20}, C^{24}, C^{25}].$$

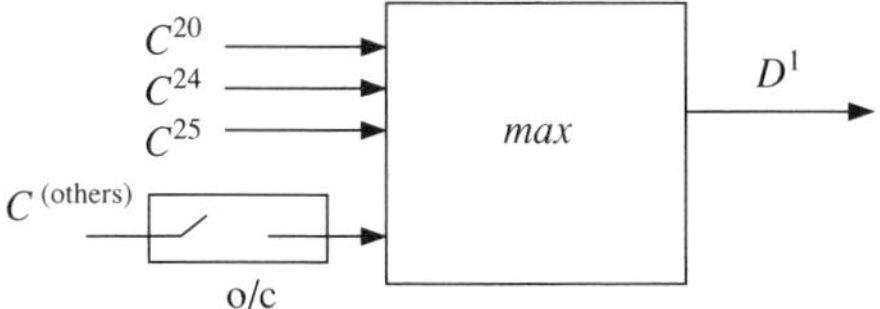


**Fig. 9.27** Model of the process of aggregating the fuzzy rules

Using the same principle, the entire set of fuzzy rules of the FVG is implemented into the VHDL component **Infer** simply by defining the association between the output conditions and the linguistic values ($D^1$ to $D^7$). The following functions are used to incorporate the rule base of the FVG into the FLC.

$$D^1 = \max[C^{20}, C^{24}, C^{25}]$$
$$D^2 = \max[C^{15}, C^{15}, C^{23}]$$
$$D^3 = \max[C^{10}, C^{14}, C^{18}, C^{22}]$$
$$D^4 = \max[C^{5}, C^{9}, C^{13}, C^{17}, C^{21}]$$
$$D^5 = \max[C^{4}, C^{8}, C^{12}, C^{16}]$$
$$D^6 = \max[C^{3}, C^{7}, C^{11}]$$
$$D^7 = \max[C^{1}, C^{2}, C^{6}]$$

#### *9.2.4.4 Defuzzifier*

The function of the component **Defuzz** is to convert the fuzzy output value of the control system into the corresponding crisp value. This is achieved using the weighted average defuzzification method. This defuzzification operation requires several multipliers and a divider. Behavioural modelling in VHDL supports multiplication and division but these operations are complicated to realise in the synthesis and implementation stages. However, in this chapter only the functional simulation is discussed. Therefore, the multiplication operator '*' and the division operator '/' are used. At subsequent stages, modifications are required, as most synthesis tools do not support the division operator '/'. In addition, using a lower level of design description has the advantage of requiring fewer gates than the '*' operator. Using the multiplication and division operators results in the defuzzification code being fairly straightforward, as shown below:

```
DEFUZZ_PROCESS:
process(CLK)
variable Dividend, Divisor : integer;
```

```
begin
  if CLK'event and CLK='1' then
    Dividend  := (E1*D1)+(E2*D2)+  (E3*D3)+(E4*D4)+(E5*D5)+
                 (E6*D6)+(E7*D7)+(E8*D8)+(E9*D9);
    Divisor   := (D1+D2+D3+D4+D5+D6+D7+D8+D9);
    if Divisor = 0 then
      – *** Avoid division by zero ***
      Y := 0;
    else
      Y := (Dividend/Divisor);
      ...
    end if;
  end if;
end process;
```

The **dividend** is obtained from the sum of the product of the linguistic values and their respective weightings while the **divisor** is simply the sum of all the linguistic values. A section of the design, marked '***Avoid division by zero***', is dedicated to checking if the value of the **divisor** is zero and taking the appropriate measures to avoid a '*division by zero*' error.

Once all the components of the FLC are successfully designed, a top hierarchy VHDL component, **Control**, is created to bind them together in the manner shown in Fig. 9.24 such as to form the complete controller. Within **Control**, the interfacing components, **Fuzzify**, **Infer** and **Defuzz** are *instantiated* and connected to one another. **Control** also features two clocking signals for synchronising its internal components. By combining the completed design of **Control** with the plant models, the performance can be analysed using computer simulations.

### 9.2.5 Simulations

A series of computer simulations is conducted on the FLC to analyse its performance before proceeding to hardware implementation. For this task, a VHDL test component, **Sim**, is created to simulate the control environment. By using the models developed already, **Engine** and **Genrect**, it is possible to simulate the relationship between the FLC and the control plant within the VHDL platform in order to study the controller. Figure 9.28 shows a simplified block diagram of the structural composition of **Sim**.

The shaded blocks represent components that have been previously described and are *instantiated* inside the structure of **Sim**. The white blocks are sections of code which perform the data type conversions between the signals of the plant models and those of the controller. A listing of the VHDL code for **Sim** is found in Appendix E – 11.5.9.

The method adopted for running the simulation is by using a VHDL test-bench. A test-bench is a VHDL design written specifically to provide a test environment for the design under study. A model of the test-bench is shown in Fig. 9.29. It is made up of two main parts, the stimulus and the observer. The former generates the necessary input signals to the design while the observer reads and records the output signals resulting from the stimulus signals.

A test-bench **TB_Sim** is subsequently created for **Sim**. The code for **TB_Sim** can be found in Appendix E – 11.5.10. The following VHDL codes are two important sections in **TB_Sim:**

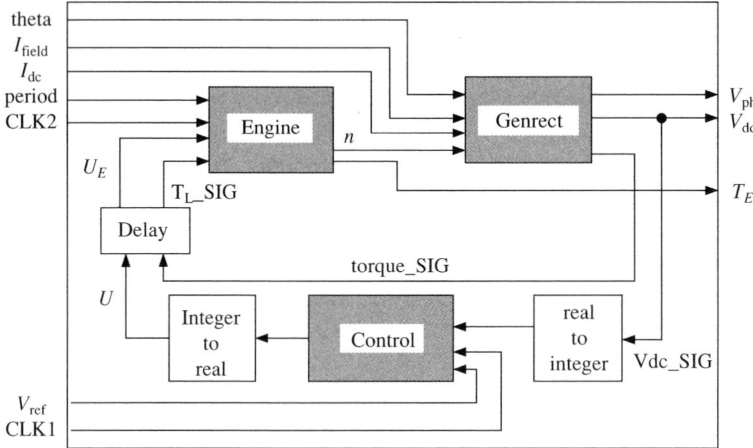


**Fig. 9.28** Block diagram of Sim

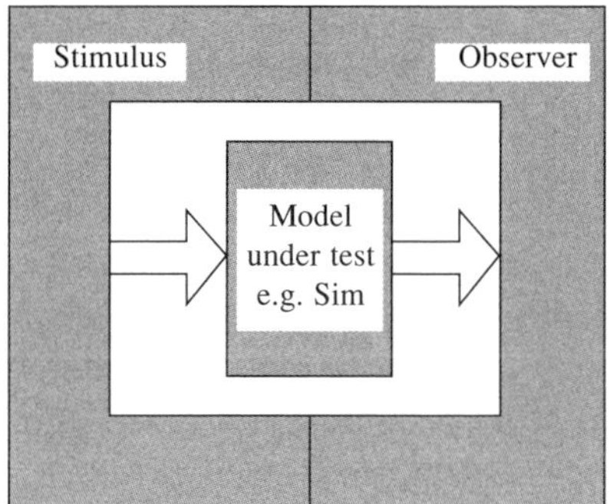


**Fig. 9.29** Model of a VHDL test-bench

**VHDL code (TB_Sim):**

```
  -- Stimulus
    CLK1 <= not CLK1 after 10 ns;
    CLK2 <= not CLK2 after 1 ns;
    Period <= 3.0;
    Theta  <= 0.0;
    Ifield <= 2.5;
    Vref <= 1000;
    Idc  <= 2.0, 6.0 after 4000 ns;
  -- Observer
  -- Write results into file
  process (CLK1)
    file outfile : text is out
    "C:\My Designs\Simulation\src\Results\Result1a.txt";
    variable out_line : line;
  begin
    write(out_line, Vdc);
```

```
    writeline(outfile, out_line);
  end process;
```

The first section, marked *Stimulus*, is responsible for assigning the appropriate values to the components in the test environment. In the given example, the load current (measured at the d.c. link) **Idc** is changed from 2.0 A to 6.0 A when $t = 4000$ ns. The other section of code (with the marking *Observer*) records the response of the system and writes the information into a text file. This is achieved using the **write** and **writeline** statements that are found in the **std.textio** library package. In this chapter, graphical representations of the system responses designed using MS Excel are shown instead of the raw numerical data. The voltage values shown in the graphs are normalised to the reference voltage.

### 9.2.5.1 Simulation 1

In this simulation, the effects of the weightings $\mathbf{E}^{1-7}$ for the defuzzification process in the component **Defuzz** are investigated. The values of the weightings in ***Simulation No.1a*** are:

$$E^1 = -8;\ E^2 = -6;\ E^3 = -4;\ E^4 = -2;\ E^5 = 0;\ E^6 = 2;\ E^7 = 4;\ E^8 = 6;\ E^9 = 8$$

Using the stimulus signals described by the VHDL code below, the d.c. voltage response of the system is simulated.

**VHDL code (TB_Sim):**

```
  -- Stimulus
    CLK1   <= not CLK1 after 10 ns;
    CLK2   <= not CLK2 after 1 ns;
    Period <= 3.0;
    Theta  <= 0.0;
    Ifield <= 2.5;
    Vref   <= 1000;
    Idc    <= 10.0;
```

The reference voltage is set to 1000 as this is a convenient value with which to analyse the performance of the controller. In the graphs shown, Figs 9.30 to 9.37, the values of

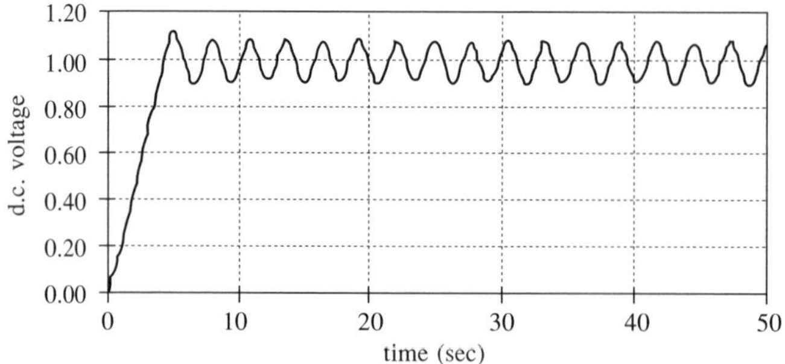


**Fig. 9.30** Voltage response in simulation No. 1a

the d.c. voltage in the graphs are normalised to the reference voltage (i.e. d.c. voltage = $V_{ds}/V_{ref}$).

It is clearly visible from Fig. 9.30 that the system is unstable. The d.c. voltage oscillates continuously with a magnitude of up to 9 per cent of the reference value. This is partly caused by overcompensation, which results from the large weighting values in the defuzzification process. In ***Simulation No. 1b***, the weighting values are reduced to:

$$E^1 = -4;\ E^2 = -3;\ E^3 = -2;\ E^4 = -1;\ E^5 = 0;\ E^6 = 1;\ E^7 = 2;\ E^8 = 3;\ E^9 = 4$$

The voltage response, plotted in the graph in Fig. 9.31, shows that the system has been stabilised by reducing the weightings in the defuzzification process. However, there is a steady-state error of almost 7.5 per cent. In order to improve the performance in this respect, it is necessary to examine the shapes of the input membership functions in the component **Fuzzify** and its effect on the control performance.

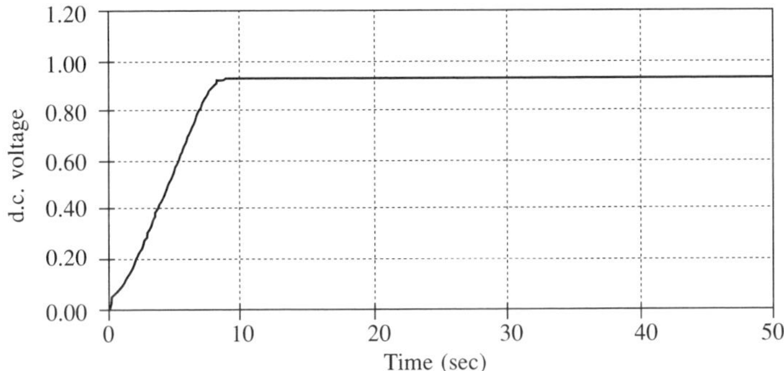


**Fig. 9.31** Voltage response in Simulation No.1b

The conditions set in the fuzzification process in ***Simulations No. 1a*** and ***No. 1b*** are as follows:

```
VHDL Code (Fuzzify):
Simulation Nos.1a and b
   constant a1  :INTEGER := -80;
   constant b1  :INTEGER := -40;
   constant a2  :INTEGER := -80;
   constant b2  :INTEGER := -40;
   constant c2  :INTEGER := 0;
   constant a3  :INTEGER := -40;
   constant b3  :INTEGER := 0;
   constant c3  :INTEGER := 40;
   constant a4  :INTEGER := 0;
   constant b4  :INTEGER := 40;
   constant c4  :INTEGER := 80;
   constant a5  :INTEGER := 40;
   constant b5  :INTEGER := 80;
```

Plotted on a graph, these conditions present a membership function diagram similar to the one shown in Fig. 9.32. The shapes of the membership functions of the three linguistic values in the middle are identical, with each of them spanning 80 units over the $x$-axis. By making the linguistic value at the centre (**Zero**) more focused around the value *zero*, it is possible to reduce the magnitude of the steady-state error. The FLC design is constructed such that this task can be achieved simply by changing some values of the 13 constants **(a1-4, b1-5, c2-5)** inside **Fuzzify.**

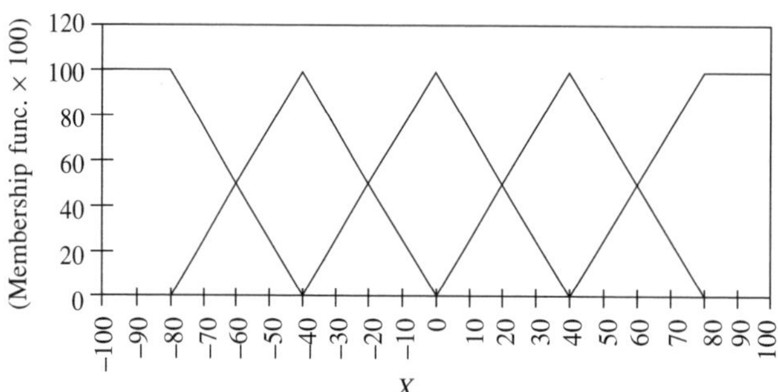


**Fig. 9.32** Input variables membership functions in Simulations No. 1a and No. 1b

### *9.2.5.2 Simulation 2*

If the membership function in Fig. 9.32 is made too narrow, the system loses its stability. This is demonstrated by ***Simulation No. 2a***, in which the membership function of the input variables is changed to that of Fig. 9.33 whereby the linguistic value **Zero** ranges between (–5) and (+5). The voltage response is plotted in the graph in Fig. 9.34. The system response is oscillatory albeit with a smaller magnitude (2.5 per cent) than that of ***Simulation No. 1a***.

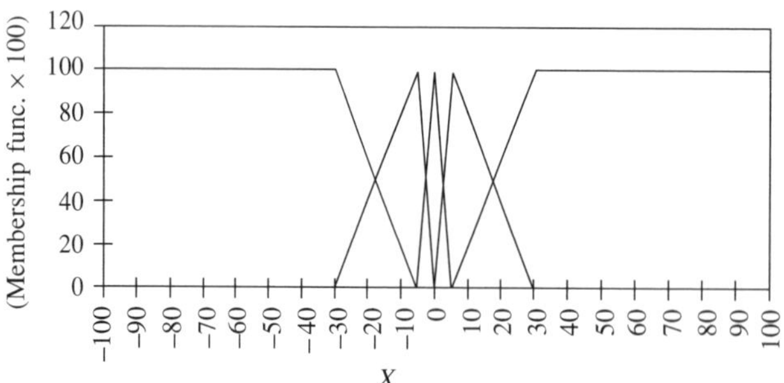


**Fig. 9.33** Input variables membership functions in Simulation No. 2a

In ***Simulation No. 2b***, the membership function of the input variables is modified to resemble the diagram in Fig. 9.35. The result, shown by the graph in Fig. 9.36 clearly proves that, when the correct balance is achieved in the choice of membership function, the FLC performance is considerably improved. The steady-state error in this case is less than 1.6 per cent.

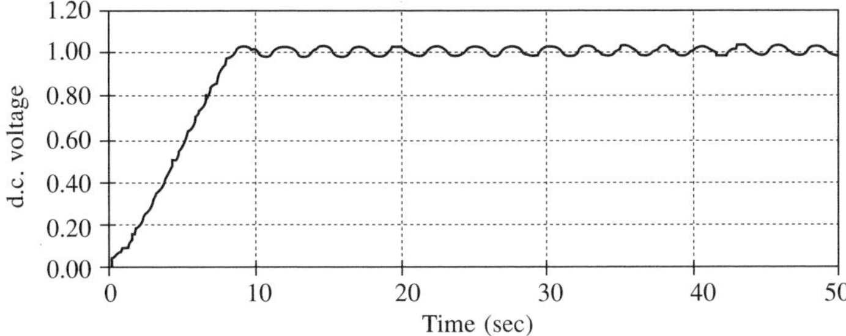


**Fig. 9.34** Voltage response in Simulation No. 2a.

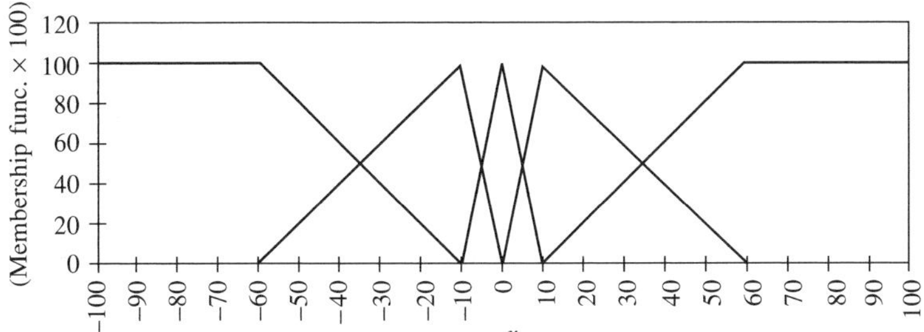


**Fig. 9.35** Input variables membership functions in Simulation No. 2b

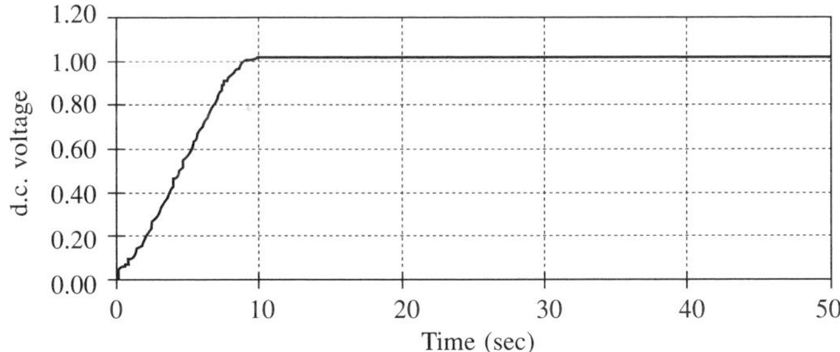


**Fig. 9.36** Voltage response in Simulation No. 2b

### *9.2.5.3 Simulation 3*

Once the parameters in **Fuzzify** and **Defuzz** are properly tuned, the system can be simulated under varying load conditions. In ***Simulation No. 3a***, the d.c. load current is changed from 2 A to 6 A. Figure 9.37 shows the voltage response whereby the change occurs when $t = 20$ seconds. There is a transient dip in the voltage of less than 14 per cent when the load current increases suddenly, but the FLC is shown to be capable of restoring the voltage to its original value in less than 5 seconds.

This section has presented an extensive example of the VHDL design of a fuzzy logic controller. A series of simulations was presented and the results provided the necessary system validation to proceed to hardware implementation with confidence.

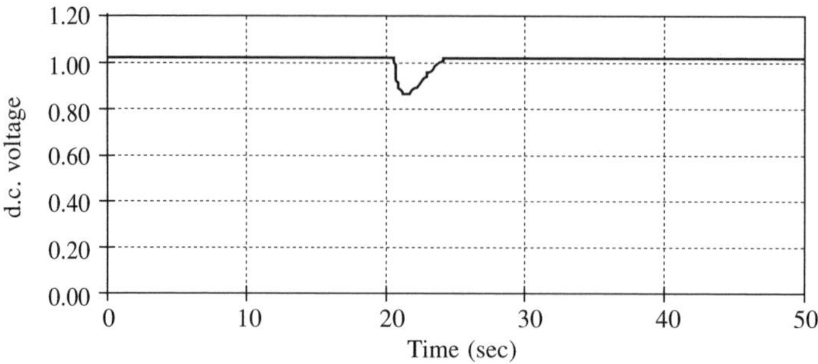


**Fig. 9.37** Voltage response in Simulation No. 3a

## 9.3 FPGA implementation

In order to transform a behavioural level VHDL design such as the one presented in the previous section into a hardware design, a number of considerations have to be made and the design must be optimised before implementation. The *synthesis* and *implementation* processes are discussed. *Synthesis* is the act of converting the VHDL code into a *netlist*. A netlist is a standard method of describing the design at a level of abstraction that is suitable for hardware implementation, either at an architectural level which consists of logic blocks or at a logic level which is made up of logic primitives. *Implementation* is the process of converting the netlist into a format that can be downloaded into the target technology. This procedure is very technology-specific since different devices require different formats. The target technology for this design is the XC4010XL-PC84, a member of the Xilinx XC4000XL Series of FPGAs [8]. These devices are fully reconfigurable and can be reprogrammed an unlimited number of times. The Xilinx XC4010 has a maximum logic gate count of 10 000. The relevance of this point will become clear later in the section when it is shown that the original FLC design discussed so far is too large to fit into a Xilinx XC4010.

The design of the FLC presented so far in this example did not take into account the synthesis and implementation considerations and the limitations of the target technology. Xilinx XC4010XL FPGA has a total of 400 CLBs and an equivalent gate count of 10 000.

Using the EDA tool, *Xilinx Foundation Series* [14], a VHDL design can be synthesised then implemented for a particular device. Figure 9.38 shows a simplified block diagram of the process. In the synthesis stage, the VHDL code is converted into a netlist, which

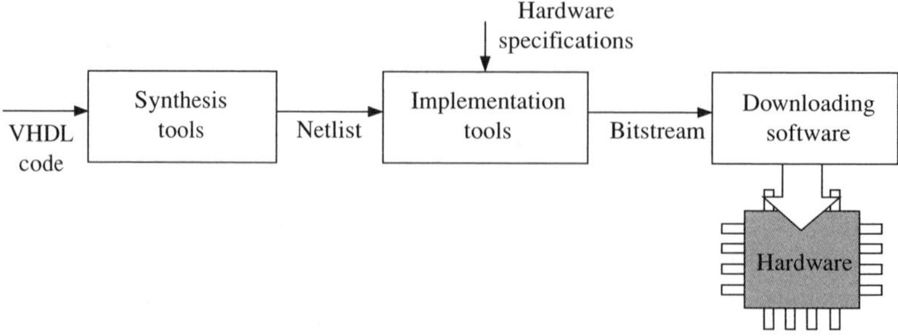


**Fig. 9.38** Simplified block diagram of the hardware design process

is a format that contains the structural hardware description of the design. A function is considered NOT to be 'synthesisable' if the synthesis tool cannot readily convert it into hardware description. Functions such as mathematical division and trigonometric operations are not synthesisable using currently available synthesis tools. Implementation in this context is the process of converting the netlist into a bitstream file. The information in the bitstream is used to configure the functionality of the target FPGA. During the implementation stage, the target technology and other hardware design specifications such as pin allocation have to be confirmed. For troubleshooting and analysis purposes, Xilinx Foundation also generates a status report at each stage of the process.

Before a VHDL design can be synthesised, all the VHDL functions which are not 'synthesisable' have to be eliminated and replaced with functions that can be translated into hardware.

### 9.3.1 The divider

The FLC's defuzzification process requires a division operation. In the behavioural design, the division is achieved with the following VHDL statements (refer to Appendix E – 11.5.6):

```
process(CLK)
  variable Dividend, Divisor : integer;
begin
  ...
  Y <= (Dividend/Divisor);
  ...
end process;
```

This process is not suitable for synthesis for two reasons:

- The variables **Dividend** and **Divisor** are declared as open-ended integers, with no meaning in hardware design. Integer variables and signals should be given a well-defined range such as

```
process(CLK)
  variable Dividend :integer range 0 to 512;
begin
```

This way of declaration informs the synthesis tool about the maximum number of bits required by the variables. By not explicitly defining a range, the synthesis tool will assign a default number of bits to model the variable. This usually results in a waste of resources. Alternatively, they can be declared as *std_logic_vector* type, which has a more relevant form in hardware terms. The declaration is written as:

```
process(CLK)
  variable Dividend :std_logic_vector(8 downto 0);
begin
```

- The division operator '/' is not supported by present synthesis tools, including Xilinx Foundation. To implement this calculation in hardware, it is necessary to design a digital divider at structural level.

### 9.3.2 Analysis of binary integer division

Before entering a discussion on binary division, it is appropriate to define some of the terms which are used. The *dividend* (divA) is defined as the number to be divided while the *divisor* (divB) is the number by which the dividend is divided. The division process can then be expressed by the following equation:

$$\frac{\text{div A}}{\text{div B}} = Y + \frac{R}{\text{div B}}$$

where $Y$ is the *quotient* and $R$ is the *remainder* whereby $R <$ divB.
In binary notation, the numbers can be written as

$$\text{div A} = 2^{(n-1)} \cdot a_{(n-1)} + 2^{(n-2)} \cdot a_{(n-2)} \ldots + 2^1 \cdot a_1 + 2^\circ \cdot a_0$$

$$\text{div B} = 2^{(m-1)} \cdot b_{(m-1)} + 2^{(m-2)} \cdot b_{(m-2)} \ldots + 2^1 \cdot b_1 + 2^\circ \cdot b_0$$

$$Y = 2^{(n-1)} \cdot y_{(n-1)} + 2^{(n-2)} \cdot y_{(n-2)} \ldots + 2^1 \cdot y_1 + 2^\circ \cdot y_0$$

Using the pencil-and-paper method, a binary division process is performed in a number of steps. In each step, a *partial dividend* $\{PD\}$ is divided by the divisor such that if $j$ is the index for the steps, then,

$$y_j = \frac{\{PD\}_j - R_j}{\text{div B}}$$

and

$$\{PD\}_{(j-1)} = 2R_j + a_{(j-1)}$$

where:
$j = (n-1), (n-2) \ldots, 0$
$R_j$ is the *partial remainder* and $R_j <$ div B.

For example, dividing the binary number '1101' by '11':

$$11\overline{)\{1\}101}\quad \text{quotient: } 0100$$

using the pencil-and-paper method,

$$\begin{aligned}
&\text{when } j = 3, \{PD\} = \{1\} \Rightarrow y_3 = 0, R_3 = 1,\\
&\text{when } j = 2, \{PD\} = \{11\} \Rightarrow y_2 = 1, R_2 = 0,\\
&\text{when } j = 1, \{PD\} = \{0\} \Rightarrow y_1 = 0, R_1 = 0,\\
&\text{when } j = 0, \{PD\} = \{01\} \Rightarrow y_0 = 0, R_0 = 1.
\end{aligned}$$

$R_0$ is also the true remainder $R$.
Therefore, the result of the division can be written as:

$$\frac{1101_2}{11_2} = 0100_2 + \frac{1_2}{11_2}$$

or, in decimal notation,

$$\frac{13}{3} = 4 + \frac{1}{3}$$

### 9.3.3 Binary integer division algorithm

There are several different algorithms, which can perform binary integer division. Some algorithms have parallel operations and require only one clock cycle to carry out the calculation while others operate in a sequential manner and may require more clock cycles, such as the one presented later in this chapter. As a general rule, the algorithms with parallel operations are larger in area size than the sequential ones when they are implemented into hardware. The division algorithm used in this example is derived from the pencil-and-paper method of binary division. Let

| | |
|---|---|
| divA | be an $n$-bit register which holds the unsigned dividend, |
| divB | be an $m$-bit register which holds the unsigned divisor, |
| $B_p$ | be a signed $(m + 1)$-bit register which holds the positive value of $B$, |
| $B_n$ | be a signed $(m + 1)$-bit register which holds the negative value of $B$, |
| $A$ | be an $n$-bit register which holds the partial dividend {PD}, |
| $A_s$ | be a flip-flop appended to the left end of $A$ to store the sign bit, and |
| $Z_sZ$ | be the temporary register used to store. |

Using the symbol '$\leftarrow$' to denote 'is assigned as', the division algorithm used in this book can be written as follows:

- Block 1:
  Initialisation:
    clear $A$;
    $B_p \leftarrow$ positive value of divB;
    $B_n \leftarrow$ negative value of divB (two's complement);
- Block 2:
  For $j = (n - 1)$:
    shift left $(A)$(divA);
    $Z_jZ \leftarrow (A_sA + B_n)$;
    $q_j \leftarrow \text{not}(Z_j)$;
- Block 3
  For $j = (n - 2), (n - 3) \ldots, 1, 0$
    shift left AP;
    if $q_{(j-1)}$ = '0' then:
    $Z_sZ \leftarrow (A_sA + B_p)$;
    else if $q_{(j-1)}$ = '1' then:
    $Z_sZ \leftarrow (A_sA + B_n)$;
    end if;
  $q_j \leftarrow \text{not}(Z_s)$;
    $A_sA \leftarrow Z_sZ$.

A VHDL code is written to implement this algorithm for a 14-bit dividend and a 9-bit divisor. The code defines the dividend as **divA**, the divisor as **divB** and the quotient is defined as **Y**. It is assumed that the effects of the Remainder on the control performance is small, therefore the Remainder can be truncated altogether. Since the algorithm is sequential, the entire operation is carried out in a VHDL *process*. A circuit is also included to avoid a division-by-zero error by clearing **Y** (assign all bits to zero) when **divB** equals zero. Using two's complement conversion to obtain the negative value of

**divB**, the 'Initialisation' procedure in Block 1 is performed by the following section of code:

```
-- Initialisation
  -- Clear A
A(13 downto 0) :="00000000000000000";
-- Assign divB(+ve) and divB(-ve)
  Bp:="00000"&divB;
  Bn:=not(Bp)+"000000000000001";
```

It can be observed that the variables are treated as binary-valued bits and bytes instead of whole numbers. This is a realistic view of digital circuits and highlights an important difference between hardware description and software programming. The next sequence of commands converts the dividend **divA** into an unsigned value. This is essential because the algorithm is not designed to cope with negative values of the dividend. From the defuzzification process, it is known that the divisor **divB** is always a positive number. Therefore, it follows that the sign status (whether positive or negative) of the quotient **Y** is always the same as the sign status of **divA**. The functions in Blocks 2 and 3 are implemented by introducing the status bits **Load** and **Ready**. '**Load** = 1' relates to the condition whereby a new set of dividend and divisor values are loaded into the divider circuit, i.e. when $j = (n - 1)$, while '**Ready** = 1' is the condition whereby the division calculation for the recent set of values has been completed, that is when $j = 0$. The VHDL code also includes the description of the output interface circuit of the controller, which implements the following function:

$$u \leftarrow u \cdot z^{-1} + y$$

where:
$y$ is the crisp output from the defuzzifier
$u$ is the actual control signal.

The section of code that describes this circuit is written as follows:

```
...
if SIGN='1' then
...
- - Negative
  U-var := U_Past - Y;
...
else
- - Positive
  U_var := U_Past + Y;
...
end if;
...
```

In any hardware code that involves continuous increment or decrement of numerical values, there is a danger of *overflow*. Therefore, the upper and lower limit of the output is set at '11111111' and '00000000', respectively. This prevents the 8-bit output register from overflowing into a 9-bit value (e.g. 11111111 + 1 = [1]00000000).

The VHDL code is subsequently configured as a component and a netlist is created. Figure 9.39 shows the symbol of the component **Divider** as viewed in the Xilinx Foundation Schematic Editor.

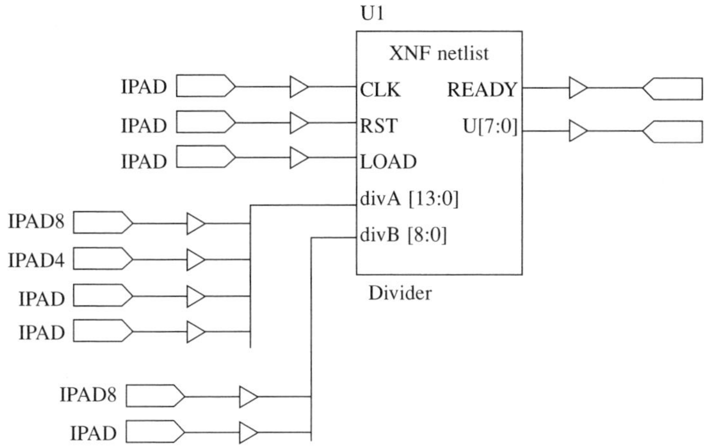


**Fig. 9.39** Symbol of the component 'Divider' with I/O buffers and pads

### 9.3.4 Design optimisation

Even after the behavioural design of the FLC is converted into a format that is fully supported for synthesis, there are still other issues to consider, particularly regarding implementation. This section looks at the question of *area optimisation*. Figure 9.40 shows an extract from the implementation status report of the FLC design targeted at the Xilinx XC4010 FPGA.

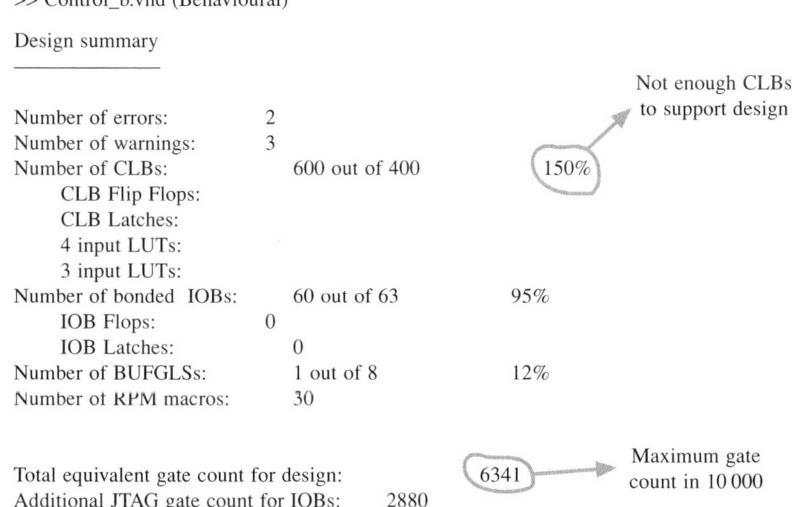

>> Control_b.vhd (Behavioural)

Design summary

| | | | |
|---|---|---|---|
| Number of errors: | 2 | | |
| Number of warnings: | 3 | | |
| Number of CLBs: | | 600 out of 400 | 150% |
| CLB Flip Flops: | | | |
| CLB Latches: | | | |
| 4 input LUTs: | | | |
| 3 input LUTs: | | | |
| Number of bonded IOBs: | | 60 out of 63 | 95% |
| IOB Flops: | 0 | | |
| IOB Latches: | | 0 | |
| Number of BUFGLSs: | | 1 out of 8 | 12% |
| Number of RPM macros: | | 30 | |

Not enough CLBs to support design

Total equivalent gate count for design: 6341

Additional JTAG gate count for IOBs: 2880

Maximum gate count in 10 000

**Fig. 9.40** Synthesis report for top hierarchy entity: Control.vhd

The following points can be noted:

- The FLC design requires more CLBs than is readily available in a Xilinx XC4010.
- The total equivalent gate count for the design is below the maximum stated in the Data Book.
- The number of bonded IOBs in the device is sufficient for the design.

From these observations, it can be seen that although the design is too large for the Xilinx XC4010 in terms of CLBs, it has a total equivalent gate count that is within its limits. This suggests that the design is not fully utilising the logic gates in each CLB. The design can be easily implemented into one of the bigger devices in the Xilinx 4000 family, but in order to implement it into the XC4010XL, a certain number of modifications have to be made. These modifications optimise the design such that it is functionally identical to the original design but requires a smaller area. However, in reducing the area space, there is a trade-off between other properties. In this case, one of the concerns is that by optimising the design, the fuzzy rule base also becomes more deeply integrated into its structure, thereby losing some of the features which make the controller generic and flexible.

The quality of a design can be measured by two main variables, *area* and *performance*. The area of a design simply refers to the sum of the area space of the circuit components. The measure of performance is more complex as it involves analysis of the structure and behaviour of the circuit. There are a number of variables that relate to circuit performance such as propagation delay, cycle-time, latency and throughput but they are not discussed in great detail here. Briefly, *propagation delay* is the delay through the critical path of a circuit (for combinational logic circuit), *cycle-time* is the fastest clock period that can be applied to the circuit (for synchronous sequential circuit), *latency* is the number of clock cycles required to execute the operation and *throughput* refers to the rate at which data is consumed and produced by the circuit. *Optimisation* is the act of minimising the area and maximising the performance. Usually, there has to be a trade-off between the two. In most cases, design optimisation is subject to constraints such as:

- minimise the area under performance constraints
- maximise the performance under area constraints.

The task in this design falls under the first category, which is to minimise the area.

#### *9.3.4.1 Structural multiplication*

It has already been shown how multiplication in VHDL programs can be carried out using the operator '*'. A simple multiplication by an arbitrary number, say five, can be performed with the following statement:

```
A<= B*5;
```

Another method of implementing this operation is to explicitly describe the step-by-step procedure of the binary multiplication. Returning to the recent example, the pencil-and-paper method of the binary multiplication when B = 9 is written as follows:

```
   1001   ←   B
  × 101       ←   5
 ---------
   1001   ←   B
  0000    ←   0
 1001     ←   B, shift left by 2 bits
 ---------
 101101   ←   A
```

This can be described in VHDL as: A<= B + shl(B,"s10");
where **shl(** B, "10" **)** is the command statement to shift the contents of **B** to the *left* by 2 bits.

The multiplication process effectively becomes a combination of an addition and a *shift-left* operation. Using this method to describe all the multiply-by-a-constant procedures, the area size of the design can be reduced significantly.

#### *9.3.4.2 Fuzzifier optimisation*

In the fuzzification process of the FLC design that was previously presented, the two inputs, $x_1$ and $x_2$, are processed using two separate fuzzification blocks. Since both of the inputs are fuzzified in exactly the same manner, it is possible to create just one fuzzification block to be shared between the two input variables. This reduces the area but it is at the expense of the performance, particularly *latency*.

While previously the entire process can be completed in just one clock cycle, this approach requires at least two clock cycles. To achieve this, there is also the need for memory elements to store the result of the first computation while the second set of data is being computed. At the end of the second computation, both sets of results are released simultaneously. In order to reduce the area size further, another method of optimisation is devised.

The fuzzification block has five outputs, one for each fuzzy value defined in the inputs' universe of discourse. However, the fuzzification process entails that, for any single crisp value of the input $x_i$, only two adjacent fuzzy values are significant (with non-zero membership values). By ignoring the insignificant fuzzy values, the number of output signals can also be reduced from five to two. The possible combinations of significant fuzzy values for an arbitrary input are:

$$B_i^1 \text{ and } B_i^2;\quad B_i^2 \text{ and } B_i^3;\quad B_i^3 \text{ and } B_i^4;\quad B_i^4 \text{ and } B_i^5$$

It is found that using just three variables, ADRi, Bi_A and Bi_B, all the combinations can be sufficiently represented for any value of $x_i$ as shown by the following statements:

| | | | |
|---|---|---|---|
| ADRi = "00" | : | Bi_A $= B_i^1$ , | Bi_B $= B_i^2$ |
| ADRi = "01" | : | Bi_A $= B_i^2$ , | Bi_B $= B_i^3$ |
| ADRi = "10" | : | Bi_A $= B_i^3$ , | Bi_B $= B_i^4$ |
| ADRi = "11" | : | Bi_A $- B_i^4$ , | Bi_B $= B_i^5$ |

Figure 9.41 illustrates how these conditions correspond with the universe of discourse.

The complete VHDL code of the fuzzifier can be found in Appendix F. An outline of the code's functional structure is described by the flowchart shown in Fig. 9.42. Management of the input and output signals is mainly controlled by a status bit, **R_sig**. When **R_sig** is '1', the input **x1** is fuzzified but the output values remain unchanged.

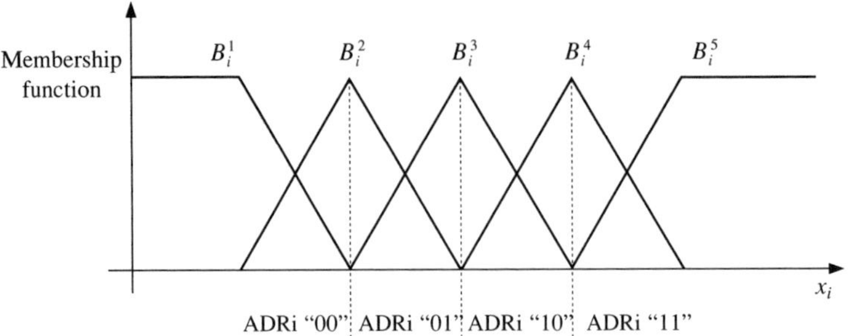


**Fig. 9.41** Definition of input fuzzy values

When **R_sig** is '0', the input **x2** is fuzzified and the new values of all the outputs are assigned. The variables **temp, temp_A** and **temp_B** represent the temporary registers used to store the fuzzy values of **x1** while the fuzzy values of **x2** are being computed. The operation takes two clock cycles to complete and all the fuzzy values are simultaneously released.

### *9.3.4.3 'Mini' FAM tables*

The FAM table of the FLC design discussed in an earlier section is shown again in Table 9.3. It was mentioned that the inference of the fuzzy rules is achieved using Mamdani's inference technique and the VHDL code presented uses an inference engine which triggers all 25 rules during every calculation. This section describes an algorithm which is developed to reduce the amount of computation required by focusing only on the relevant rules and ignoring those which are irrelevant to the conditions in question. From the previous section, it is known that for every set of inputs, only four fuzzy values (two for each input) are significant. This means that only four fuzzy rules are relevant at any one time.

An easier way of explaining the technique is to imagine the entire FAM table to be covered from view. Access to the content of the FAM table is only allowed through a small window and only four adjoining rules can be viewed through this window at a time. Therefore, instead of having to access 25 rules, the inference engine only has to access four rules during every computation. The window can move around the FAM table and its position is identified by an index $j$ defined as:

ADR1 = "00" & ADR2 = "00" $\rightarrow j = 0$
ADR1 = "00" & ADR2 = "01" $\rightarrow j = 1$
ADR1 = "00" & ADR2 = "10" $\rightarrow j = 2$
ADR1 = "00" & ADR2 = "11" $\rightarrow j = 3$
ADR1 = "01" & ADR2 = "00" $\rightarrow j = 4$
ADR1 = "01" & ADR2 = "01" $\rightarrow j = 5$
ADR1 = "11" & ADR2 = "11" $\rightarrow j = 15$

There are 16 'window positions' altogether and the first six are shown in Fig. 9.43. The shaded blocks are the rules which are considered relevant for the input conditions corresponding to the index $j$. To distinguish the 'windowed' view of the FAM table from the original table, the first is referred to as the *'Mini' fuzzy associative memory (FAM) table*.

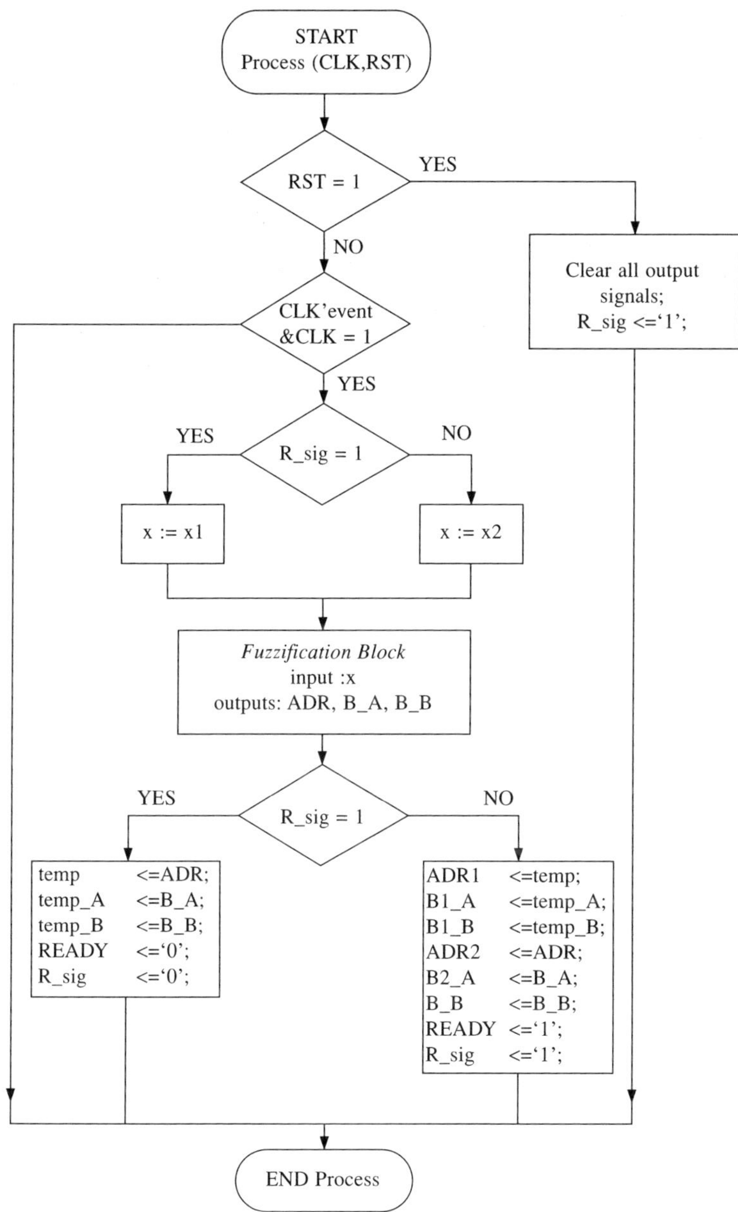


**Fig. 9.42** Flowchart of the fuzzifier

When the window technique is applied to the FAM table in Table 9.3, it is observed that a number of the *mini-FAM* tables are identical (e.g. $j = 1$ and $j = 4$). Out of the 16 mini-FAM tables, there are only seven unique tables as shown in Fig. 9.44.

If **WIN** is the index for the new set of tables, then the tables can be arranged using the following:

**Table 9.3** FAM table of the FLC design

| $E$ \ $\Delta E$ | NB | N | Z | P | PB | | |
|---|---|---|---|---|---|---|---|
| **NB** | $R^1$ | $R^2$ | $R^3$ | $R^4$ | $R^5$ | | |
| | $u$ = PVB | $u$ = PB | $u$ = P | $u$ = PS | $u$ = Z | **NVB** | Negative Very Big |
| **N** | $R^6$ | $R^7$ | $R^8$ | $R^9$ | $R^{10}$ | **NB** | Negative Big |
| | $u$ = PB | $u$ = P | $u$ = PS | $u$ = Z | $u$ = NS | **N** | Negative |
| **Z** | $R^{11}$ | $R^{12}$ | $R^{13}$ | $R^{14}$ | $R^{15}$ | **NS** | Negative Small |
| | $u$ = P | $u$ = PS | $u$ = Z | $u$ = NS | $u$ = N | **Z** | Zero |
| **P** | $R^{16}$ | $R^{17}$ | $R^{18}$ | $R^{19}$ | $R^{20}$ | **PS** | Positive Small |
| | $u$ = PS | $u$ = Z | $u$ = NS | $u$ = N | $u$ = NB | **P** | Positive |
| **PB** | $R^{21}$ | $R^{22}$ | $R^{23}$ | $R^{24}$ | $R^{25}$ | **PB** | Positive Big |
| | $u$ = Z | $u$ = NS | $u$ = N | $u$ = NB | $u$ =NVB | **PVB** | Positive Very Big |

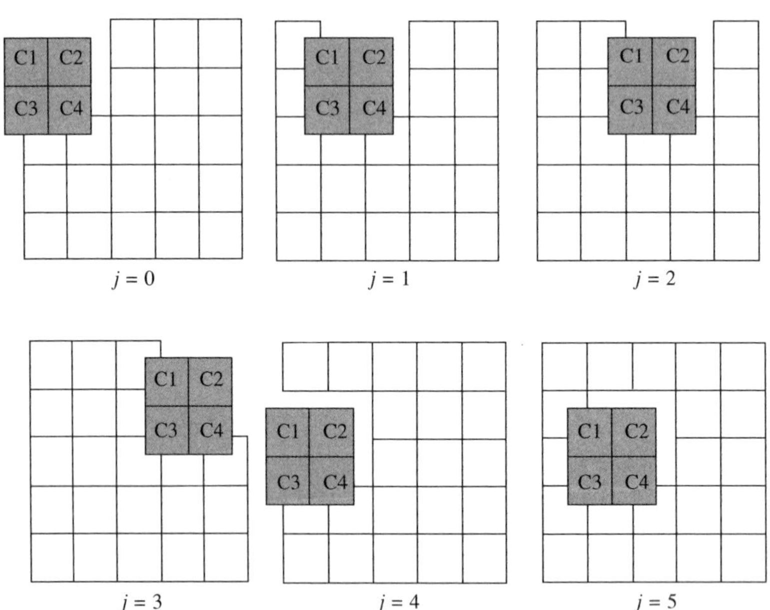


**Fig. 9.43** 'Mini' FAM tables

```
IF j=0 THEN WIN="0000"
IF j=1 OR j=4 THEN WIN="0001"
IF j=2 OR j=5 OR j=8 THEN WIN="0010"
...
F j=11 OR j=14 THEN WIN="0101"
IF j=15 THEN WIN="0110"
```

This algorithm requires a considerable number of IF-THEN operations and is not necessarily an efficient way to implement the design into hardware. By observing the pattern in the original FAM table, it can be shown that the mini-FAM tables are identical when the sum of **ADR1** and **ADR2** is the same. Therefore, instead of using numerous IF-THEN operations, the arrangement of the mini-FAM tables is achieved using a single addition

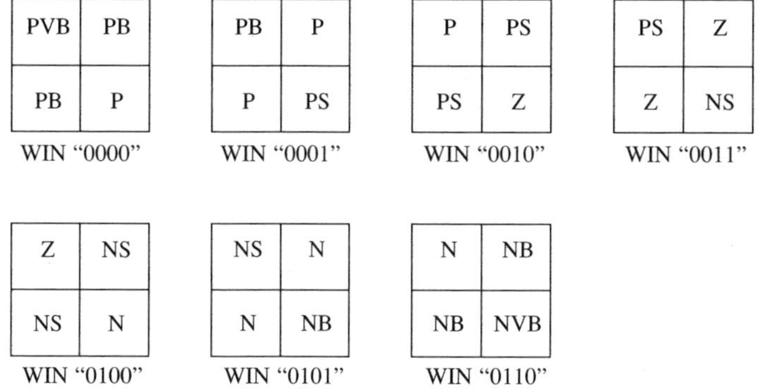


**Fig. 9.44** Mini-FAM tables for the FLC design

operation as shown by the following statement in the VHDL code (see Appendix F – 11.6.4):

```
WIN <= ("00"&ADR1) + ADR2;
```

where **ADR1** and **ADR2** are signals from the component **Fuzzify**. The function of the statement **("00"&ADR1)** is to expand the value of **ADR1** from 2 bits to 4 bits such that it is compatible with the 4-bit signal **WIN**. The variables inside the mini-FAM table are subsequently processed in the section of the code that is marked '*Mini-Fuzzy Inference Engine*'. In the original code, the inference engine contains 25 MIN-operations. The modified code consists of only four MIN-operations, which is a notable reduction.

### 9.3.4.4 Defuzzification algorithm

The original algorithm for the MAX-operation and defuzzification process contains two important computations: the aggregation of 25 rule-consequents into nine output (fuzzy) values, and the multiplication of each output value by a constant weighting.

By incorporating the modifications discussed in the previous sections, only four significant rule-consequents are considered. Therefore, the number of rule-consequents to be aggregated is reduced but the allocation of correct weightings for the significant output values becomes slightly more complicated. From the tables in Fig. 9.44 it is obvious that regardless of the **WIN** value, the consequents C2 and C3 always point to the same fuzzy value (e.g. when **WIN** = "0000": C1→PVB, C2→**PB**, C3→**PB**, C4→P). This implies that only C2 and C3 have to be aggregated, hence:

$$DA = C1$$
$$DB = \max[C2, C3]$$
$$DC = C4$$

where DA, DB and DC represent the membership function of the output fuzzy values.

The actual fuzzy values referred to by DA, DB and DC are determined by the value of **WIN**. If $E^i$ is the weighting while VA, VB and VC are the weighted values, then

WIN = "0000" : $VA = DA * E^{PVB}$, $VB = DB * E^{PB}$, $VC = DC * E^{P}$ ;
WIN = "0001" : $VA - DA * E^{PB}$, $VB = DB * E^{P}$, $VC - DC * E^{PS}$ ;

WIN = "0010" : VA = DA* $E^P$, VB = DB* $E^{PS}$, VC = DC* $E^Z$ ;
WIN = "0011" : VA = DA* $E^{PS}$, VB = DB* $E^Z$, VC = DC* $E^{NS}$ ;
WIN = "0100" : VA = DA* $E^Z$, VB = DB* $E^{NS}$, VC = DC* $E^N$ ;
WIN = "0101" : VA = DA* $E^{NS}$, VB = DB* $E^N$, VC = DC* $E^{NB}$ ;
WIN = "0110" : VA = DA* $E^N$, VB = DB* $E^{NB}$, VC = DC* $E^{NVB}$ ;

Although the functions above can be implemented with seven IF-THEN statements it is preferable to adopt a less space consuming method. Figure 9.45 shows a flowchart of the modified **Defuzz** design. The complete VHDL code can be found in Appendix F. In this design, instead of seven IF-THEN operations, only three are required to accomplish the main task (not counting the reset and clocking circuits). The command statements in *Block 0* implement the reset conditions whereby all the outputs (except **READY**) and internal variables are cleared. The condition **LOAD**='1' and **READY**='1' indicates that the component is ready to initialise the defuzzification process and proceeds to execute the commands in *Block 1* and *Block 2*. *Block 1* performs the aggregation of C2 and C3 using a MAX-operator:

***Block 1:***
DA = C1
DB = max[ C2, C3]
DC = C4

In *Block 2,* it is assumed that **WIN** = "0000", and the fuzzy values are then multiplied by the appropriate weightings for PVB, PB and P. If the weightings are defined as:

$E^{PVB}$ = 40; $E^{PB}$ = 30; $E^P$ = 20; $E^{PS}$ = 10; $E^Z$ = 0;
$E^{NS}$ = –10; $E^N$ = –20; $E^{NB}$ = –30; $E^{NVB}$ = –40;

then *Block 2* can be written as:

***Block 2:***
VA = DA * 40
VB = DB * 30
VC = DC * 20

The subsequent commands check to see if **WIN** is in fact "0000". If the case is true (i.e. **COUNT** = **WIN**), then the computation is complete and the output signals **divA**, **divB** and **READY** are assigned with the appropriate values. Otherwise, the computation is incomplete (**READY** = '0') and *Block 3* is executed in the next clock cycle.

The strategy of this algorithm relies on the fact that the values of VA, VB and VC decrease steadily as **WIN** increases. In other words,

IF
WIN = "0000": $VA^0$ = DA*40, $VB^0$ = DB*30, $VC^0$ = DC*20
THEN,
WIN = "0001": $VA^1 = VA^0$ – (DA*10), $VB^1 = VB^0$ – (DB*10), $VC^1 = VC^0$ – (DC*10)
WIN = "0010": $VA^2 = VA^1$ – (DA*10), $VB^2 = VB^1$ – (DB*10), $VC^2 = VC^1$ – (DC*10)
WIN = "0011": $VA^3 = VA^2$ – (DA*10), $VB^3 = VB^2$ – (DB*10), $VC^3 = VC^2$ – (DC*10)
....

By taking advantage of the recurring pattern, *Block 3* can be described with just four logical statements:

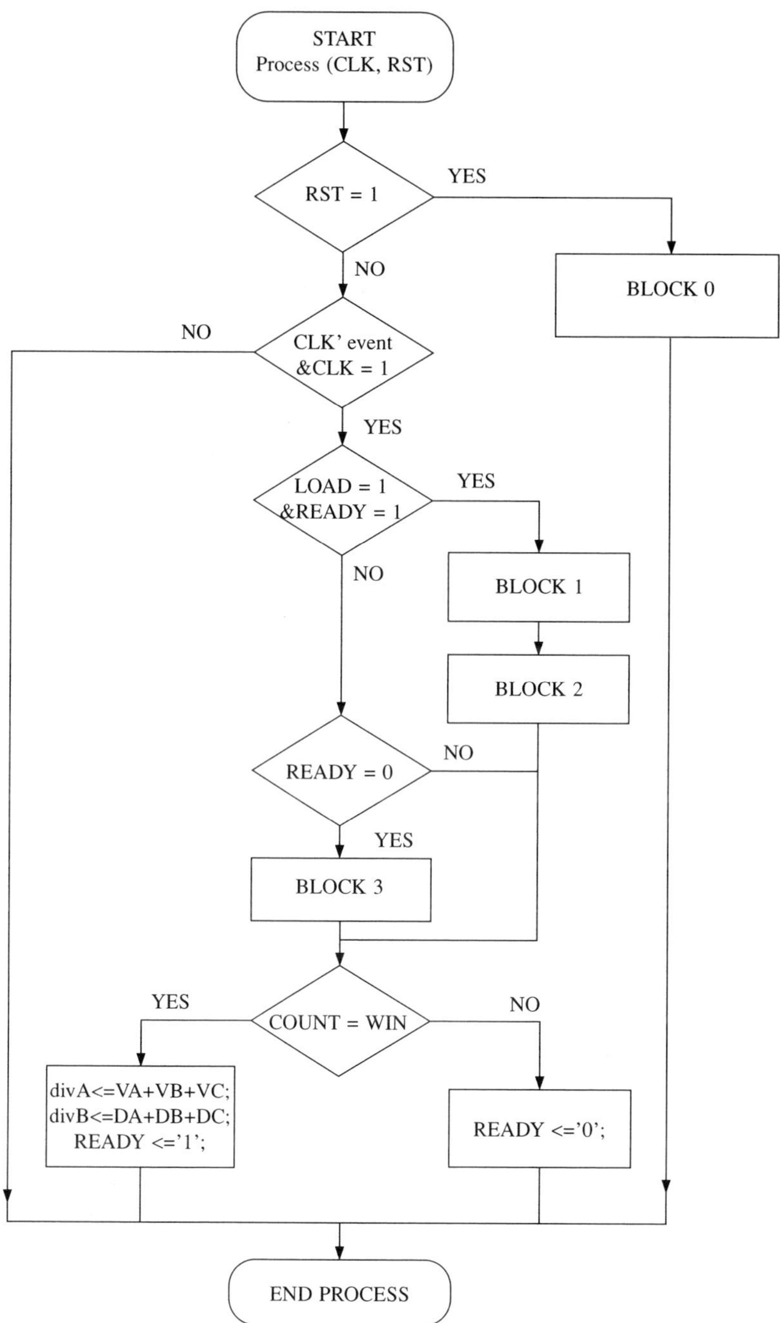


**Fig. 9.45** Flowchart of the code for defuzzification

***Block 3:***

COUNT = COUNT + 1

VA = VA – (DA*10)

VB = VB – (DB*10)
VC = VC – (DC*10)

Once the optimised structural-level design is successfully completed, it is implemented using Xilinx Foundation HDL Editor.

### 9.3.5 Implementation

Each element of the FLC is designed and carefully optimised for synthesis. Five VHDL components **DERIV**, **FUZZIFY**, **INFER**, **DEFUZZ** and **DIVIDER** make up the core of the fuzzy controller. The nature of the components' connection and the functionality of the processes are described by an upper hierarchy VHDL code **Control.vhd**. These components are wired to each other to form the complete control system. In addition to the components, two synchronous processes **Process1** and **Process2** are used to synchronise various signals. The code is subsequently synthesised to generate a netlist of the upper hierarchy component **Control**.

In creating a single top hierarchy component, it is easier to proceed into the implementation stage using the Xilinx Foundation Schematic Editor whereby the design can be developed in a graphical form. The '**Create Macro symbol from netlist**' function allows the component **Control** to be converted into a Macro symbol that can exist in the Schematic Editor. Then, using Xilinx Foundation Implementation tools, the design can be compiled into a bitstream file.

Xilinx XC4010 FPGA is available in several packages and the one used for this design is the PC-84 package which has 84 I/O pins in total. During the generation of the bitstream, the inputs and outputs of the design are mapped to the physical I/O pins of the FPGA. The allocation of pin numbers can either be automatically performed by the implementation tools or explicitly specified by the user. In this design, all the pins are specified by manually assigning the appropriate pin numbers to the IPADs and OPADs in the Schematic Editor. This enables the designer to have full control over the function of the physical pins in the FPGA.

The allocation of pin numbers to the I/O pads is shown by the schematic diagram (as seen in the Xilinx Foundation Schematic Editor) of the design in Fig. 9.46. The numbers (preceded by the letter 'p') in the I/O pads are the allocated pin numbers. It can be seen from the diagram that the clock input of the design is allocated to pin 35, and as it is the main clock signal, a global buffer IBUFG is used instead of the normal input buffer.

Once the hardware specifications have been confirmed, the netlist is compiled into a bitstream file using the *Implementation* procedure in Xilinx Foundation Project Manager. The implementation report shows that the modifications were effective in utilising the logic blocks in the FPGA in a more efficient manner and confirms that the bitstream file is successfully generated without errors and is ready for downloading.

The circuit board used to house the target FPGA during downloading is the XS40 Board, a Xilinx FPGA evaluation board from X Engineering Software Systems (XESS) Corp. Figure 9.47 shows a picture of the XS40. The XS40 Board connects to the parallel port of a personal computer (PC) via a cable with DB-25 connectors. Having set up the board appropriately, the bitstream file can then be downloaded into the FPGA using the XSTOOLS software. The downloading of the FLC design in FPGA represents a major milestone in the hardware realisation of the control system. The next section shows some experimental test results of the complete system using the FPGA controller.

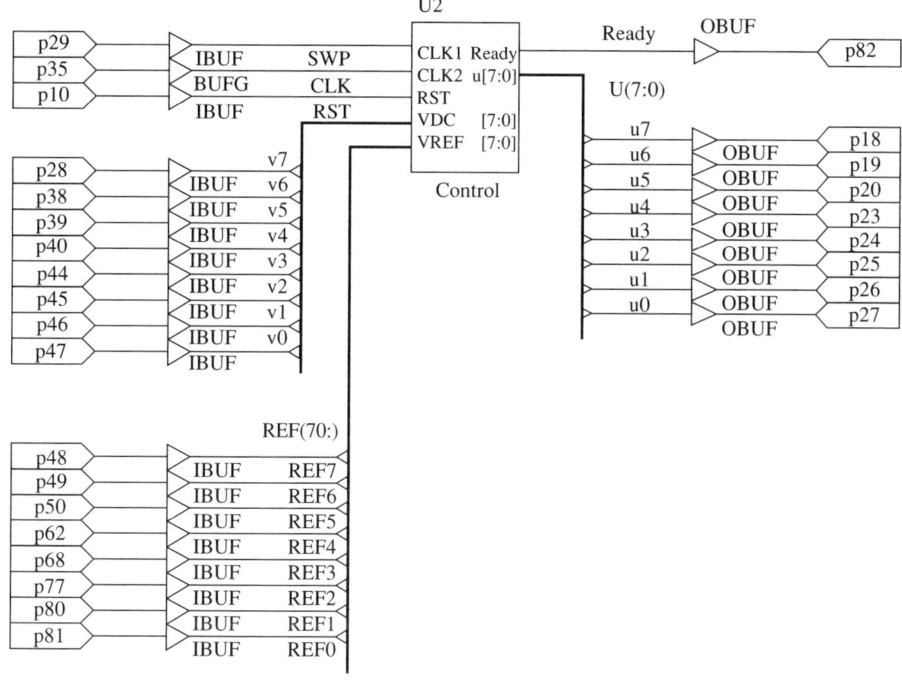


**Fig. 9.46** Xilinx foundation schematic editor design

**Fig. 9.47** Picture of XS40 Board (reproduced with permission of XESS Corporation, USA)

## 9.4 System assembly and experimental tests

A block diagram of the complete test system is shown in Fig. 9.48.

The system can be subdivided into three main sections:

- an *electromechanical system* which comprises the generator, the diesel engine and the electromagnetic actuator

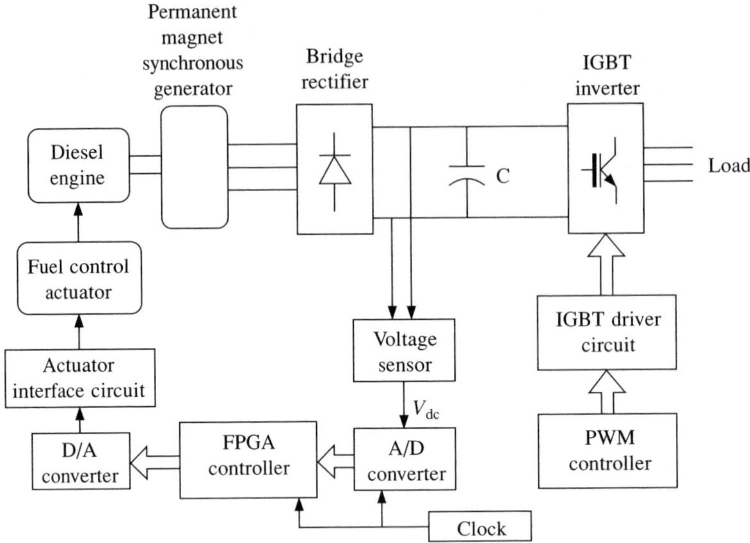


**Fig. 9.48** Block diagram of engine–generator set and controller

- a *power electronic system* consisting of the diode bridge rectifier, d.c. link and IGBT inverter
- an *electronic control system.*

The *electronic control system* can be further divided into two components, namely the PWM controller and a control loop consisting of the sensing and interfacing circuits as well as the FPGA (fuzzy) controller.

### 9.4.1 The fuzzy logic controller (FLC)

The integration of the FLC in the system is illustrated by Fig. 9.49 whereas the main VHDL components of the FLC are shown in Fig. 9.50.

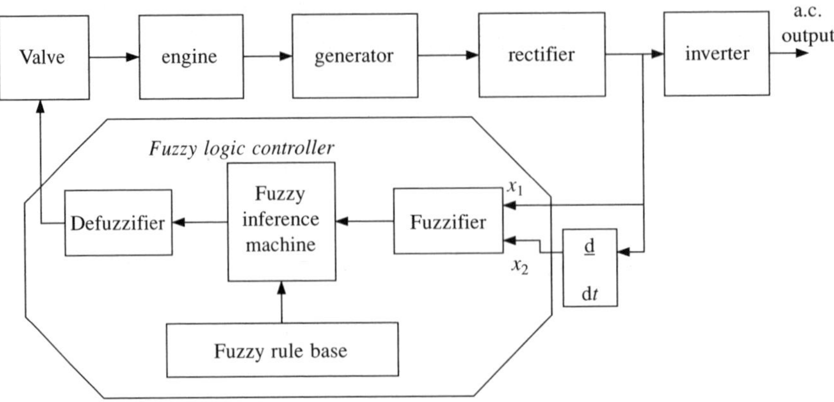


**Fig. 9.49** Block diagram of fuzzy variable speed governor and generator system

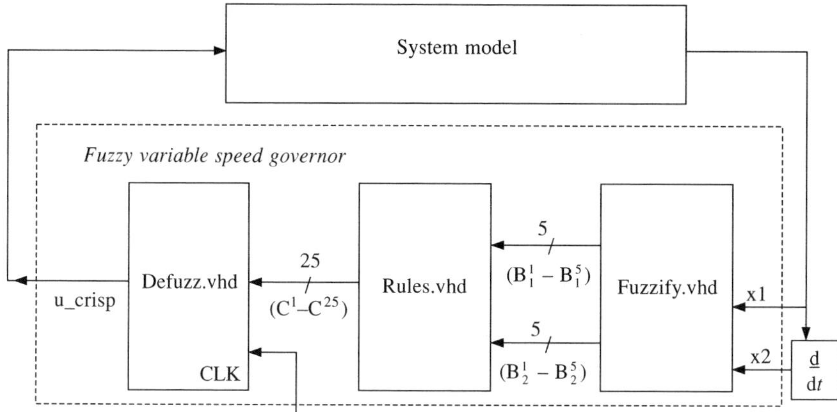


**Fig. 9.50** VHDL entity diagram of FVSG

### 9.4.2 The PWM controller

The PWM control circuit used for testing the system does not form a specific objective for presentation as part of this example because it uses a PWM controller commercially available and not a controller specifically designed for this application. However, this will be briefly presented in this section.

The PWM IGBT inverter is controlled using SA828, a three-phase PWM waveform generator from GEC Plessey. The SA828 has six TTL level PWM outputs that control the six switches in a three-phase inverter via a driver circuit.

#### *9.4.2.1 Functional description*

The PWM fundamental waveform (a standard sinusoidal reference waveform) is stored in an on-chip ROM, storing only the shape of the power waveform and not the actual switching pattern. Calculation of the switching pattern is performed on the chip. The advantage of this approach is the ability to change PWM parameters such as the modulation ratios during the generation of the switching pattern. Using the maximum clock frequency of 12.5 MHz, the triangular carrier frequency can be selected up to 24 kHz. The SA828 is designed to operate in conjunction with an external microprocessor. Values of two groups of registers have to be programmed into the device in order to set the parameters for the desired output waveform. The SA828 is controlled by loading data into two 24-bit registers, the *initialisation register* and the *control register*, via a microprocessor interface. The initialisation register contains all the information on the parameters, which are constant. The parameters set in the initialisation register are:

- Carrier frequency.
- Power frequency range.
- Pulse delay time.
- Pulse deletion time.
- Counter reset.

The data in the initialisation register is loaded only once, during power up, and does not change in the course of the operation. Data in the control register, however, can be modified during the operation of the chip, allowing real-time control of the PWM waveform. The parameters set in the control register are:

- Power frequency.
- Overmodulation.
- Forward/Reverse.
- Output inhibit.
- Power waveform amplitude.

The choices for the PWM parameters are based on the desired output waveform as well as the overall system requirements. The following calculations are used to determine the necessary data for the initialisation and control registers. The clock frequency used is 12 MHz.

#### *9.4.2.2 Initialisation register*

The initialisation register controls five PWM parameters. They are:

***The carrier frequency $f_{carr}$ is set at 11.7 kHz***
This is the frequency of the triangular waveform that is compared to the reference waveform to obtain the PWM switching pattern.

$$f_{\text{carr}} = \frac{f_{\text{clk}}}{512 \times n}$$

$$\Rightarrow n = \frac{f_{\text{clk}}}{512 \times f_{\text{carr}}} = \frac{12 \times 10^6}{512 \times 11.7 \times 10^3} = 2$$

where $f_{\text{carr}}$ = carrier frequency and $f_{\text{clk}}$ = clock frequency.

***The reference frequency range $f_{range}$ is set for 61 Hz***
This parameter sets the maximum reference frequency of the waveform. In variable speed motor applications, this feature prevents the machine from being operated outside its design parameters.

$$f_{\text{range}} = \frac{f_{\text{carr}} \times m}{384}$$

$$\Rightarrow m = \frac{f_{\text{range}} \times 384}{f_{\text{carr}}} = \frac{61 \times 384}{11.7 \times 10^3} = 2$$

***The pulse delay time $t_{delay}$ is set at 5.83 μs***
The pulse delay time is the blanking time between two complementary switching waveforms. Setting a large delay time increases low harmonic distortion to the output waveform. At the same time, the delay has to be large enough to prevent a 'shoot-through' between two complementary power switches. The pulse delay time is determined by a 6-bit word, PDY.

$$t_{\text{delay}} = \frac{pdy}{f_{\text{carr}} \times 512}$$

$$\Rightarrow pdy = t_{\text{delay}} \times f_{\text{carr}} \times 512 = 5.83 \times 10^{-6} \times 11.7 \times 10^3 \times 512 = 35$$

where $pdy$ is the value of the PDY word.

***The pulse deletion time $t_{pd}$ is set at 10 μs***

Theoretically, the pulses in a PWM waveform can be infinitesimally narrow. However, in practice, a narrow pulse may cause problems to the power switches due to storage effects. The pulse deletion time sets a minimum pulse width time. Pulses with widths narrower than the minimum time are eliminated altogether. From the data sheet:

$$t_{pd} = \frac{pdt}{f_{carr} \times 512}$$

$$\Rightarrow pdt = t_{pdt} \times f_{carr} \times 512 = 10 \times 10^{-6} \times 11.7 \times 10^{3} \times 512 = 60$$

PDT is the 7-bit word which determines the pulse deletion time and *pdt* is the value of PDT. From the data sheet, when *pdt* = 60, *PDT* =1001110.

***Counter reset (active low)***

This facility allows the internal reference frequency counter to be set to zero. When the counter reset is active LOW, the internal reference frequency phase counter is set to 0° for the red phase. The counter reset is released when HIGH. For this application the counter is set HIGH all the time.

### 9.4.2.3 Control register

The control register provides on-line control over four parameters. They are:

***Reference frequency (power frequency)***

The reference frequency is the frequency of the inverter's output voltage. The present application does not require variable frequency output. The reference frequency is set at a constant value of 50 Hz by the 8-bit word PFS in the control register. The decimal value of the word is denoted by *pfs*.

$$f_{power} = \frac{f_{range} \times pfs}{4096}$$

$$pfs = \frac{f_{power} \times 4096}{f_{range}} = \frac{50 \times 4096}{61} = 3357$$

From the data sheet, when *pfs* = 3357, *PFS* = 00011011.

***Overmodulation***

Overmodulation is not required, therefore the overmodulation switch OM is set to 0.

***Reference waveform amplitude (power waveform amplitude)***

The reference waveform amplitude is set in percentage of the maximum value. For this application, the reference waveform amplitude is required to be maintained constant even if disturbances cause the d.c. line voltage to change. This means that on-line control of the reference waveform amplitude setting is required to maintain a constant output. However, for initial testing purposes, the amplitude setting of this open-loop control circuitry is set at 80 per cent. Hence the amplitude in percentage is:

$$A_{power} = \frac{A}{255} \times 100\%$$

$$\Rightarrow A = \frac{A_{power} \times 255}{100} = \frac{80 \times 255}{100} = 204$$

where $A$ is the decimal value of the 7-bit word AMP which determines the reference waveform amplitude. From the data sheet, when $A$ = 204, $AMP$ =1001100.

***Inhibit***

Output inhibit is set to '0' during the first control sequence and set to '1' during subsequent control sequences.

From the preceding set of calculations, a table containing the register data can be obtained as shown in Table 9.4. The first column shows the register while the last column shows the hexadecimal value of the register. The present design employs the use of a PIC16x00 series microcontroller to control the SA828. The microcontroller is programmed to feed the necessary signals into the PWM SA828. A listing of the program code is included in Appendix G. This control circuit was successfully implemented and used in the practical experiments.

**Table 9.4** Values for initialisation and control registers

| | | | | | | | | | |
|---|---|---|---|---|---|---|---|---|---|
| Initialisation | | | | | | | | | |
| | CR | $PDT_6$ | $PDT_5$ | $PDT_4$ | $PDT_3$ | $PDT_2$ | $PDT_1$ | $PDT_0$ | |
| R0 | 1 | 1 | 0 | 0 | 1 | 1 | 1 | 0 | CE |
| | $FRS_2$ | $FRS_1$ | $FRS_0$ | X | X | $CFS_2$ | $CFS_1$ | $CFS_0$ | |
| R1 | 0 | 0 | 1 | 1 | 1 | 0 | 0 | 1 | 39 |
| | X | X | $PDY_5$ | $PDY_4$ | $PDY_3$ | $PDY_2$ | $PDY_1$ | $PDY_0$ | |
| R2 | 1 | 1 | 1 | 0 | 0 | 0 | 0 | 1 | E1 |
| Control 1 | | | | | | | | | |
| | $PFS_7$ | $PFS_6$ | $PFS_5$ | $PFS_4$ | $PFS_3$ | $PFS_2$ | $PFS_1$ | $PFS_0$ | |
| R0 | 0 | 0 | 0 | 1 | 1 | 0 | 1 | 1 | 1B |
| | F/R | OM | INH | X | $PFS_{11}$ | $PFS_{10}$ | $PFS_9$ | $PFS_8$ | |
| R1 | 0 | 0 | 0 | 1 | 1 | 1 | 0 | 1 | 1D |
| | $AMP_7$ | $AMP_6$ | $AMP_5$ | $AMP_4$ | $AMP_3$ | $AMP_2$ | $AMP_1$ | $AMP_0$ | |
| R2 | 1 | 1 | 0 | 0 | 1 | 1 | 0 | 0 | CC |
| Control 2 | | | | | | | | | |
| | $PFS_7$ | $PFS_6$ | $PFS_5$ | $PFS_4$ | $PFS_3$ | $PFS_2$ | $PFS_1$ | $PFS_0$ | |
| R0 | 0 | 0 | 0 | 1 | 1 | 0 | 1 | 1 | 1B |
| | F/R | OM | INH | X | $PFS_{11}$ | $PFS_{10}$ | $PFS_9$ | $PFS_8$ | |
| R1 | 0 | 0 | 1 | 1 | 1 | 1 | 0 | 1 | 3D |
| | $AMP_7$ | $AMP_6$ | $AMP_5$ | $AMP_4$ | $AMP_3$ | $AMP_2$ | $AMP_1$ | $AMP_0$ | |
| R2 | 1 | 1 | 0 | 0 | 1 | 1 | 0 | 0 | CC |

## 9.4.3 Experimental results

The system was initially tested under a 'step change in rectifier load' condition. The objective of this test was to check if the control system is capable of maintaining the d.c. voltage at a desired level under a step change in loading condition. The output of the generator is connected to a rectifier, which is loaded with a set of resistive banks. The load is changed from 119 Ω to 37.5 Ω and the voltage response is monitored. The clock frequency of the controller in this test is 1.2 kHz. For the purpose of comparison, an initial test is carried out without the control system, whereby the actuator is set at a fixed position. The result shown in Fig. 9.51 confirms that the generator system is not intrinsically

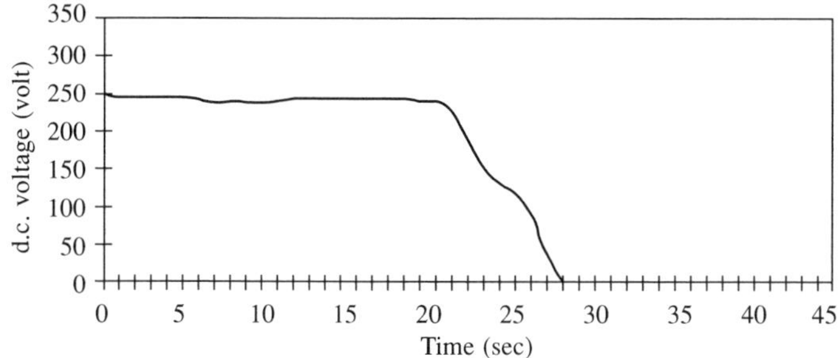


**Fig. 9.51** Voltage response to d.c. load step variation without controller

stable under changing load; when the load changes, it can be observed that the operational speed drops steadily until the engine finally stalls. Figure 9.52 shows the voltage response when the controller is connected to the system. The desired d.c. voltage is set at 250 V. The graph shows that the controller is successful in stabilising the generator system. Although there is a voltage drop of about 14 per cent when the load changes, this effect is soon counteracted by the controller and the voltage level recovers to a steady value.

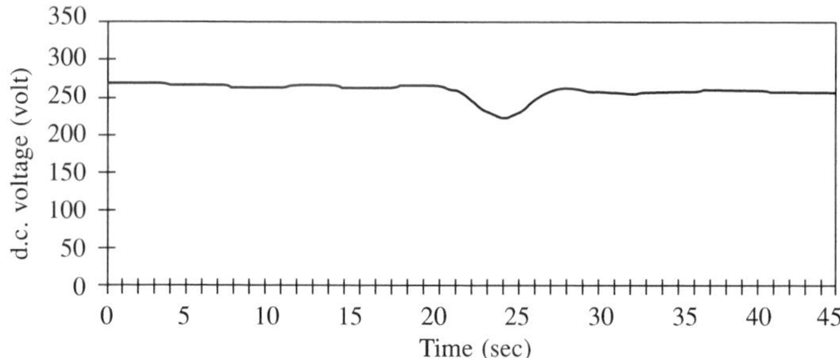


**Fig. 9.52** Voltage response to d.c. load step variation with control system

Another test was that of a step variation of inverter load. The inverter was loaded with a star connected resistive load at time $t = 10$ s. Figure 9.53 shows the response of the d.c. link voltage when the controller is connected. The results show that the control system manages to maintain the d.c. voltage level constant despite load changes. The clocking frequency value used for test is 1.2 kHz. The load current measured on the d.c. link is 5 A. It was observed that the output frequency of the inverter is maintained at 50 Hz during the entire operation, while there are variations in the generator speed, depending on the load and the actions taken by the controller. These speed variations do not affect the output frequency.

From the experimental results presented it can be concluded that the controller is successful at governing the d.c. voltage level over different loading conditions. The

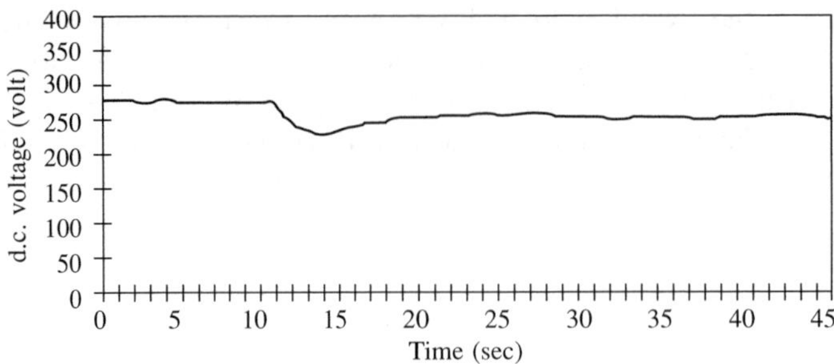


**Fig. 9.53** Voltage response to a.c. load step variation with control system

controller coped well with rectifier load changes as well as inverter load changes. When the complete system is considered, the a.c. output frequency is not affected by the variations of the generator speed and the desired output frequency is maintained constant. This feature allows the variable speed operation of the generator set.

### 9.4.4 Conclusions

A fuzzy controller for the diesel driven stand-alone generator system was designed using VHDL, implemented into a Xilinx FPGA XC4010E device and comprehensively tested in conjunction with the power system. The main achievements of the system are:

- The output voltage and frequency are independent of the rotational speed of the generator, thus allowing the optimum speed operation of the diesel engine.
- Fuel economy is achieved by the use of the fuzzy logic controller.
- The PWM control system maintains the output voltage at the desired magnitude and frequency against changes in $V_{dc}$ which arise from changes in speed and/or load.
- The FPGA controller provides a reliable hardware framework for design verification.
- The system provides a suitable platform for extensive study and development of efficient diesel engine driven variable speed generator systems.

The reconfigurability of the controller design is achieved at two levels: the easy modification of the rule base by simple changes of the VHDL code, and the reprogramming capability of the FPGA. These features have enabled extensive practical tests to be carried out for a number of different operating conditions at no extra cost.

# Final notes

The VHDL approach to the holistic modelling, simulation, design and controller implementation of complex power systems including digital controllers enables a unique working environment, a short development time, multiple choices for the implementation target technology and universal compatibility of the design with respect to a range of existing modern CAD tools.

Specific digital electronic modules, implementing fast and effective neural and fuzzy logic control elements, can be included in complex control systems as intellectual property (IP) blocks and reused in accordance with modern principles of design reuse.

Due to these advantages, the authors estimate that such methodologies, based on hardware description languages, will be used on a larger scale in the future for digitally controlled power systems modelling, design and analysis.

# 10

# References


[1] *Analogue and Mixed Signal Catalogue*, Texas Instruments, copyright 1999.

[2] *Designing with Powerview/WorkviewPLUS*, Viewlogic Systems, Inc., 293 Boston Post Road West, Marlboro, Massachusetts 01752-4615, copyright 1993.

[3] *Digital Analysis with Powerview/WorkviewPLUS*, Viewlogic Systems, Inc., 293 Boston Post Road West, Marlboro, Massachusetts 01752-4615, copyright 1993.

[4] *FH2/3 MkIV Electrical Machines Teaching System*, TecQuipment, Bonsall Street, Long Eaton, Nottingham, NG10 2AN, UK, 1997.

[5] *Foundation Series Quick Start Guide 1.4*, Xilinx, Inc., 1997.

[6] *Technical Reference Manual*, Newage Intern. Ltd, Section 4131, Jan. 1990, p.1.

[7] *Semikron Innovation + Service*, Semikron International, Sigmundstr. 200, D-90431, Nuremberg, Germany, copyright 1998.

[8] *The Programmable Logic Data Book*, Xilinx, Inc., 2100 Logic Drive, San José, California 95124, USA, copyright 1999.

[9] *TMS320C5x – User's Guide*, Texas Instruments, Inc., 1992.

[10] *IEEE Standard VHDL Language Reference Manual*, IEEE standard 1076-1993.

[11] *Energy Savings with Electric Motors and Drives*, guide provided by Energy Efficiency Enquiries Bureau, Crown copyright 1998.

[12] *Intel 80170NX ETANN Data Sheets*, Feb. 1991.

[13] *Sensorless Control with Kalman Filter on TMS320 Fixed-Point DSP*, Literature Number: BPRA057, Texas Instruments Europe, July 1997 (www.ti.com).

[14] *Foundation Series Quick Start Guide 1.4*, Xilinx, Inc., 1991–1997.

[15] *VHDL SelfStart_Kit*, Topdown Design Solutions, 1995.

[16] Adler, M.S. *et al.*: 'The Evolution of Power Device Technology', *IEEE Trans. Electron Devices*, vol. 31, no. 11, Nov. 1984, pp. 1570–1591.

[17] Akmese, R., Chalmers, B.J.: 'Permanent Magnet Alternators with Constant Output Voltage', *Proc. Universities Power Eng. Conf. UPEC'97*, vol. 1, 10–12 Sept. 1997, pp. 34–36.

[18] Alolah, A.I.: 'Steady State Operating Limits of Three Phase Self-excited Reluctance Generator', *IEE Proc. C*, vol. 139, no. 3, May 1992, pp. 261–268.

[19] Alspector, J., Allen, R.B., Jayakumar, A., Zeppenfeld, T., Meir, R.: 'Relaxation Networks for Large Supervised Learning Problems', in *Advances in Neural Information Processing Systems 3* (Lippmann, R.P., Moody, J.E., Touretzky, D.S., eds), Morgan Kaufmann, San Mateo, CA, 1991, pp. 1015–1021.

[20] Al-Tayie, J.K., Acarnley, P.P. 'Estimation of Speed, Stator Temperature and Rotor Temperature in Cage Induction Motor Drive using the Extended Kalman Filter Algorithm', *IEE Proceedings on Electric Power Applications*, vol. 144, no. 5, Sept. 1997, pp. 301–309.

[21] Andreou, A.G., Boahen, K.A., Pouliqueen, P.O., Jenkins, R.E., Strohbehn, K. 'Current-mode Subthreshold MOS Circuits for Analogue VLSI Neural Systems', *IEEE Transactions on Neural Networks*, vol. 2, 1991, pp. 205–213.

[22] Armstrong, W.W., Gecsei, J.: 'Adaption Algorithms for Binary Tree Networks', *IEEE Transactions on System, Man, Cybernetics*, no. 9, 1979, pp. 276–285.

[23] Atkinson, D.J., Hopfensperger, B., Lakin, R.A.: 'Field Oriented Control of a Doubly-fed Induction Machine using Coupled Microcontrollers', EPE'99 Lausanne (CD).

[24] Aubebart, F., Girerd, C., Chapuis, Y.A., Poure, P., Braun, F.: 'ASIC Implementation of Direct Torque Control for Induction Machines: Functional Validation by Power and Control Simulation', PCIM'98, Power Converter and Intelligent Motion Conference, Nuremberg, 25–28 May 1998, pp. 251–260.

[25] Ayestaran, H.E., Prager, R.W.: 'The Logical Gates Growing Network', Technical Report CUED/F-INFENG/TR 137, Engineering Dept., Cambridge University, 1993.

[26] Baliga, B.J. et al.: 'The Insulated Gate Transistor (IGT) – A New Power Switching Device', *IEEE/IAS Ann. Meet. Conf. Rec.*, 1983, pp. 794–803.

[27] Bartos, F.J.: 'Sensorless Vector Drives Strive for Recognition', *Control Engineering*, Sept. 1996.

[28] Bashagha, A.E., Ibrahim, M.K.: 'A New Digit-serial Divider Architecture', *International Journal on Electronics*, vol. 75, no. 1, pp. 133–140.

[29] Bashagha, A.E., Ibrahim, M.K.: 'A New High Radix Non-restoring Divider Architecture', *International Journal on Electronics*, vol. 79, no. 4, 1995, pp. 455–470.

[30] Bashagha, A.E., Ibrahim, M.K.: 'Design of a Square-root Architecture: Digit Serial Approach', *International Journal on Electronics*, vol. 76, no. 1, 1994, pp. 15–25.

[31] Bashagha, A.E., Ibrahim, M.K.: 'Digit-serial Squaring Architecture', *Journal of Circuits, Systems and Computers*, vol. 4, no. 1, 1994, pp. 99–108.

[32] Bashagha, A.E., Ibrahim, M.K.: 'Nonrestoring Radix-$2^k$ Square Rooting Algorithm', *Journal of Circuits, Systems and Computers*, vol. 6, no. 3, 1996, pp. 267–285.

[33] Baum, E.B.: 'On the Capabilities of Multilayer Perceptrons', *J. Compl.*, no. 4, 1988, pp. 193–215.

[34] Beierke, S., Konigbauer, R., Krause, B., Altrock, C.V.: 'Fuzzy Logic Enhanced Control of AC Motor Using DSP', *Embedded Systems Conference California*, 1995, pp. 101–106.

[35] Beiu, V.: 'Entropy, Constructive Neural Learning and VLSI Efficiency', in *Proceedings of NEUROTOP'97: Neural Priorities on Data Transmission and EDA*, Brasov Romania, 22–30 May 1997, pp. 38–74.

[36] Beiu, V.: *VLSI Complexity of Discrete Neural Networks*, Gordon and Breach & Harwood Academics Publishing, 1998.

[37] Beiu, V., Taylor, J.G.: 'Optimal Mapping of Neural Networks onto FPGAs', 'Natural to Artificial Neural Computation', Lecture Notes in CS930, Springer-Verlag, Berlin, June 1995.

[38] Ben-Brahim, L.: 'Motor Speed Identification Via Neural Networks', *IEEE Industry Applications Magazine*, vol. 1, no. 1, Jan./Feb. 1995, pp. 28–32.

[39] Binns, K.J., Kurdali, A.: 'Permanent Magnet A.C. Generator', *Proc. IEE*, vol. 126, no. 7, July 1979, pp. 690–696.

[40] Blayliss, J.: '*Electronic Design Automation Report*,' Cambridge Market Intelligence, 1994.

[41] Bleijs, J.A.M., Wong, K.J.: 'Control of a Variable Speed Wind Turbine with a Synchronous Generator and Boost Rectifier,' *Proc. 37th Universities Power Eng. Conf. UPEC'97*, 10–12 Sept. 1997, pp. 722–725.

[42] Bleijs, J.A.M.: 'Improvement in Performance of a Passive Pitch Wind Turbine with Variable Speed Operation', *Proc. of 5th Eur. Wind Energy Assoc. Conf. EWEC*, Thessaloniki, Greece, vol. 1, Oct. 1994, pp. 588–592.

[43] Boldea, I., Nasar, S.A.: '*Vector Control of A.C. Drives*', CRC Press, Boca Raton, 1992.

[44] Boost, M.A., Ziogas, P.D.: 'State-of-the-art Carrier PWM. Techniques: A Critical Evaluation', *IEEE Trans. Ind. Applicat.*, vol. 24, no. 24, March/April 1988, pp. 271–280.

[45] Bose, B.K.: 'Expert System, Fuzzy Logic and Neural Network Applications in Power Electronics and Motion Control', *Proc. IEEE*, vol. 82, Aug. 1994, pp.1303–1323.

[46] Bose, B.K.: 'High Performance Control of Induction Motor Drives', *IEEE Industrial Electronics Society Newsletter*, Sept. 1999, pp. 7–11.

[47] Bose, N.K., Garga, A.K.: 'Neural Network Design Using Voronoi Diagrams', *IEEE Transactions on Neural Networks*, vol. 4, no. 5, Sept. 1993, pp. 778–787.

[48] Bowes, S.R., Mount, M.J.: 'Microprocessor Control of PWM Inverters', *IEEE Trans. Ind. Applicat.*, vol. 128, no. 6, 1981, pp. 293–305.

[49] Brod, D.M., Novotny, D.W.: 'Current Control of VSI-PWM Inverters', *IEEE-IAS Conf. Rec.*, 1984, pp. 418–425.

[50] Butcher, J.C.: *The Numerical Analysis of Ordinary Differential Equations – Runge–Kutta and General Linear Methods*, John Wiley & Sons, 1987.

[51] Carter, D.E., Stilwell-Baker, B.: *Concurrent Engineering: The Product Development Environment for the 1990s*, Mentor Graphics, 1991.

[52] Carter, D.E., Sullivan, T.N.: *Concurrent Engineering: Best Practices for Global Success*, Mentor Graphics, 1994.

[53] Cecati, C., Rotondale, N.: 'Torque and Speed Regulation of Induction Motors Using the Passivity Theory Approach', *IEEE Transactions on Industrial Applications*, vol. 46, no.1, Feb. 1999, pp. 119–127.

[54] Chan, T.F.: 'Self-excited Induction Generators Driven by Regulated and Unregulated Turbines', *IEEE Trans. on Energy Conversion*, vol. 11, no. 3, Sept. 1996, pp. 338–343.

[55] Chapuis, Y.A., Poure, P., Braun, F.: 'Torque Dynamic Correction of Direct Torque Control for Induction Machine Using a DSP', *PCIM'98, Power Converter and Intelligent Motion Conference*, Nuremberg, 25–28 May 1998, pp. 241–250.

[56] Chapuis, Y.A., Roye, D.: 'Optimization of Square Wave Transition for Direct Torque Control of Induction Machine', *Proceedings of the Intelligent Motion Conference*, Nuremberg, 10–12 June 1997.

[57] Chen, Z., Spooner, E.: 'Grid Interface Options for Variable Speed Permanent Magnet Generators', *IEE Proc. Electr. Power Appl.*, vol. 145, no. 4, July 1998, pp. 273–282.

[58] Churcher, D., Baxter, D.J., Hamilton, A., Murray, A.F., Reekie, H.M.: 'Generic Analog Neural Computation – The EPSILON chip', in *Advances in Neural Information Processing Systems*, (Hanson, S.J., Cowan, D.J., Giles, C.L. eds.), Morgan Kaufmann 1993, San Mateo, CA: pp. 773–780.

[59] Cilia, J., Asher, G.M., Bradley, K.J.: 'Sensorless Position Detection for Vector Controlled Induction Motor Drives Using an Asymmetric Outer-section Cage', *IEEE Trans. on IAS*, vol. 33, no. 5, Sept./Oct. 1997, pp. 1162–1169.

[60] Cirstea, M.N.: 'An investigation into ASIC Control of a 6-pulse Cycloconverter Quad-winding Induction Motor', PhD Thesis, Nottingham Trent University, 1996.

[61] Cirstea, M.N., Dinu, A., Nicula, D.: *A Practical Guide to VHDL Design*, Editura Tehnica, Bucharest, Romania, 2001, ISBN: 973–31-1539–8.

[62] Cirstea, M.N., Patterson, E.B., Morley, D., Koczara, W., Przybylski, J.: 'Electronic Design Automation Techniques Transferred to Central/Eastern Europe', 'Advances in Design and Manufacturing' – Concerted Efforts for Europe – European Commission DG iii, 1995, vol. 6, Ch. 43, pp. 85–92.

[63] Cirstea, M.N., Patterson, E.B., Holmes, P.G., Morley, D.: 'A New Technique to Model and Simulate Divided Winding Induction Motors', *Proc. of IEE Int. Conf. on Computation in Electromagnetics (CEM'96)*, Bath, UK, April 1996, pp. 194–199.

[64] Cirstea, M.N., Patterson, E.B., Morley, D., Holmes, P.G.: 'A Complete ASIC Controlled Electric Drive System', *Proc. of the International Symposium on Circuits and Systems* (ISCAS'96 – IEEE sponsored), Atlanta, Georgia, USA, May 1996, vol. 1, pp. 561–564.

[65] Cirstea, M.N., Giamusi, M., McCormick, M.: 'A Modern ASIC Controller for a 6-pulse Rectifier', *Proceedings of the ASIC'97 IEEE Conference*, Portland, Oregon, USA, 7–10 Sept., 1997, pp. 335–338.

[66] Cirstea, M.N.: 'ASIC Control System for Electric Drives', *Proc. of IEE/IEEE International Symposium of Integrated Circuits – (ISIC'97)*, Singapore, 11–12 Sept. 1997, pp. 621–624.

[67] Cirstea, M.N., McCormick, M.: 'Intelligent ASIC Control of Power Electronics', *Proc. of the*

*International Conference on Power Conversion and Intelligent Motion (PCIM'98)*, Nuremberg, Germany, 26–28 May 1998, pp. 95–99.

[68] Cirstea, M.N., McCormick, M.: 'Control Systems for Power Electronics', *Proc. of the IEE International Conf. on Simulation (SIMULATION'98)*, York, UK, Oct. 1998, pp.182–186.

[69] Cirstea, M.N., Khor, J.G., McCormick M., Haydock, L.: 'VHDL Design of an Intelligent Fuzzy Logic Controller for Synchronous Generator Sets Implemented in FPGA', *Proc. of the International HDL Conference*, Santa Clara, California, USA, April 1999, pp. 43–47.

[70] Cirstea, M.N., Khor, J.G., Hu, Y., McCormick, M., Haydock, L.: 'Intelligent Fuzzy Logic Controller for Power Generation Systems', *Proc. of the 9th International Conference on Electrical Machines and Drives*, Canterbury, UK, Sept. 1999, pp. 321–324.

[71] Cirstea, M.N., Aounis, A., Dinu, A., McCormick, M.: 'A VHDL Approach to Induction Motor Modelling', *Proc. of IEE CONTROL '00*, Cambridge, UK, Sept. 2000, CDROM.

[72] Cirstea, M.N., Dinu, A., McCormick, M., Nicula, D.: 'A VHDL Success Story: Electric Drive System using Neural Controller', *Proc. of the IEEE VIUF/BMAS Fall Conference*, Orlando, Florida, Oct. 2000, pp. 118–122.

[73] Cirstea, M.N., Dinu, A., McCormick, M.: 'A New Neural Networks Approach to Induction Motor Speed Control', *Proc. of the IEEE Power Electronics Specialists Conference (PESC)*, Vancouver, Canada, 18–22 June 2001, CD-ROM.

[74] Cirstea, M.N., Khor, J.G., McCormick, M.: 'FPGA Fuzzy Logic Controller for Variable Speed Generators', *Proc. of IEEE International Conference on Control Applications (CCA)*, Mexico City, 4–7 Sept. 2001, CDROM.

[75] Comnac, V., Cirstea, M.N., Giamusi, M., Cernat, R.M.: 'Torque and Speed Control of Inverter-fed Interior Permanent Magnet Synchronous Motor Using Sliding Mode', *Proc. of Int. European Conf. on Power Electronics & Applications (EPE)*, Lausanne, Switzerland, 1999, CDROM.

[76] Conradi, P.: *Reuse in Electronic Design: from Information Modelling to Intellectual Properties*, John Wiley & Son Ltd, 1999.

[77] Cybenko, G.: 'Approximations by Superposition of a Sigmoidal Function', *Mathematics of Control, Signal and Systems*, vol. 2, 1989, pp. 303–314.

[78] De Mello, F.P., Hannet, L.N.: 'Large Scale Induction Generators for Power Systems', *IEEE Trans. on Power Apparatus & Systems*, PAS 100, 1981, pp. 2610–2618.

[79] DeDoncker, R., Novotny, D.W.: 'The Universal Field Oriented Controller', *IEEE-IAS Trans.*, vol. 30, no. 1, Jan./Feb. 1994, pp. 92–100.

[80] Den Bout, D.V.: *The Practical Xilinx Designer Lab Book*, Prentice Hall, 1998.

[81] Dinu, A.: 'FPGA Neural Controller for Three Phase Sensorless Induction Motor Drive Systems', PhD Thesis, De Montfort University, 2000.

[82] Dinu, A., Cirstea, M.N., McCormick, M., Ometto, A., Rotondale, N.: 'An Adaptive Control Strategy for Electric Drives', *Proc. of the International Conference on Power Conversion and Intelligent Motion (PCIM'98)*, Nuremberg, Germany, 26–28 May 1998, pp. 101–107.

[83] Dinu, A., Cirstea, M.N., McCormick, M.: 'Virtual Prototyping of a Digital Neural Current Controller', *Proc. of the IEEE International Workshop on Rapid Systems Prototyping*, Leuven, Belgium, 2–5 June 1998, pp. 176–181.

[84] Dinu, A., Cirstea, M.N., McCormick, M., Ometto, A., Rotondale, N.: 'Load Independent Current Control Strategy for PWM Inverters', *Proc. of the IEE CONTROL'98*, Swansea, UK, 1–4 Sept. 1998, vol. 2, pp.1118–1122.

[85] Dinu, A., Cirstea, M.N., McCormick, M., Ometto, A., Rotondale, N.: 'Neural ASIC Controller for PWM Power Systems', *Proc. of the 11th IEEE International ASIC Conference*, Rochester, USA, 13–16 Sept. 1998, pp. 29–33.

[86] Dinu, A., Cirstea, M.N., McCormick, M., Haydock, L., Al-Khayat, N.: 'Neural Current Controller for Induction Motor Applications', *Proc. of Int. Conf. on Optimization of Electric & Electronic Equipment (OPTIM-IEE, IEEE)*, Brasov, Romania, May 2000, pp. 665–670. Award: 3rd ABB prize!

[87] Dinu, A., Cirstea, M.N., McCormick, M., Ometto, A., Rotondale, N.: 'Sensorless Induction

Motor Control Strategy Optimised for FPGA Hardware Implementation', *Journal of Electrical Engineering*, vol.1, no. 1, 2001, pp. 26–31, ISSN 1582–4594.

[88] Driankov, D., Hellendoorn, H., Reinfrank, M.: *An Introduction to Fuzzy Control*, Springer-Verlag, Berlin Heidelberg, 1993.

[89] Dwyer, R.A.: 'High-dimensional Voronoi Diagrams in Linear Expected Time', *Discr. Comput. Geom.*, vol. 6, 1991, pp. 343–367.

[90] Eichmann, G., Caulfield, H.J.: 'Optical Learning (Inference) Machine', *Applied Optics*, no. 24, 1985, pp. 378.

[91] Eldridge, S.E., Walter, C.D.: 'Hardware Implementation of Montgomery's Modular Multiplication Algorithm', *IEEE Transactions on Computers*, vol. 42, 1993, pp. 693-699.

[92] EL-Sharkawi, M., Neibur, D. (eds): 'Artificial Neural Networks Applied to Power Systems', IEEE Power Engineering Society tutorial course, IEEE catalogue number 96 TP 112–0, 1996.

[93] Evans, W.R.: *Graphical Analysis of Control Systems*, Trans. AIEE, 1948.

[94] Farhat, N.H., Psaltis, D., Prata, A., Paek, E.: 'Optical Implementation of the Hopfield Model', *Applied Optics*, no. 24, 1985, p. 339.

[95] Farsi, M., Zachariah, K.J., Finch, J.W.: 'Implementation of a Self-tuning AVR', *IEE Proc. – Control Theory Appl.*, vol. 144, no. 1, Jan. 1997, pp. 32–39.

[96] Fodor, D., Vas, J., Katona, Z.: 'Fuzzy Logic Based Vector Control of AC Motor Using Embedded DSP Controller Board', *Proc. of PCIM'98 – Intelligent Motion*, pp. 235–240.

[97] Foussier, P., Calmon, F., Carrabina, J., Fathallah, M., Grennerat, V., Jorda, X., Gontrand, C., Retif, M.J., Chante, J.-P.: 'Practical Example of Algorithm Integration for Electrical Drives', *EPE'99 Lausanne* (CDRom).

[98] Ghosh, A., Ledwich, G., Malik, O.P., Hope, G.S.: 'Power Systems Stabilisers Based on Adaptive Control Techniques', *IEE Trans. on Power Apparatus and Systems*, vol. PAS-103, no. 8, 1984, pp. 1983–1989.

[99] Godhwani, A., Basler, M.J.: 'A Digital Excitation Control System for Use on Brushless Excited Synchronous Generators', *IEEE Trans. on Energy Conversion*, vol. 11, no. 3, Sept. 1996. pp. 616–619.

[100] Goslin, G.R.: 'Using Xilinx FPGAs to Design Custom Digital Signal Processing Devices', *Proceedings of the 1995 DSP Technical Program*, pp. 595–604.

[101] Goslin, G., Newgard, B.: '*16-Tap, 8-bit FIR Filter Application Guide*', Xilinx Inc., Nov., 1994.

[102] Gottlieb, I.M.: '*Power Supplies, Switching Regulators, Inverters and Converters*', 1st edition, Blue Ridge Summit, PA, 1984.

[103] Gottlieb, I.M.: *Practical Power-control Techniques*, 1st edition, Indianapolis, IN, USA: H.W. Sams, 1987.

[104] Gottlieb, I.M.: *Regulated Power Supplies*, 4th edition, Blue Ridge Summit, PA, 1992.

[105] Grzesiak, L., Beliczynski, B.: 'Simple Neural Cascade Architecture for Estimating of Stator and Rotor Flux', *EPE'99* Lausanne (CDRom).

[106] Habetler, T.G., Divan, D.M.: 'Acoustic Noise Reduction in Sinusoidal PWM Drives Using a Randomly Modulated Carrier', *IEEE Trans. on Power Electronics*, vol. 6, July 1991, pp. 356–363.

[107] Habetler, T.G.: 'A Space Vector-based Regulator for AC/DC/AC Converters', *IEEE Trans. on Power Electronics*, vol 8, no. 1, 1993, pp. 30–36.

[108] Hammerstrom, D.: 'The Connectivity Analysis of Simple Association – or – How Many Connections Do You Need', *Proceedings NIPS'87* (Denver, USA), Amer. Inst. Phys., 1987, pp. 338–347.

[109] Harrer, H., Nossek, J.A., Stelzl, R.: 'An Analog Implementation of Discrete-time Cellular Neural Networks', *IEEE Transactions on Neural Networks*, vol. 3, no. 3, May 1992, pp. 466–476.

[110] Haykin, S.: *Neural Networks – a Comprehensive Foundation*, Macmillan College Publishing Company Inc., 1994.

[111] Hazen, H.L.: *Theory of Servomechanism*, J. Franklin Inst., 1934.

[112] Heht-Nielsen, R.: *Neurocomputing. Reading*, Addison-Wesley Publishing Co., 1990.

[113] Hilloowala, R.M., Sharaf, A.M.: 'A Rule-based Fuzzy Logic Controller for a PWM Inverter in a Stand Alone Wind Energy Conversion Scheme', *IEEE Trans. on Industry Appl.*, vol. 32, no. 1, 1996, pp. 57–65.

[114] Hirayama, T.: 'Robustness of Fuzzy Logic Power System Stabilisers Applied to Multimachine Power System', *IEEE Trans. Energy Conversion*, vol. 9, no. 3, 1994, pp. 451–459.

[115] Hirayama, K. *et al.*: 'Digital A.V.R. Application to Power Plants', *IEEE Trans. on Energy Conversion*, vol. 8, no. 4, Dec. 1993.

[116] Hmaifar, A., McCormick, E.: 'Simultaneous Design of Membership Functions and Rule Sets for Fuzzy Controllers Using Genetic Algorithms', *IEEE Trans. Fuzzy Syst.*, vol. 3, no. 2, May 1995, pp. 129–139.

[117] Holtz, J.: 'Pulsewidth Modulation – A Survey', *IEEE Trans. Ind. Electron.*, vol. 39, no. 5, 1992, pp. 410–420.

[118] Holtz, J.: 'Speed Estimation and Sensorless Control of AC Drives', *Proceedings of IECON'93*, IEEE-Industrial Electronics Society, 1993, pp. 649–654.

[119] Holtz, J., Stadtfeld, S.: 'A Predictive Controller for the Stator Current Vector of AC Machines Fed from Switched Voltage Source', *JIEE IPEC-Tokyo Conf.*, 1983, pp. 1665–1675.

[120] Hopcroft, J.E., Mattson, R.L.: 'Synthesis of Minimal Threshold Logic Networks', *IEEE Trans. on Electr. Comp., EC-6*, 1965, pp. 552–560.

[121] Hornik, K., Stinchcombe, M., White, H.: 'Multilayer Feedforward Networks are Universal Approximators', *Neural Networks*, no. 2, 1989, pp. 359–366.

[122] Hu, W.Y., Zhong, L., Rahman, M.F., Lim, K.W.: 'A Fuzzy Observer for Stator Resistance for Application in Direct Torque Control of Induction Motor Drives', *Proc. of the Second International Conference on Power Electronics and Drives (PEDS'97)*, 26–29 May, 1997 Singapore, vol. 1, pp. 91–96.

[123] Hu, Y.W., Rahman, M.F. *et al.* 'Direct Torque Control of Induction Motor Using Fuzzy Logic', *Canadian Conference on Electrical and Computer Engineering*, St John's, Newfoundland, Canada, 25–28 May 1997, vol. 2, pp. 767–772.

[124] Huber, D.W., Runtz, K.J., Malik, O.P., Hope, G.S.: 'Analytical and Experimental Studies of a Digital A.V.R.', *IEEE 8th PICA Conf. Proc.*, 1973, pp 195–203.

[125] Hui, S.Y.R., Oppermann, I., Sathiakumar, S.: 'Microprocessor-based Random PWM Schemes for D.C.–A.C. Power Conversion', *IEEE Trans. on Power Electronics*, vol. 12, no. 2, March 1997, pp. 253–260.

[126] Hwang, B., Saif, M., Jamshidi, M.: 'A Neural Network Based Fault Detection and Identification (FDI) for a Pressurized Water Reactor', *Proceedings of the 12th IFAC World Congress on Automatic Control*, Sydney, Australia, 1993.

[127] Hwang, B.C., Saif, M., Jamshidi, M.: 'Fault Detection and Diagnosis of a Nuclear Power Plant Using Artificial Neural Networks', *Journal of Intelligent and Fuzzy Systems*, vol. 3, no. 3, 1995, pp. 197–213.

[128] Hwang, H.S.: 'Automatic Design of Fuzzy Rule Base for Modelling and Control Using Evolutionary Programming', *IEE Proc. Contr. Theory Appl.*, vol. 146, no. 1, Jan. 1999, pp. 9–16.

[129] Ibrahim, A.S., Hogg, B.W., Sharaf, M.M.: 'Self-tuning Automatic Voltage Regulators for a Synchronous Generator', *IEE Proc. Pt. D*, vol. 136, no. 5, Sept. 1989, pp. 252–260.

[130] Ibrahim, M.K., Bashagha, A.E.: 'Area-Time Efficient Two's Complement Square Rooting', *International Journal on Electronics*, vol. 86, no. 2, 1999, pp. 127–140.

[131] Irwin, G.W., Warwick, K., Hunt, K.J.: '*Neural Network Applications in Control*', Institution of Electrical Engineers, London, 1995.

[132] Javurek, J: 'Possibilities of Improving the Method of Direct Control of Asynchronous Machine Torque', *Automatizace Journal* (Czech Republic), no. 12, 1997, pp. 789–794.

[133] Jeong, S.G., Myung-Ho, W.: 'DSP Based Active Power Filter with Predictive Current Control', *IEEE Trans. on Industrial Electronics*, vol. 44, No. 3, June 1997

[134] Jones, R., Smith, G.A.: 'High Quality Mains Power Form Variable Speed Wind Turbines', *Int. Conf. on Renewable Energy – Clean Power*, 2001, IEE Conf. Publ. No. 385, 17–19 Nov. 1993, pp. 202–206.

[135] Jönsson, R.: 'Natural Field Orientation (NFO) Provides Sensorless Control of AC Induction Servo Motors', *PCIM Magazine*, June 1995, pp. 10–17.

[136] Jönsson, R., Leonhard, W.: 'Control of an Induction Motor without a Mechanical Sensor, based on the Principle of "Natural Field Orientation" (NFO)', *International Power Electronics Conference IPEC*, Yokohama, 1995, pp. 101–106.

[137] Jung, S.L., Tzou, Y.Y.: 'Discrete Sliding Mode Control of a PWM Inverter for Sinusoidal Output Waveform Synthesis with Optimal Sliding Curve', *IEEE Trans. on Power Electronics*, vol. 11, no. 4, July 1996, pp. 567–577.

[138] Kalman, R.E.: 'A New Approach to Linear Filtering and Prediction Problems', *Trans. ASME (J. Basic Engineering)*, vol. 82D, no. 1, March 1960, pp. 35–45.

[139] Kalman, R.E., Bucy, R.S.: 'New Results in Linear Filtering and Prediction Theory', *Trans. ASME (J. Basic Engineering)*, vol. 83D, no. 1, March 1961, pp. 95–108.

[140] Kanmachi, T., Takahashi, I: 'Sensor-less Speed Control of an Induction Motor', *IEEE Industry Applications Magazine*, vol. 1, no. 1, Jan./Feb. 1995, pp. 22–27.

[141] Karr, C.L., Gentry, E.J.: 'Fuzzy Control of pH Using Genetic Algorithm', *IEEE Trans. Fuzzy Syst.*, vol. 1, no. 1, Jan. 1993, pp. 46–53.

[142] Kawamura, A., Hoft, R.G.: 'Instantaneous Feedback Controlled PWM Inverters with Adaptive Hysteresis', *IEEE Trans. Ind. Appl.*, vol. IA-20, 1984, pp. 769–775.

[143] Kazmierkowski, M.P., Dzieniakowski, M.A.: 'Review of Current Regulation Techniques for Three-phase PWM Inverters', *IEEE IECON Conf. Rec.*, 1994, pp. 567–575.

[144] Kelemen, A., Panã, T.: 'Simultaneous Speed and Rotor Resistance Estimation for Sensorless Vector-controlled Induction Motor Drives', *Proceedings of the Twenty-Seventh International Intelligent Motion Conference*, 20–22 June, Nuremberg, Germany, 1995, pp. 523–530.

[145] Kennedy, D.C., Quintana, V.H.: 'Neural Network Regulators for Synchronous Machines', *Proc. of ESAP*, 1993, pp. 531–535.

[146] Khor, J.: 'An Intelligent Controller for Synchronous Generators', PhD Thesis, De Montfort University, 1999.

[147] Khor, J.G., Cirstea, M.N., McCormick M., Low, W.F.: 'PWM Control Techniques for Variable Speed Generator Set', *Proc. of Int. Conf. on Optimization of Electric and Electronic Equipment* (OPTIM'98-IEE, IEEE), vol. II, Brasov, Romania, May 1998, pp. 379–382.

[148] Khor, J.G., Cirstea, M.N., McCormick M.: 'Control of Stand Alone Synchronous Generators at Optimum Speed', *Proc. of the 33rd IEEE Intersociety Energy Conversion Engineering Conference* (IECEC'98), Colorado Springs, USA, 2–6 Aug. 1998, CD-ROM.

[149] Knapp, S.K., Xilinx Corporate Applications Manager: *Using Programmable Logic to Accelerate DSP Functions*, Xilinx Inc., Dec. 1996.

[150] Kobayashi, T., Yokoyama, A.: 'An Adaptive Neuro-control System of Synchronous Generator for Power System Stabilisation', *IEEE Trans. on Energy Conversion*, vol. 11, no. 3, Sept. 1996, pp. 621–630.

[151] Koc, C.K., Johnson S.: 'Multiplication of Signed-digit Numbers', *Electronics Letters*, May 1994, pp. 840–841.

[152] Kolar, J.K., Ertl, H., Zach, F.C.: 'Analysis of On- and Off-line Optimised Predictive Current Controllers for PWM Converter Systems', *IEEE Trans. on Power Electronics*, vol. 6, July 1991, pp. 451–462.

[153] Krzeminski, Z.: 'An Observer System for Induction Motor without Speed Sensor', *Proceedings of the Twenty-Seventh International Intelligent Motion Conference*, 20–22 June 1995, Nuremberg, Germany, pp. 143–153.

[154] Krzeminski, Z., Guzinski, J.: 'DSP Based Sensorless Control System of the Induction Motor', *Proc. of PCIM'98 – Intelligent Motion*, pp. 137–146.

[155] Kung, S.Y.: '*Digital Neural Networks*', Prentice Hall, 1993.

[156] Lanser, J.A., Lehmann, T.: 'An Analog CMOS Chip Set for Neural Networks with Arbitrary Topologies', *IEEE Trans. on Neural Networks*, vol. 4, no 3, May 1993, pp. 441–444.

[157] Lee, J.: 'On Methods for Improving Performance of PI-Type Fuzzy Logic Controllers', *IEEE Trans. Fuzzy Systems*, vol. 1, no. 4, Nov. 1993, pp. 298–302.

[158] Le-Huy, H., Dessiant, L.A.: 'An Adaptive Current Control Scheme for PWM Synchronous Motor Drives: Analysis and Simulation', *IEEE Trans. on Power Electronics*, vol. 4, Oct. 1989, pp. 486–495.

[159] Leonhard, W.: *Control of Electrical Drives*, 2nd edition, Springer, Berlin, 1996.

[160] Lin, F.-J., Shyu, K.-K., Wai, R.-J.: 'DSP-based Minmax Speed Sensorless Induction Motor Drive with Sliding Mode Model-following Speed Controller', *IEE Proceedings – Electric Power Applications*, Nov. 1999, vol. 146, Issue 6, p. 471.

[161] Lindsey, C.S., Lindblad, Th.: 'Review of Hardware Neural Networks: A User's Perspective', *Proceedings of the 3rd Workshop on Neural Networks: From Biology to High Energy Physics*, Isola d'Elba, Italy, 26–30 Sept. 1994.

[162] Lüdtke, I., Jayne, M.G.: 'A New Direct Torque Control Strategy', *IEE Colloquium on Advances in Control Systems for Electric Drives*, 24 May 1995, London, UK, Digest No. 1995, pp. 114–120.

[163] Macq, D., Verleysen, M., Jespers, P., Legat, J.D.: 'Analog Implementation of a Kohonen Map with On-Chip Learning', *IEEE Transactions on Neural Networks*, vol. 4, no. 3, May 1993, pp. 456–461.

[164] Malik, O.P., Hope, G.S., Huber, D.W.: 'Design and Test Results of a Software Based Digital A.V.R', *IEEE Trans. on Power Appliances and Systems*, vol. PAS-92, no. 2, March/April 1976, pp. 634–642.

[165] Manes, C., Parasiliti, F., Tursini, M.: 'Comparative Study of Rotor Flux Estimation in Induction Motors with a Nonlinear Observer and the Extended Kalman Filter', *Proceedings of the 20th International Conference on Industrial Electronics Control and Instrumentation IECON*, Bologna, vol. 3, 1994, p. 2149.

[166] Manes, C., Parasiliti, F., Tursini, M.: 'DSP Based Fied-oriented Control of Induction Motor with a Non-Linear State Observer', *27th Annual IEEE Power Electronics Specialists Conference*, vol. 2, 1996, p. 1254.

[167] Marino, R.: 'An Example of Non-linear Regulator', *IEEE Trans. Automatic Contr.*, vol. 29, no. 3, 1984, pp. 276–279.

[168] Massengill, L.W., Mundie, D.B.: 'An Analog Neural Hardware Implementation Using Charge-injection Multipliers and Neuron-specific Gain Control', *IEEE Transactions on Neural Networks*, vol. 3, no. 3, May 1992, pp. 354–362.

[169] Matsuo, T., Lipo, T.A.: 'A Rotor Parameter Identification Scheme for Vector Controlled Induction Motor Drives', Rec. *IEEE-IAS Annual Meeting*, pp. 538–545.

[170] Maxwell, J.C.: 'On Governors', *Proc. Roy. Soc.* (*London*), 1868.

[171] Mead, C.A., Ismail, M.: '*Analogue VLSI Implementation of Neural Systems*', MA: Kluwer, Boston, 1989.

[172] Miller, A., Muljadi, E., Zinger, D.S.: 'A Variable Speed Wind Turbine Power Control', *IEEE Trans. on Energy Conversion*, vol. 12, no. 2, 1997, pp. 181–186.

[173] Mirchandini, G., Cao, W.: 'On Hidden Nodes in Neural Nets', *IEEE Transactions on Circuits and Systems*, no. 36, May 1989, pp. 661–664.

[174] Misawa, E.A., Hedrick, J.K.: 'Non-linear Observers – a State of the Art Survey', *ASME J. Dyn. Syst. Meas. Contr.*, vol. 111, 1989, pp. 344–352.

[175] Morioka, Y. *et al.*: 'Application of Multivariable Optimal Controller to Real Power Systems', *IEEE Trans. on Power Systems*, vol. 9, no. 4, 1994, pp. 1949–1955.

[176] Mortara, A., Vittoz, E.A.: 'A Communication Architecture Tailored for Analog VLSI Artificial Neural Networks: Intrinsic Performance and Limitations', *IEEE Transactions on Neural Networks*, vol. 5, no. 3, May 1994, pp. 459–466.

[177] Mudi, R.K., Pal, N.R.: 'A Robust Self-tuning Scheme for PI- and PD-type Fuzzy Controllers', *IEEE Trans. Fuzzy Systems*, vol. 7, no. 1, Feb. 1999, pp. 2–16.

[178] Murray, A.F., Del Corso, D., Tarassenko, L.: 'Pulse-stream VLSI Neural Networks Mixing Analog and Digital Techniques', *IEEE Transactions on Neural Networks*, vol. 2, no. 2, March 1991, pp. 193–204.

[179] Nabae, S., Ogasawara, M., Akagi, H.: 'A New Control Scheme for Current Controlled PWM Inverters', *IEEE Trans. Ind. Appl.*, vol. IA–22, no. 4, July/Aug. 1986, pp. 697–701.

[180] Navabi, Z.: 'VHDL – Analysis and Modeling of Digital Systems', *Electrical and Computer Engineering Series*, McGraw-Hill International Editors, 1993.

[181] New, B.: 'A Distributed Arithmetic Approach to Designing Scaleable DSP Chips', *EDN*, 17 August 1995, pp. 107–114.

[182] Nicula, D., Cirstea, M.N.: 'Successful CAD Tools Application to FPGA/ASIC Design', *International Journal of Engineering Education*, vol. 15, no. 1, 1999, pp. 72–76.

[183] Novotny, D.W., Lipo, T.A.: '*Vector Control and Dynamics of AC Drive*', Oxford Science Publications, Clarendon Press, Oxford, 1996.

[184] Ogata, K.: '*Discrete-time Control System*', Prentice-Hall, 1995.

[185] Pahalawaththa, N.C., Hope, G.S., Malik, O.P.: 'Multivariable Self Tuning Power System Stabiliser Simulation and Implementation Studies', *IEEE Trans. on Energy Conversion*, vol. 6, no. 2, 1991, pp. 310–319.

[186] Patterson, E.B.: 'Electronic Design Automation for Power Electronic Drives', PhD Thesis, Nottingham Trent University, 1993.

[187] Patterson, E.B., Holmes, P.G., Morley, D.: 'Electronic Design Automation (EDA) Techniques for the Design of Power Electronic Control Systems', *IEE Proc. – G*, vol. 139, no. 2, April 1992, pp. 191–198.

[188] Pena, R.S., Asher, G.M., Clare, J.C.: 'A Doubly-fed Induction Generator Using Back-to-Back PWM Converters Supplying an Isolated Load from a Variable Speed Wind Turbine', *IEE Proc. – B*, vol. 143, no. 5, Sept. 1996, pp. 380–387.

[189] Perry, D.L.: *VHDL*, 2nd edition, McGraw-Hill Series on Computer Engineering, McGraw-Hill Inc., 1994.

[190] Profumo, F., Griva, G., Tenconi, A., Abrate, M., Ferraris, L.: 'Stability Analysis of Luenberger Observers for Speed Sensorless High Performance Spindle Drives', *EPE'99*, Lausanne (CD-Rom).

[191] Rahman, M.A., Osheiba, A.M., Radwan, T.S.: 'Modelling and Controller Design of an Isolated Diesel Engine Permanent Magnet Synchronous Generator', *IEEE Trans. on Energy Conversion*, vol. 11, no. 2, June 1996, pp. 324–329.

[192] Rajashekara, K., Kawamura, A., Matsuse, K.: '*Sensorless Control of AC Drives*', IEEE Press, 1996.

[193] Ramacher, U., Wesseling, M.: 'A Geometrical Approach to Neural Network Design', *Proceedings IJCNN'89* (Washington USA), IEEE Press, vol. 2, January 1989, pp. 147–153.

[194] Ross, R.J.: '*Fuzzy Logic with Engineering Applications*', McGraw-Hill, 1995.

[195] Säkinger, E., Boser, B.E., Jackel, L.D.: 'A Neurocomputer Board Based on the ANNA Neural Network Chip', in '*Advances in Neural Information Processing Systems*', (Moody, J.E., Hanson, S.J., Lippmann, R.P., eds.), Morgan Kaufmann 1992, San Mateo, CA, pp.773–780.

[196] Sankara Narayanan, E.M., De Souza, M.M., Qin, Z.: 'Devices and Technologies for High Voltage Integrated Circuits', *Proc. Int. Conf. on Semiconductor Materials and Technologies*, New Delhi, India,16–21 Dec. 1996.

[197] Satieo, S., Torrey, D.A.: 'Fuzzy Logic Control of a Space–Vector PWM Current Regulator for Three-phase Power Converters', *IEEE Transactions on Power Electronics*, vol. 13, no. 3, 1998, pp. 419–426.

[198] Savaresi, S.M.: 'Exact Feedback Linearisation of a Fifth Order Model of Synchronous Generators', *IEE Proc. Contr. Theory Appl.*, vol. 146, no. 1, Jan. 1999, pp. 53–57.

[199] Scharf, A.: 'Twenty Years of Innovation', *PCIM Magazine*, issue 6, 1999, pp. 10–12.

[200] Schwartz, M., Shaw, L: 'Signal Processing – Discrete Spectral Analysis, Detection and Estimation', McGraw-Hill Book Company, 1975.

[201] Seepold, R., Kunzmann, A. (eds): '*Reuse Techniques for Vlsi Design*', Kluwer Academic Publishers, 1999.

[202] Simoes, M.G., Bose, B.K.: 'Application of Fuzzy Logic in the Estimation of Power Electronic Waveforms', *IEEE/IAS Annual Meet. Conf. Rec.*, 1993, pp. 853–861.

[203] Simoes, M.G. *et al.*: 'Fuzzy Logic Based Intelligent Control of a Variable Speed Cage Machine Wind Generation System', *IEEE Trans. Power Electronics*, vol. 12, no. 1, Jan. 1997, pp. 87–94.

[204] Sjoholm, S., Lindh, L.: *VHDL for Designers*, Prentice-Hall, 1997.

[205] Smieja, F.J.: 'Neural Network Constructive Algorithm: Trading Generalisation for Learning Efficiency?', *Circuits, Systems, Signal Processing*, vol. 12, no. 2, 1993, pp. 331–374.

[206] Smith, G.A., Stephens, R.G., Marshall, P., Kansara, M.: 'A Sinewave Interface for Variable Speed Wind Turbines', *5th European Conf. on Power Electronics & Applications*, IEE Conf. Publ. No. 377, 13–16 Sept. 1993, pp. 97–102.

[207] Smith, F.S., Shen, Q.: 'Selecting Inference and Defuzzification Techniques for Fuzzy Logic Control', *Proc. of UKACC Intern. Conference on Control*, 1–4 Sept. 1998, pp. 54–59.

[208] Song, Y., Johns, A.T.: 'Applications of Fuzzy Logic in Power Systems, Part 1', *IEE Power Engineering Journal*, Oct. 1997.

[209] Song, Y., Johns, A.T.: 'Applications of Fuzzy Logic in Power Systems, Part 2', *IEE Power Engineering Journal*, Aug. 1998, pp. 185–190.

[210] Sousa, G.C.D., Bose, B.K.: 'A Fuzzy Set Theory Based Control of a Phase-Controlled Converter dc Machine Drive', *IEEE Trans. Ind. Appl.*, vol. 30, 1994, pp. 34–44.

[211] Sousa, G.C.D. *et al.*: 'Fuzzy Logic Based On-line Efficiency Optimisation Control of an Indirect Vector Controlled Induction Motor Drive', *Proc. IEEE/IECON Conf.*, 1993, pp. 1168–1174.

[212] Summer, M., Campbell, J., Curtis, M.: 'A Stator Resistance Estimator for Sensorless Vector Controlled Drives using Artificial Neural Networks', *EPE'99*, Lausanne (CD Rom).

[213] Tajima, H., Hori, Y.: 'Speed Sensorless Field Orientation Control of the Induction Machine', *IEEE Trans. Ind. Appl.*, vol. 29, 1993, pp. 175–180.

[214] Tan, S., Vandewalle, J.: 'Efficient Algorithm for the Design of Multilayer Feed-forward Neural Networks', *Proc. IJCNN'92* (Baltimore, USA), IEEE Press, vol. 2, 1992, pp. 190–195.

[215] Tenti, P., Malesani, L., Rossetto, L.: 'Optimum Control of N-input K-output Matrix Converters', *IEEE Trans. Power Electron.*, vol. 7, no. 4, 1992, pp. 707–713.

[216] Teske, N., Asher, G.M., Bradley, K.J., Summer, M.: 'Sensorless Position Control of Induction Machines using Rotor Saliencies under Load Conditions', *EPE'99*, Lausanne (CD Rom).

[217] Thomas, M.J.: 'Designing Intelligent Muscle into Industrial Motion Control', *IEEE Trans. Ind. Electron.*, vol. 37, no. 5, 1990, pp. 329–341.

[218] Trynadlowski, A.M., Legowski, S.: 'Minimum-loss Vector PWM Strategy for Three-phase Inverters', *IEEE Trans. Power Electron.*, vol. 9, no. 1, 1994, pp. 26–34.

[219] Tzou, Y.-Y., Tsai, M.-F., Lin, Y.F., Wu, H.: 'Dual-DSP Fully Digital Control of an Induction Motor', *IEEE ISIE Conf. Rec.*, Warsaw, Poland, 17–20 June 1996, pp. 673–678.

[220] Tzou, Y.-Y., Hsu, H.-J.: 'FPGA Realization of Space–Vector PWM Control IC for Three-phase PWM Inverters', *IEEE Transactions on Power Electronics*, vol. 12, no. 6, Nov. 1997, pp. 953–963.

[221] Tzou, Y.-Y., Yeh, S.-T., Wu, H.: 'DSP-based Rotor Time Constant Identification and Slip Gain Auto-tuning for Indirect Vector-controlled Induction Drives', *IECON Proc.*, vol. II, Taiwan, 1996, p. 1228.

[222] Vadivel, S., Bhuvaneswari, G., Rao, G.S.: 'A Unified Approach to Real Time Implementation of DSP Based PWM Waveforms', *IEEE Trans. Power Electron.*, vol. 6, no. 4, 1991, pp. 565–575.

[223] Van der Broeck, H.W., Skudelny, H.C., Stanke, G.V.: 'Analysis and Realisation of a Pulsewidth Modulator Based on Voltage Space vectors', *IEEE Transactions on Industry Applications*, vol. 24, no. 1, Jan./Feb. 1988, pp. 142–150.

[224] Vanlandingham, H.F.: '*Introduction to Digital Control Systems*', MacMillan Press, NewYork, 1992.

[225] Vas, P., Li J., Stronach, A.F.: 'Artificial Neural Network-based Control of Electromechanical Systems', *Proceedings of 4th European Conference on Control IEE*, Coventry, 1994.

[226] Vas, P.: 'Artificial-intelligence-based Electrical Machines and Drives. Application of Fuzzy, Neural, Fuzzy-neural and Genetic Algorithms', *Monographs in Electrical and Electronic Engineering*, Oxford University Press, 1999.

[227] Vas, P.: 'Electrical Machines and Drives, A Space-Vector Theory Approach', *Monographs in Electrical and Electronic Engineering*, Oxford University Press, 1992.

[228] Vas, P.: 'Sensorless Vector and Direct Torque Control', *Monographs in Electrical and Electronic Engineering*, Oxford University Press, 1998.

[229] Vas, P.: 'Vector Control of AC Machines', *Monographs in Electrical and Electronic Engineering*, Oxford University Press, 1990.

[230] Vas, P., Stronach, A.F., Neuroth, M.: 'DSP-based Speed-sensorless High-performance Torque Controlled Induction Motor Drives', *Proc. of PCIM'98 Intelligent Motion*, pp. 225–234.

[231] Vas, P., Stronach, A.F., Neuroth, M., Du, T.: 'A Fuzzy Controlled Speed-Sensorless Induction Motor Drive with Flux Estimators', *IEE EMD*, Durham, 1995, pp. 315–319.

[232] Venturini, M.G.B., Alesina, A.: 'A New Sine Wave in, Sine Wave out, Conversion Technique Eliminates Reactive Elements', *Proc. Powercon 7*, 1980, pp. E3.1–E3.15.

[233] Walter, C.D.: 'Fast Modular Multiplication using 2-Power Radix', *International Journal of Computer Mathematics*, no. 3, 1991, pp. 21–28.

[234] Walter, C.D.: 'Systolic Modular Multiplication', *IEEE Transactions on Computers*, vol. 42, 1993, pp. 376–378.

[235] Weedy, B.M.: *Electric Power Systems*, 3rd Edition, John Wiley & Sons, 1995.

[236] Wheeler, P.W., Grant, D.A.: 'A Low Loss Matrix Converter for A.C. Variable Speed Drives', *Proc. EPE'93*, UK, vol. 5, 1993, pp. 27–32.

[237] White, D.A., Sofge, D.A.: *Handbook of Intelligent Control: Neural, Fuzzy, and Adaptive Approaches*, Multiscience Press, Inc., 1992.

[238] Wu, Q.H., Hogg, B.H.: 'Robust Self-tuning Regulator for a Synchronous Generator', *IEE Proc. – D*, vol. 135, no. 6, Nov. 1988, pp. 463–473.

[239] Zadeh, L.A.: 'Fuzzy Sets', *Information and Control*, vol. 8, 1965, pp. 338–353.

[240] Zadeh, L.A.: 'Outline of a New Approach to the Analysis of Complex Systems and Decision Processes', *IEEE Trans. Syst., Man & Cybern.*, vol. 3, 1973, pp. 28–44.

[241] Zhang, L., Watthanasarn, C., Shepherd, W.: 'Analysis and Comparison of Control Techniques for A.C.–A.C. Matrix Converters', *IEE Proc. Electr. Power Appl.*, vol. 145, no. 4, July 1998, pp. 284–294.

[242] Zhang, Y., Malik, O.P., Chen G.P.: 'Artificial Neural Network Power System Stabilisers in Multimachine Power System Environment', *IEEE Trans. on Energy Conversion*, vol. 10, no. 1, 1995, pp. 147–153.

[243] Zurada, J.M.: *Introduction to Artificial Neural Systems*, West Publishing Company, 1992.

# 11

# Appendices

The electronic format of the program codes (ASCII text) is downloadable from the publisher's website at: http://www.bh.com/newnes

## 11.1 Appendix A – C++ code for ANN implementation

### 11.1.1 CONV_NET.CPP

```
#include <iostream.h>
#include <fstream.h>
#include <string.h>
#include <stdlib.h>
#include <memmanag.h>
#define MaximumDepth 100
#define LengthInputTab 2000
#define AND 1
#define ANY 0
ofstream output_file;
int *out_layer, *index, *node, *inverter, *used;
// Index specifies the initial position of the weights in the matrix
// row before they were rearranged according to their descending
// values.Node is a vector which stores the node numbers corresponding
// to the weights. The node number is not generally speaking corelated
// with the input numbers because there are inversor gates and because
// some neurones are located in other layers than the first.
double *w; //The weights vector for one neurone
int no_w; //The number of weights per neurone
int current_node,output_node,no_gates=0,max_depth,depth=0;
int no_max_inputs;
void add_gate(int ind_init, double threshold, char and_gate);
void arrange(void)
{
  int i,i_max,i_aux,j;
  double max,max_aux;
  for(i=0;i<no_w;i++)
  {
    max=w[i];
```

```
        i_max=i;
        for(j=i+1;j<no_w;j++)
          if (max<w[jl)
          {
        max=w[j];
        i_max=j;
          }
        if(i_max != i)
        {
          max_aux=w[i_max];
          w[i_max]=w[i];
          w[i]=max_aux;
          i_aux=index[i_max];
          index[i_max]=index[i];
          index[i]=i_aux;
        }
    }
}
int add_inverter(int i)
{
    no_gates++;
    ++current_node;
    output_file<<".NOT ";
    output_file<<node[index[i]]<<" "<<current_node<<"\n";
    if(!output_file.good())
    {
      cout<<"\n\aError on file writing";
      exit(1);
    }
    return current_node;
}
double* measure_matrix(int& no_neu,int& no_values, istream& input_file)
{
    double temp;
    char buffer[2];
    no_neu=no_values=0;
    while(input_file.read(buffer,1),!input_file.eof())
    {
      if(buffer[0]=='\n')
      no_neu++;
    }
    input_file.clear();
    input_file.seekg(0,ios::beg);
    do
    {
      input_file>>temp;
      if(!input_file.eof())
        no_values++;
    }
    while(!input_file.eof());
    if(no_neu==0)
      no_neu++;
```

```
  if((no_values/no_neu)*no_neu == no_values)
    no_values=no_values/no_neu;
  else
    if((no_values/(no_neu+1)) == no_values)
    {
      no_values=no_values/(no_neu+1);
      no_neu++;
    }
    else
    {
      cout<<"\n\aError in input file!";
      exit(1);
    }
  input_file.clear();
  input_file.seekg(0,ios::beg);
  w=alloc_double(no_values);
  return w;
}
void read_line_matrix(istream& input_file)
{
  int i;
  for(i=0;i<no_w+1;i++)
    input_file>>w[i];
  if(!input_file.good())
  {
    cout<<"\n\aError when reading the input file";
    exit(1);
  }
}
int convert_neuron(void)
{
  int i;
  double threshold;
inverter=alloc_int(no_w); //Shows if the corresponding weight was
    // negative thereby requiring an inversor gate
  used=alloc_int(no_w);   //Shows if the node has been already used
    // in the past so that the inverter has already been put
  index=alloc_int(no_w+1);

  for(i=0;i<no_w;i++)
  {
    index[i]=i;
    inverter[i]=0;
    used[i]=0;
  }
  for(i=0;i<no_w;i++)
  {
    if(w[i]<0)
    {
      w[i]=-w[i];
      inverter[index[i]]=1;
    }
    w[no_w]-=w[i];
```

```
    w[i]=2*w[i];
  }
  threshold=-w[no_w];
  if(threshold<=0)
  {
    output_node=-1; //Output value is constantly 1
    cout<<"\n\aWarning: The output of a neurone is constantly 1";
  }
  else
  {
    arrange();
    add_gate(0,threshold,ANY);
  }
  delete index;
  delete used;
  delete inverter;
  return output_node;
}
int port_number(int i)
{
  int rez;
  if (inverter[index[i]]==0)
  {
    rez=node[index[i]];
    if(depth>max_depth)
      max_depth=depth;
  }
  else
  {
    if(depth+1>max_depth)
      max_depth=depth+1;
    if (used[index[i]])
      rez=used[index[i]];
    else
    {
      used[index[i]]=add_inverter(i);
      rez=used[index[i]];
    }
  }
  return rez;
}
int det_num_internal_gate_layers(int no_inp)
{
  int no_inp_top,no_layers=1;
  if(no_inp<=no_max_inputs)
    return 1;
  else
  while(no_inp>no_max_inputs)
  {
    no_inp_top=no_inp/no_max_inputs;
    if(no_inp>no_inp_top*no_max_inputs)
      no_inp_top++;
```

```
    no_inp=no_inp_top;
    no_layers++;
  }
  return no_layers;
}
int cursor=0;
int input[LengthInputTab];
void write_gate(char *name,int no_inputs, int local_cursor)
{
  int i,no_inp_top_gate,no_last_inputs;
  if(no_inputs<=no_max_inputs)
  {
    output_file<<name<<no_inputs;
    for(i=0;i<no_inputs;i++)
      output_file<<" "<<input[local_cursor+i];
    current_node++;
    output_file<<" "<<current_node<<"\n";
    output_node=current_node;
    no_gates++;
    if(!output_file.good())
    {
      cout<<"\n\aError on file writing";
      exit(1);
    }
  }
  else
  {
    no_inp_top_gate=no_inputs/no_max_inputs;
    if(no_inputs>no_inp_top_gate*no_max_inputs)
      no_inp_top_gate++;
    if(cursor+no_inputs+no_inp_top_gate>=LengthInputTab)
    {
      cout<<"\n\aError: Input table is full";
      exit(1);
    }
    for(i=0;i<no_inp_top_gate-1;i++)
    {
      local_cursor=cursor+i*no_max_inputs;
      write_gate(name,no_max_inputs,local_cursor);
      input[cursor+no_inputs+i]=output_node;
    }
    local_cursor=cursor+(no_inp_top_gate-1)*no_max_inputs;
    no_last_inputs=no_inputs-(no_inp_top_gate_1)*no_max_inputs;
    if(no_last_inputs>1) //It is possible to have only one remaining
      input
    {
      write_gate(name,no_last_inputs,local_cursor);
      input[cursor+no_inputs+no_inp_top_gate-1]=output_node;
    }
    else
      input[cursor+no_inputs+no_inp_top_gate-1]=input[cursor+no_inputs-
1];
```

```
    for(i=0;i<no_inp_top_gate;i++)
      input[cursor+i]=input[cursor+no_inputs+i];
    write_gate(name,no_inp_top_gate,cursor);
  }
}
double sum;
int j;
void add_gate(int ind_init, double threshold, char and_gate)
{
  int i,no_inputs,no_big_weights,ind_for_AND;
  //These are local variables because they need to be preserved during
    the
  //recursive calls of the function
  if(threshold<=0)
  {
    cout<<"\n\aError: The output of a subneurone is constantly 1";
    exit(1);
  }
  if(!and_gate)
  {
    sum=0;
    no_inputs=0;
    no_big_weights=0;
    ind_for_AND=ind_init;
    for(i=ind_init; i<no_w;i++)
      if(w[i]>=threshold)
      {
    no_inputs++;
    no_big_weights++;
    ind_for_AND=i+1;
      }
    for(i=ind_for_AND;i<no_w;i++)
    {
      sum=0;
      for(j=i;j<no_w;j++)
      sum+=w[j];  //The sum will be the result of several cumulated
if(sum>=threshold) //inputs anyway because these are not big weights.
      no_inputs++;
    }
  }
  if((no_inputs>1) && (!and_gate))
  {
    depth=depth+det_num_internal_gate_layers(no_inputs);
//'-1' because there is one gate anyway
    if(depth>MaximumDepth)
    {
      cout<<"\n\aError: Too many recursive calls!";
      exit(1);
    }
    for(i=0;i<no_inputs;i++)
      if(i>=no_big_weights)
      {
```

```
cursor+=no_inputs;
if(cursor>=LengthInputTab)
{
  cout<<"\n\aError: Input table is full. Enlarge the input table";
  exit(1);
}
add_gate(ind_init+i,threshold,AND);
cursor-=no_inputs;                    //necessarily be an AND gate
input[cursor+i]=output_node;
    }
    else
input[cursor+i]=port_number(ind_init+i);
    write_gate(".OR",no_inputs,cursor); //the gate is written
    depth=depth-det_num_internal_gate_layers(no_inputs);
}
else if((no_inputs==1) && (no_big_weights==1) && (!and_gate))
  output_node=port_number(ind_init); //It is just a straightforward
                                     //input-output connection
else if(((no_inputs==1) && (no_big_weights==0)) || and_gate)
                                     //An AND gate will be used
{
  no_inputs=1; //When is just a simple AND gate, it corresponds to a
  //single combination of inputs. Variable 'no_inputs' is
    //used for economy of space in the stack. The first input
      //is compulsory to be used which is why no_inputs=1.
  sum=0;
  for(i=ind_init;i<no_w;i++)
    sum+=w[i];
  for(i=ind_init+1;i<no_w;i++)
    if(sum-w[i]<threshold)
  no_inputs++;
    else
  break;
  sum=0;
  for(i=ind_init;i<ind_init+no_inputs;i++)
    sum+=w[i];
  if(threshold-sum>0)
    no_inputs++;   //A further subneurone is required
  if(no_inputs<2)
  {
    cout<<"\n\aError in algorithm! An AND gate has less than 2
      inputs!";
    exit(1);
  }
  depth+=det_num_internal_gate_layers(no_inputs);
//'-1' because there is one gate anyway
  if(depth>MaximumDepth)
  {
     cout<<"\n\aError: Too many recursive calls!";
     exit(1);
  }
  for(i=0;i<no_inputs;i++)
```

```
    {
      if((i<no_inputs-1) || (threshold-sum<=0))
  input[cursor+i]=port_number(i+ind_init);
      if((i==no_inputs-1) && (threshold-sum>0))
      {
  cursor+=no_inputs;   //The supplementary subneurone is added
  if(cursor>=LengthInputTab)
  {
    cout<<"\n\aError: Input table is full. Enlarge the input table";
    exit(1);
  }
  add_gate(ind_init+no_inputs-1,threshold-sum,ANY);
  cursor-=no_inputs;
  input[cursor+i]=output_node;
      }
    }
    write_gate(".AND",no_inputs,cursor);
    depth-=det_num_internal_gate_layers(no_inputs);
  }
  else if(no_inputs==0)
  {
    output_node=0;  //Output value is constantly 0
    cout<<"\n\aWarning: The output of a neurone is constantly 0";
  }
  else
  {
    cout<<"\n\aError in conversion algorithm";
    exit(1);
  }
}
void main(int no_par, char** par)
{
  ifstream input_file;
  int no_neu=0, no_values_per_line, previous_no_neu,gate_layers=0;
  int i,no_file;
  if(no_par<4)
  {
    cout<<"\nToo few parameters";
    exit(1);
  }
  no_max_inputs=atoi(par[no_par-1]);
  output_file.open(par[no_par-2],ios::out);
  if(!output_file.good())
  {
    cout<<"\n\aError: The output file cannot be opened!";
    exit(1);
  }
  cout<<"\n----- Start conversion -----";
  for(no_file=1;no_file<no_par-2;no_file++)
  {
    input_file.open(par[no_file],ios::in);
    if(!input_file.good())
```

```
{
  cout<<"\n\aError: The input file cannot be opened!";
  exit(1);
}
cout<<"\nProcessing file "<<no_file;
previous_no_neu=no_neu;
w=measure_matrix(no_neu,no_values_per_line, input_file);
no_w=no_values_per_line-1;
if((no_file>1)&&(previous_no_neu != no_values_per_line-1))
{
  cout<<"\n\aError: Wrong number of neurones in layer"<<no_file;
  exit(1);
}
out_layer=alloc_int(no_neu);
if(no_file==1)
{
  node=alloc_int(no_w);
  for(i=0;i<no_w;i++)
  {
node[i]=i+1;
output_file<<".INPUT "<<node[i]<<"\n";
if(!output_file.good())
{
  cout<<"\n\aError on file writing";
  exit(1);
}
  }
  current_node=no_w;
}
max_depth=0;
for(i=0;i<no_neu;i++)
{
  read_line_matrix(input_file);
  out_layer[i]=convert_neuron();
}
gate_layers+=max_depth;
if(no_file==no_par-3)  //It was the last input file so the output
  //ports must be written
  for(i=0;i<no_neu;i++)
  {
output_file<<".OUTPUT "<<out_layer[i]<<"\n";
if(!output_file.good())
{
  cout<<"\n\aError on file writing";
  exit(1);
  }
  }
delete node;
if(no_file<no_par-3)
              //File no 'no_par-2' is the output file so this is
{                      //not the last input file
  node=alloc_int(no_neu);
```

```
      for(i=0;i<no_neu;i++)
      {
  node[i]=out_layer[i];  //The outputs of the previous layer are
          //the inputs for the next one
  if(node[i]<=0)
  {
    cout<<"\n\aWarning: The output of the hidden neuron "<<(i+1);
    cout<<" in layer "<<no_file<<" is constant!";
  }
      }
    }
    input_file.close();
    delete w;
    delete out_layer;
  }
  input_file.close();
  cout<<"\nThe output file contains "<<no_gates<<" logic gates on ";
  cout<<gate_layers<<" gate layers\n";
}
```

### 11.1.2 OPTIM.CPP

```
#include <iostream.h>
#include <fstream.h>
#include <stdlib.h>
#include <string.h>
#include <memmanag.h>
#define NOT 3
#define AND 1
#define OR 4
#define INPUT 2
#define OUTPUT 5
#define NO_WORDS 5
#define CANCELLED -2
//'0' is already defined in CONV_NET!.CPP as 'ground' and '-1' as 'Vcc'.

char* words[5]={".AND",".INPUT",".NOT",".OR",".OUTPUT"};
int length[5]={4,6,4,3,7};
struct gate_string
{
  char name[8];
  int no_gates;
  int cursor;
  int* gate_nodes;
};
gate_string *and_s, *or_s;
int *input_s, *output_s, *not_s;
int max_input=0,max_output=0,max_and=0,max_or=0,max_not=0;
int cursor_input=0,cursor_output=0,cursor_not=0;
int no_init_gates,no_fin_gates;
gate_string* alloc_gate_string(int no_gate_strings)
{
```

```
  gate_string *pointer;
  if(no_gate_strings>0)
  {
    if(!(pointer=new gate_string[no_gate_strings]))
    {
      cout<<alloc_err;
      exit(1);
    }
    return pointer;
  }
  else
    return NULL;
}
void init_structures(void)
{
  int i;
  for(i=0;i<max_and-1;i++)
  {
    and_s[i].no_gates=0;
    and_s[i].cursor=0;
    and_s[i].name[0]=0;
  }
  for(i=0;i<max_or-1;i++)
  {
    or_s[i].no_gates=0;
    or_s[i].cursor=0;
    or_s[i].name[0]=0;
  }
}
int check_word(char* buffer)
{
  int i,j,found;
  for(i=0;i<NO_WORDS;i++)
  {
    found=1;
    for(j=0;j<length[i];j++)
      if(buffer[j] != words[i][j])
      {
  found=0;
  break;
      }
    if(found==1)
      return i+1;
  }
  return 0;
}
int det_inputs(char *buffer)
{
  int i=0;
  while(((buffer[i]>='A') && (buffer[i]<='Z')) || (buffer[i]=='.'))
    i++;
  return atoi(buffer+i);
```

```
}
void first_scan(ifstream& in_file)
{
  char buffer[10];
  int i, node,ind_word,no_inputs;
  cout<<"First scan\n";
  while(!in_file.eof())
  {
    in_file>>buffer;
    ind_word=check_word(buffer);
    switch (ind_word)
    {
      case INPUT: in_file>>node;
                                max_input++;
                                break;
      case OUTPUT:in_file>>node;
                       max_output++;
                       break;
      case AND: no_inputs=det_inputs(buffer);
                    if(no_inputs>max_and)
                      max_and=no_inputs;
                    for(i=0;i<=no_inputs;i++)
                      in_file>>node;
                      break;
      case OR: no_inputs=det_inputs(buffer);
                                     if(no_inputs>max_or)
                                     max_or=no_inputs;
          for(i=0;i<=no_inputs;i++)
                                in_file>>node;
          break;
      case NOT: in_file>>node;
                                in_file>>node;
                                max_not++;
                                break;
      default: if(buffer[0]==0)
                     break;
          else
          {
            cout<<"\n\aSyntax error in input file";
            exit(1);
          }
    }
  }
  and_s=alloc_gate_string(max_and-1);
  or_s=alloc_gate_string(max_or-1);
  input_s=alloc_int(max_input);
  output_s=alloc_int(max_output);
  not_s=alloc_int(2*max_not);
  init_structures();
  no_init_gates=max_not;
}
void second_scan(ifstream& in_file)
```

```
{
  char buffer[10];
  int i, node,ind_word,no_inputs;
  cout<<"Second scan\n";
  in_file.seekg(0,ios::beg);
  in_file.clear();
  while(!in_file.eof())
  {
    in_file>>buffer;
    ind_word=check_word(buffer);
    switch (ind_word)
    {
      case INPUT: in_file>>node;
                  input_s[cursor_input]=node;
                  cursor_input++;
                  break;
      case OUTPUT:in_file>>node;
                  output_s[cursor_output]=node;
                  cursor_output++;
                  break;
      case AND: no_inputs=det_inputs(buffer);
                and_s[no_inputs-2].no_gates++;
                strcpy(and_s[no_inputs-2].name,buffer);
                for(i=0;i<=no_inputs;i++)
                in_file>>node;
                break;
      case OR: no_inputs=det_inputs(buffer);
          or_s[no_inputs-2].no_gates++;
          strcpy(or_s[no_inputs-2].name,buffer);
          for(i=0;i<=no_inputs;i++)
                in_file>>node;
          break;
      case NOT: in_file>>node;
                not_s[cursor_not++]=node;
                in_file>>node;
                not_s[cursor_not++]=node;
                break;
      default: if(buffer[0]==0)
                break;
          else
          {
            cout<<"\n\aSyntax error in input file";
            exit(1);
          }
    }
  }
  for(i=0;i<max_and-1;i++)
  {
    and_s[i].gate_nodes=alloc_int((and_s[i].no_gates)*(i+3));
    no_init_gates+=and_s[i].no_gates;
  }
  for(i=0;i<max_or-1;i++)
```

```
  {
    or_s[i].gate_nodes=alloc_int((or_s[i].no_gates)*(i+3));
    no_init_gates+=or_s[i].no_gates;
  }
}
void third_scan(ifstream& in_file)
{
  char buffer[10];
  int i,node,ind_word,no_inputs,curs;
  cout<<"Third scan\n";
  in_file.seekg(0,ios::beg);
  in_file.clear();
  while(!in_file.eof())
  {
    in_file>>buffer;
    ind_word=check_word(buffer);
    switch (ind_word)
    {
      case AND: no_inputs=det_inputs(buffer);
                    curs=and_s[no_inputs-2].cursor;
                    for(i=0;i<=no_inputs;i++)
                    in_file>>and_s[no_inputs-2].gate_nodes[curs+i];
                    and_s[no_inputs-2].cursor+=(no_inputs+1);
                    break;
      case OR: no_inputs=det_inputs(buffer);
                    curs=or_s[no_inputs-2].cursor;
                    for(i=0;i<=no_inputs;i++)
                    in_file>>or_s[no_inputs-2].gate_nodes[curs+i];
                    or_s[no_inputs-2].cursor+=(no_inputs+1);
                    break;
      case NOT:  in_file>>node; //If it is an inverter gate there are
      case INPUT:  //two nodes to be read. If it is just
      case OUTPUT: in_file>>node; //a port, there is only one node to
                    break;  //be read;
      default: if(buffer[0]==0)
                    break;
        else
        {
            cout<<"\n\aSyntax error in input file";
            exit(1);
          }
    }
  }
}
void write_file(ofstream& out_file)
{
  int i,j,k;
  for(i=0;i<max_input;i++)
    out_file<<".INPUT "<<input_s[i]<<"\n";
  for(i=0;i<max_not;i++)
    if(not_s[2*i] != CANCELLED)
      out_file<<".NOT "<<not_s[2*i]<<" "<<not_s[2*i+1]<<"\n";
```

```
  for(i=0;i<max_and-1;i++)
    if(and_s[i].name[0])
      for(j=0;j<and_s[i].no_gates;j++)
  if(and_s[i].gate_nodes[j*(i+3)]!=CANCELLED)
  {
    out_file<<and_s[i].name;
    for(k=0;k<i+3;k++)
        out_file<<" "<<and_s[i].gate_nodes[j*(i+3)+k];
    out_file<<"\n";
  }
for(i=0;i<max_or-1;i++)
    if(or_s[i].name[0])
      for(j=0;j<or_s[i].no_gates;j++)
  if(or_s[i].gate_nodes[j*(i+3)]!=CANCELLED)
  {
    out_file<<or_s[i].name;
    for(k=0;k<i+3;k++)
      out_file<<" "<<or_s[i].gate_nodes[j*(i+3)+k];
    out_file<<"\n";
  }
for(i=0;i<max_output;i++)
    out_file<<".OUTPUT "<<output_s[i]<<"\n";
}
void arrange_inputs(int *begin, int length)
{
  int i,j,i_min,min,aux;
  for(i=0;i<length-2;i++)          //The last is the output node and the
  {                                //second-last  doesn't  need  to  be
                                     exchanged
    min=begin[i];                  //with itself
    i_min=i;
    for(j=i+1;j<length-1;j++)      //The last is the output node which is
                                     not
      if(begin[j]<min)             //to be modified
      {
  min=begin[j];
  i_min=j;
      }
    if(i != i_min)
    {
      aux=begin[i];
      begin[i]=begin[i_min];
      begin[i_min]=aux;
    }
  }
}
void arrange_all_inputs(void)
{
  int i,j;
  for(i=0;i<max_and-1;i++)
    for(j=0;j<and_s[i].no_gates;j++)
      arrange_inputs(and_s[i].gate_nodes+j*(i+3),i+3);
```

```
  for(i=0;i<max_or-1;i++)
    for(j=0;j<or_s[i].no_gates;j++)
      arrange_inputs(or_s[i].gate_nodes+j*(i+3),i+3);
}
int check_inputs(int *begin1, int *begin2, int length)
{
  int i, rez=1;
  if((*begin1 == CANCELLED) || (*begin2 == CANCELLED))
    return 0;
  else
  {
    for(i=0;i<length-1;i++)
      if(begin1[i] != begin2[i])
      {
  rez=0;
  break;
      }
    return rez;
  }
}
void replace_all(int dest, int source)
{
  int i,j;
  for(i=0;i<max_not;i++)
    if(not_s[2*i]==dest)
      not_s[2*i]=source;
  for(i=0;i<max_and-1;i++)
    for(j=0;j<(and_s[i].no_gates)*(i+3);j++)
      if(and_s[i].gate_nodes[j]==dest)
  and_s[i].gate_nodes[j]=source;
  for(i=0;i<max_or-1;i++)
    for(j=0;j<(or_s[i].no_gates)*(i+3);j++)
      if(or_s[i].gate_nodes[j]==dest)
    or_s[i].gate_nodes[j]=source;
  for(i=0;i<max_output;i++)
    if(output_s[i]==dest)
      output_s[i]=source;
}
void optimise_structure(void)
{
  int i,j,k,replacement;
  do
  {
    replacement=0;
    arrange_all_inputs();
    for(i=0;i<max_not-1;i++)
      for(j=i+1;j<max_not;j++)
  if((not_s[2*i]==not_s[2*j]) && (not_s[2*i] != CANCELLED))
  {
    not_s[2*j]=CANCELLED;
    replace_all(not_s[2*j+1],not_s[2*i+1]);
    replacement=1;
```

```
    no_fin_gates--;
  }
    for(i=0;i<max_and-1;i++)
      for(j=0;j<and_s[i].no_gates-1;j++)
  for(k=j+1;k<and_s[i].no_gates;k++)
    if(check_inputs(and_s[i].gate_nodes+j*(i+3),
      and_s[i].gate_nodes+k*(i+3),i+3))
    {
      and_s[i].gate_nodes[k*(i+3)]=CANCELLED;
      replace_all(and_s[i].gate_nodes[k*(i+3)+i+2],
                and_s[i].gate_nodes[j*(i+3)+i+2]);
      replacement=1;
      no_fin_gates--;
    }
  for(i=0;i<max_or-1;i++)
      for(j=0;j<or_s[i].no_gates-1;j++)
  for(k=j+1;k<or_s[i].no_gates;k++)
    if(check_inputs(or_s[i].gate_nodes+j*(i+3),
              or_s[i].gate_nodes+k*(i+3),i+3))
    {
      or_s[i].gate_nodes[k*(i+3)]=CANCELLED;
      replace_all(or_s[i].gate_nodes[k*(i+3)+i+2],
                or_s[i].gate_nodes[j*(i+3)+i+2]);
      replacement=1;
      no_fin_gates--;
    }
  }
  while(replacement);
}
void dealloc_everything(void)
{
  int i;
  for(i=0;i<max_and-1;i++)
    delete and_s[i].gate_nodes;
  for(i=0;i<max_or-1;i++)
    delete or_s[i].gate_nodes;
  delete and_s;
  delete or_s;
  delete not_s;
  delete input_s;
  delete output_s;
}
void main(int no_par, char** par)
{
  ifstream in_file;
  ofstream out_file;
  if(no_par<3)
  {
    cout<<"\nToo few parameters";
    exit(1);
  }
  in_file.open(par[1],ios::in);
```

```
  if(!in_file.good())
  {
    cout<<"\n\aError: The input file cannot be opened!";
    exit(1);
  }
  cout<<"\n---------- Start optimisation ----------\n";
  first_scan(in_file);
  second_scan(in_file);
  third_scan(in_file);
  cout<<no_init_gates<<" gates in the input file\n";
  no_fin_gates=no_init_gates;
  optimise_structure();
  out_file.open(par[2],ios::out);
  if(!out_file.good())
  {
    cout<<"\n\aError: The output file cannot be opened!";
    exit(1);
  }
  write_file(out_file);
  dealloc_everything();
  in_file.close();
  out_file.close();
  cout<<no_fin_gates<<" gates in the output file\n";
  if(no_fin_gates<no_init_gates)
  {
    cout<<"The structure has been compressed at ";
    cout<<((100.0*no_fin_gates)/no_init_gates)<<"% from the initial
                                                              size\n";
  }
  else
    cout<<"The structure could not be compressed\n";
}
```

### 11.1.3 VHDL_TR.CPP

```
#include <iostream.h>
#include <fstream.h>
#include <stdlib.h>
#include <string.h>
#include <ctype.h>
#include <memmanag.h>
#define NO_WORDS 5
#define BUFFER_SIZE 30
#define NO_INPUTS_MAX 25
#define NO_PORTS 300
#define NO_MAX_FILES 5
#define _and 1
#define _or 2
#define _inv 3
#define _input 4
#define _output 5
```

```
#define INPUT_TYPE 1
#define NODE_TYPE 2
#define OUTPUT_TYPE 3

typedef char standard_list[2][NO_INPUTS_MAX];
char *dict[NO_WORDS]={"AND","OR","NOT","INPUT","OUTPUT"};
char buffer[BUFFER_SIZE];
char* node_list;
int input_list[NO_PORTS],output_list[NO_PORTS];
int max_node_number,internal_nodes,input_cursor,output_cursor;
int gate_count[NO_MAX_FILES];
int no_first_net;
ofstream output_file;
void init_list_gate_count(void)
{
  int i;
  for(i=0;i<NO_MAX_FILES;i++)
    gate_count[i]=0;
}
int search_word(char* name, int length)
{
  int i,answer=0;
  char* temp=alloc_char(length+1);
  for(i=0;i<length;i++)
    temp[i]=name[i];
  temp[length]=0;
  for(i=0;i<NO_WORDS;i++)
    if(!strcmp(dict[i],temp))
    {
      answer=i+1;
      break;
    }
  delete temp;
  return answer;
}
int word_limit(char* buffer, int& w_beg, int& w_end)
{
  int i;
  w_beg=w_end=-1;
  for(i=0;i<BUFFER_SIZE && buffer[i]!=0;i++)
  {
    if(buffer[i]>='A' && buffer[i]<='Z' && w_beg==-1)
      w_beg=i;
    if(buffer[i]<'A' || buffer[i]>'Z' && w_beg!=-1)
      w_end=i-1;
    if(w_beg!=-1 && w_end!=-1)
      return 1;
  }
  if(buffer[i]==0 && w_beg>-1)
  {
    w_end=i-1;
    return 1;
  }
```

```
  return 0;
}
void count_ports_and_nodes(int no_file, char** par)
{
  ifstream input_file;
  int w_beg,w_end,index;
  int input_number=0,output_number=0,current_node_number;
  input_cursor=0;
  output_cursor=0;
  input_number=0;
  output_number=0;
  cout<<"\n Processing file "<<no_file;
  input_file.open(par[no_file],ios::in);
  if(!input_file.good())
  {
    cout<<"\n\aError: The input file cannot be opened";
    exit(1);
  }
  input_file>>buffer;
  while(!input_file.eof() || buffer[0])
  {
    if(word_limit(buffer,w_beg,w_end))
      if(index=search_word(&buffer[w_beg],w_end-w_beg+1))
      {
  if(index==_input)
      input_number=1;
  if(index==_output)
      output_number=1;
      }
      else
      {
  cout<<"\n\aError: Syntax error in input file";
  exit(1);
      }
    else
    {
      current_node_number=atoi(buffer);
      if(max_node_number<current_node_number)
    max_node_number=current_node_number;
      if(input_number==1)
      {
    input_list[input_cursor++]=current_node_number;
    input_number=0;
      }
      if(output_number==1)
      {
    output_list[output_cursor++]=current_node_number;
    output_number=0;
      }
    }
    input_file>>buffer;
  }
```

```
  input_file.close();
  if(input_cursor>=NO_PORTS)
  {
    cout<<"\n\aError: Too many input ports";
    exit(1);
  }
  if(output_cursor>=NO_PORTS)
  {
    cout<<"\n\aError: Too many output ports";
    exit(1);
  }
}
int find_node_in_vector(int* vector, int node_num, int total_num)
{
  int i;
  for(i=0; i<NO_PORTS;i++)
  {
    if(vector[i]==node_num)
      return total_num-i-1;
  }
  cout<<"\n\aSerious internal error in the algorithm";
  exit(1);
  return 0;
}
void write_network_entity(int no_file)
{
  output_file<<"LIBRARY ieee;\nUSE ieee.std_logic_1164.all;\n\n";
  output_file<<"ENTITY network"<<(no_file+no_first_net-1)<<" IS\n";
  output_file<<" PORT(d_in : IN std_logic_vector("<<(input_cursor-1);
  output_file<<" DOWNTO 0);\n d_out: OUT std_logic_vector(";
  output_file<<(output_cursor-1)<<" DOWNTO 0));\nEND network";
  output_file<<(no_file+no_first_net-1)<<";\n\n";
}
void write_logic_exp(ifstream& input_file,int no_file)
{
  #define NO_GATE_INP 30
  char gate_name[5];
  int i,no_inputs,w_beg,w_end,index;
  int gate_nodes[NO_GATE_INP];
  input_file>>buffer;
  while(!input_file.eof())
  {
    if(word_limit(buffer,w_beg,w_end))
    {
      index=search_word(&buffer[w_beg],w_end-w_beg+1);
      if(index<=3)
      {
  for(i=0;*(dict[index-1]+i)!=0;i++)
    gate_name[i]=(*(dict[index-1]+i));
  gate_name[i]=0;
  gate_count[no_file-1]++;
  if(index<3)
```

```
    no_inputs=atoi(&buffer[w_end+1]);
  else
    no_inputs=1;
  if(no_inputs>NO_GATE_INP)
  {
    cout<<"\n\aError: One of the gates has too many inputs";
    exit(1);
  }
  for(i=0;i<no_inputs+1;i++)
    input_file>>gate_nodes[i];
  if(node_list[gate_nodes[no_inputs]-1]==NODE_TYPE)
    output_file<<" n"<<gate_nodes[no_inputs];
  else
  {
    output_file<<" d_out("<<find_node_in_vector(output_list,
          gate_nodes[no_inputs],output_cursor);
    output_file<<")";
  }
  output_file<<"<=";
  if (no_inputs>1)
  {
    if(node_list[gate_nodes[0]-1]==NODE_TYPE)
      output_file<<" n"<<gate_nodes[0];
    else
    {
      output_file<<" d_in("<<find_node_in_vector(input_list,
       gate_nodes[0],input_cursor);
      output_file<<")";
    }
    for(i=1;i<no_inputs;i++)
    {
      output_file<<" "<<gate_name;
      if(node_list[gate_nodes[i]-1]==NODE_TYPE)
        output_file<<" n"<<gate_nodes[i];
      else
      {
        output_file<<" d_in("<<find_node_in_vector(input_list,

gate_nodes[i],input_cursor);
        output_file<<")";
      }
    }
    output_file<<";\n";
  }
  else
  {
    output_file<<" "<<gate_name;
    if(node_list[gate_nodes[0]-1]==NODE_TYPE)
      output_file<<" n"<<gate_nodes[0];
    else
    {
      output_file<<" d_in("<<find_node_in_vector(input_list,
```

```
gate_nodes[0],input_cursor);
        output_file<<")";
      }
      output_file<<";\n";
  }
        }
      }
      input_file>>buffer;
  }
}
void write_network_architecture(char* file_name,int no_file)
{
  ifstream input_file;
  int i,it_is_input,it_is_output,w_beg,w_end,index,current_node_number;
  int no_signals;
  node_list=alloc_char(max_node_number+1); //+1 is for safety
  for(i=0;i<max_node_number;i++)
    node_list[i]=0;
  cout<<"\n Reprocessing file "<<no_file;
  output_file<< "ARCHITECTURE arch_network"<<(no_file+no_first_net-1);
  output_file<<" OF network"<<(no_file+no_first_net-1)<<" IS\n";
  input_file.open(file_name,ios::in);
  if(!input_file.good())
  {
    cout<<"\n\aError: The input file cannot be opened";
    exit(1);
  }
  input_file>>buffer;
  while(!input_file.eof()|| buffer[0])
  {
    if(word_limit(buffer,w_beg,w_end))
    {
      index=search_word(&buffer[w_beg],w_end-w_beg+1);
      if(index==_input)
  it_is_input=1;
      else
  it_is_input=0;
      if(index==_output)
  it_is_output=1;
      else
  it_is_output=0;
  }
  else
  {
  current_node_number=atoi(buffer);
  if(it_is_input)
    node_list[current_node_number 1]-INPUT_TYPE;
  else if(it_is_output)
    node_list[current_node_number-1]=OUTPUT_TYPE;
  else if(node_list[current_node_number-1]!=INPUT_TYPE &&
    node_list[current_node_number-1]!=OUTPUT_TYPE)
  {
```

```
    node_list[current_node_number-1]=NODE_TYPE;
    internal_nodes=1;
    }
}
input_file>>buffer;
}
  no_signals=1;
  if (internal_nodes)
{
    output_file<<" SIGNAL n";
    for(i=0;i<max_node_number;i++)
      if(node_list[i]==NODE_TYPE)
      {
  output_file<<(i+1);
  break;
      }
  for(i++;i<max_node_number;i++)
  {
    if(node_list[i]==NODE_TYPE)
    {
    no_signals++;
  if(no_signals%10==0)
    output_file<<",\n              n";
  else
    output_file<<",n";
  output_file<<(i+1);
      }
    }
    output_file<<": std_logic;\n";
  }
  output_file<<"BEGIN\n";
  input_file.seekg(0,ios::beg);
  input_file.clear();
  write_logic_exp(input_file, no_file);
  output_file<<"END arch_network"<<(no_file+no_first_net-1)<<";\n\n";
  input_file.close();
  delete node_list;
}
void write_network_configuration(int no_file)
{
  output_file<<"CONFIGURATION conf_network" <<(no_file+no_first_net-
    1);
  output_file<<" OF network";
  output_file<<(no_file+no_first_net-1)<<" IS\n FOR arch_network";
  output_file<<(no_file+no_first_net-1);
  output_file<<"\nEND FOR;\nEND conf_network" <<(no_file+no_first_net-
    1);
  output_file<<";\n\n";
}
void write_networks(int no_file, char** par)
{
  write_network_entity(no_file);
```

```
  write_network_architecture(par[no_file],no_file);
  write_network_configuration(no_file);
}
void main(int no_par, char** par)
{
  ifstream input_file;
  int i,j,total_gate_count=0;
  init_list_gate_count();
  if(no_par<4)
  {
    cout<<"\n\aError: Too few parameters!";
    exit(1);
  }
  if(no_par>NO_MAX_FILES+3) //3 is for prog.name+out file+ no. first
      net.
  {
    cout<<"\n\aError: Too many files!";
    exit(1);
  }
  no_first_net=atoi(par[no_par-1]);
  output_file.open(par[no_par-2],ios::out);
  if(!output_file.good())
  {
    cout<<"\n\aError: Output file cannot be opened!";
    exit(1);
  }
  for(i=1;i<no_par-2;i++)
  {
    for(j=0;j<NO_PORTS;j++)
    {
      input_list[j]=0;
      output_list[j]=0;
    }
    max_node_number=0;
    internal_nodes=0;
    count_ports_and_nodes(i, par);
    write_networks(i,par);
    cout<<"\nArchitecture no. "<<i<<" contains "<<gate_count[i-1]<<"
      gates";
    total_gate_count+=gate_count[i-1];
  }
  cout<<"\nTotal gate count: "<<total_gate_count;
  output_file.close();
}
```

### 11.1.4 MEMMANAG.H

```
//This is a header file used by the three universal programs
#if !defined( __STDLIB_H )
#include <stdlib.h>
#endif
```

```
char* alloc_err="\n\aError: Insufficient RAM memory for dinamic
    allocation!";
float* alloc_float(int mem_length)
{
  float *pointer;
  if(mem_length>0)
  {
    if(!(pointer=new float[mem_length]))
    {
      cout<<alloc_err;
      exit(1);
    }
    return pointer;
  }
  else
    return NULL;
}
double* alloc_double(int mem_length)
{
  double *pointer;
  if(mem_length>0)
  {
    if(!(pointer=new double[mem_length]))
    {
      cout<<alloc_err;
      exit(1);
    }
    return pointer;
  }
  else
    return NULL;
}
int* alloc_int(int mem_length)
{
  int *pointer;
  if(mem_length>0)
  {
    if(!(pointer=new int[mem_length]))
    {
      cout<<alloc_err;
      exit(1);
    }
    return pointer;
  }
  else
    return NULL;
}
char* alloc_char(int mem_length)
{
  char *pointer;
  if(mem_length>0)
  {
```

```
    if(!(pointer=new char[mem_length]))
    {
      cout<<alloc_err;
      exit(1);
    }
    return pointer;
  }
  else
    return NULL;
}
```

### 11.1.5 MATRIX.H

```
//This is a header file used by the three universal programs
#include <iostream.h>
#include <process.h>
char* alloc_error="\n\aError: Not enough memory for dinamic allocation!";
class vector
{
  private:
    int length;
    double* no;
  public:
    vector(void);
    vector(int);
    ~vector(void);
    double& operator[](int);
    void resize(int);
};
class matrix
{
  private:
    int rows,columns;
    vector* val;
  public:
    matrix(int,int);
    ~matrix();
    vector& operator[](int);
    int no_rows(void);
    int no_columns(void);
};
vector::vector(void)
{
  length=0;
  no=NULL;
}
vector::vector(int nlength)
{
  length=nlength;
  if(length>0)
  {
```

```
      if(!(no=new double[length]))
      {
        cout<<alloc_error;
        exit(1);
      }
    }
    else
      no=NULL;
}
vector::~vector(void)
{
    if(no!=NULL)
      delete no;
}
double& vector::operator[](int index)
{
    if(index<0 || index>=length)
    {
      cout<<"\n\aError: Index value is outside limits";
      exit(1);
    }
    return no[index];
}
void vector::resize(int nlength)
{
    if (no!=NULL)
      delete no;
    length=nlength;
    if(length>0)
    {
      if(!(no=new double[length]))
      {
        cout<<alloc_error;
        exit(1);
      }
    }
    else
      no=NULL;
}
matrix::matrix(int length1, int length2)
{
    int i;
    if(length1<=0 || length2<=0)
    {
      cout<<"\n\aError: The matrix dimensions must be positive!";
      exit(1);
    }
    rows=length1;
    columns=length2;
    if(!(val=new vector[rows]))
    {
      cout<<alloc_error;
```

```
    exit(1);
  }
  for(i=0;i<rows;i++)
    val[i].resize(columns);
}
matrix::~matrix(void)
{
  delete [] val;
}
vector& matrix::operator[](int index)
{
  if(index<0 || index>=rows)
  {
    cout<<"\n\aError: Index value is outside limits";
    exit(1);
  }
  return val[index];
}
int matrix::no_rows(void)
{
  return rows;
}
int matrix::no_columns(void)
{
  return columns;
}
```

## 11.2 Appendix B – C++ Programs for PWM generation

### 11.2.1 ANGLES.CPP

```
#include <iostream.h>
#include <fstream.h>
#include <math.h>
#include <stdlib.h>

int main(int no_par, char** par)
{
  int no_bits, no_sectors;
  int no_lines,i,j,k;
  double val[2],val_dig,step,threshold;
  ofstream f;
  if(no_par<4)
  {
    cout<<"\nError: Too few pparameters!";
    exit(1);
  }
  no_bits=atoi(par[1]);
  no_sectors=atoi(par[2]);
  no_lines=no_sectors/2;
```

```
  step=M_PI/no_lines;
  f.open(par[3],ios::out);
  if(!f.good())
  {
    cout<<"\n\aError: The output file cannot be opened!";
    exit(1);
  }
  f.precision(12);
  f.setf(ios::fixed);
  for(i=1;i<=no_lines;i++)
  {
    if(i*step-step/2!=M_PI_2)
    {
      val[0]=-tan(i*step-step/2);
      if(val[0]<0)
  val[1]=1;
      else
      {
  val[0]=-val[0];
  val[1]=-1;
      }
    }
    else
    {
      val[0]=-1;
      val[1]=0;
    }
    for(j=0;j<2;j++)
      for(k=no_bits-1;k>=0;k--)
      {
  val_dig=val[j]*pow(2,k)/(pow(2,no_bits)-1);
  if(k==no_bits-1)
    val_dig=-val_dig; //In order to have C2 codification for inputs
  f<<val_dig<<" ";
      }
    threshold=(val[0]+val[1])/(pow(2,no_bits)-1);
    f<<(-threshold)<<endl;
  }
  f.close();
  return 0;
}
```

### 11.2.2 REGIONS.CPP

```
#include <iostream.h>
#include <fstream.h>
#include <process.h>
#include <matrix.h>
#include <math.h>
#include <stdlib.h>

void write_matrix(matrix& mat,char* file_name)
```

```
{
  int i,j;
  ofstream f(file_name,ios::out);
  f.precision(12);
  if(!f.good())
  {
    cout<<"\n\aError:Output file could not be open in regions.exe";
    cout<<"\nFile name:"<<file_name;
    exit(1);
  }
  for(i=0;i<mat.no_rows();i++)
  {
    for(j=0;j<mat.no_columns()-1;j++)
      f<<mat[i][j]<<"  ";
    f<<mat[i][mat.no_columns()-1]<<"\n";
  }
  f.close();
}
int main(int no_par, char** par)
{
  int no_bits, no_int_stripes,no_ext_stripes,no_boundaries;
  int i,j,k,ind,outer_limit,matrix_lines,third_vertex_j,line,no_regions;
  double vertical_step;
  if(no_par<6)
  {
    cout<<"\n\aError:Too few parameters";
    exit(1);
  }
  no_bits=atoi(par[1]);
  no_int_stripes=atoi(par[2]);
  no_ext_stripes=atoi(par[3]);
  outer_limit=(no_ext_stripes+no_int_stripes)/2-1;
  matrix_lines=3*(2*outer_limit+1);
  {
    matrix w1(matrix_lines,3);
    for(j=0,i=-outer_limit;i<=outer_limit;i++,j++)
    {
      w1[j][0]=0;
      w1[j][1]=1;
      w1[j][2]=-vertical_step*i-1.0/(pow(2,no_bits)-1);
    }
    for(i=-outer_limit;i<=outer_limit;i++,j++)
    {
      w1[j][0]=-sqrt(3);          //positive slopes y=-w1*x-w3
      w1[j][1]=1;
      w1[j][2]=-2*vertical_step*i-(1-sqrt(3))/(pow(2,no_bits)-1);
    }
    for(i=-outer_limit;i<=outer_limit;i++,j++)
    {
      w1[j][0]=sqrt(3);           //negative slopes y=-w1*x-w3
      w1[j][1]=1;
      w1[j][2]=-2*vertical_step*i-(1+sqrt(3))/(pow(2,no_bits)-1);
```

```
    }
    matrix w_dig(matrix_lines,2*no_bits+1);
    for(i=0;i<matrix_lines;i++)
    {
      for(j=0;j<no_bits;j++)
      {
  w_dig[i][j]=w1[i][0]*pow(2,(no_bits-j-1))/(pow(2,no_bits)-1);
  w_dig[i][no_bits+j]=w1[i][1]*
  pow(2,(no_bits-j-1))/(pow(2,no_bits)-1);
  if(j==0)                //Condition necessary in order to obtain
  {                       //C2 representation for the input values
    w_dig[i][j]=-w_dig[i][j];
    w_dig[i][no_bits+j]=-w_dig[i][no_bits+j];
  }
      }
      w_dig[i][2*no_bits]=w1[i][2];
    }
    write_matrix(w_dig,par[4]);
  }
  ofstream f2(par[5],ios::out);
  if(!f2.good())
  {
    cout<<"\n\aError:Output file could not be opened by regions.exe";
    cout<<"\nFile name:"<<par[5];
    exit(1);
  }
  no_regions=6*(no_int_stripes+no_ext_stripes)*
          (no_int_stripes+no_ext_stripes)/4;
  matrix points(no_regions,2);
  line=0;
  for(i=-outer_limit-1;i<=outer_limit;i++)          //i & j are vertex
                                                       coordinates
    for(j=-outer_limit-1;j<=outer_limit+1;j++)
      if(j-i<=outer_limit+1 && j-i>=-outer_limit)
  //j-i=index of positive slope boundary
  for(k=-1;k<=1;k+=2)
  {
    third_vertex_j=j+k;
    if(third_vertex_j>=-outer_limit-1 && third_vertex_j<=outer_limit+1)
    {
      no_boundaries=0;
      if(k>0)
      {
        for(ind=-outer_limit;ind<=outer_limit;ind++)
                if(j==ind) //horizontal boundary
                {
                f2<<"1 ";
                no_boundaries++;
                }
                else
                f2<<"0 ";
    for(ind=-outer_limit;ind<=outer_limit;ind++)
```

```
            if(j-i==ind)  //positive slope boundary
            {
            f2<<"-1 ";
            no_boundaries++;
            }
            else
            f2<<"0 ";
  for(ind=-outer_limit;ind<=outer_limit;ind++)
            if(i+1==ind) //negative slope boundary
            {
            f2<<"-1 ";
            no_boundaries++;
            }
            else
            f2<<"0 ";
  }
  else
  {
  for(ind=-outer_limit;ind<=outer_limit;ind++)
            if(j==ind) //horizontal boundary
            {
            f2<<"-1 ";
            no_boundaries++;
            }
            else
            f2<<"0 ";
  for(ind=-outer_limit;ind<=outer_limit;ind++)
            if(j-(i+1)==ind) //positive slope boundary
            {
            f2<<"1 ";
            no_boundaries++;
            }
            else
            f2<<"0 ";
  for(ind=-outer_limit;ind<=outer_limit;ind++)
            if(i==ind) //negative slope boundary
            {
            f2<<"1 ";
            no_boundaries++;
            }
            else
            f2<<"0 ";
    }
    f2<<(-no_boundaries+0.5)<<"\n";
    points[line][0]=2.0/no_int_stripes*(i-j/2.0+0.5);
    points[line][1]=2.0/no_int_stripes*(j+k/3.0)*sqrt(3)/2;
    line++;
  }
 }
 f2.close();
 write_matrix(points,par[6]);
 return 0;
}
```

## 11.2.3 CTRL.CPP

```
#include <iostream.h>
#include <fstream.h>
#include <string.h>
#include <stdlib.h>
#include <math.h>
#include <memmanag.h>
int no_regions,no_angles;
double xh[7]={1,0.5,-0.5,-1,-0.5,0.5,0.0},yh[7]={0,sqrt(3)/2, sqrt(3)/
              2,0,
       -sqrt(3)/2,-sqrt(3)/2,0.0};
char* there_is_neu;
double* read_matrix(double *w,ifstream& in_file)
{
  int i;
  char buffer[40];
  no_regions=0;
  in_file>>buffer;
  while(buffer[0])
  {
    no_regions++;
    in_file>>buffer;
  }
  no_regions/=2;
  in_file.seekg(0,ios::beg);
  in_file.clear();
  w=alloc_double(2*no_regions);
  for(i=0;i<2*no_regions;i++)
    in_file>>w[i];
  return w;
}
double minimal_difference(double ang1, double ang2)
{
  double dif;
  dif=fabs(ang2-ang1);
  while(dif>M_PI)
    dif=fabs(2*M_PI-dif);
  return dif;
}
int right_direction(double x, double y, int k, int index)
{
  int i,i_min;
  double step=2*M_PI/no_angles;
  double ang[7],min;
  for(i=0;i<7;i++)
  {
    ang[i]=atan2(yh[i]-y,xh[i]-x);
    ang[i]=minimal_difference(ang[i],k*step);
  }
  min=ang[0];
  i_min=0;
  for(i=1;i<7;i++)
```

```
    if(ang[i]<=min)         //The condition <= instead of < helps including
    {                       //the zero inverter voltage in the considered
                              set
      min=ang[i];           //of possibilities. Zero voltage is the last in
                              the
      i_min=i;              //lists xh and yh.
    }
  if (i_min==index)
    return 1;
  else
    return 0;
}
void write_first_layer(double *w, ofstream& out_file)
{
  int i,j,k,l,num,not_null;
  double step=2*M_PI/no_angles;
  int *neu_o=alloc_int(no_angles/2);
  int *neu_i=alloc_int(no_angles/2);
  for(i=0;i<7;i++)
    for(j=0;j<no_regions;j++)
    {
      num=0;
      for(k=0;k<no_angles/2;k++)
  neu_o[k]=0;
      for(k=0;k<no_angles;k++)      //Checking each possible direction
  if(right_direction(w[2*j], w[2*j+1],k,i))
  {
    for(l=0;l<no_angles/2;l++) //To determine the corresponding
                            //result from angle calculation network
      if((k*step>=l*step+step/2) && (k*step<l*step+step/2+M_PI))
        neu_i[l]=1;
      else
        neu_i[l]=-1;
    for(l=0;l<no_angles/2;l++)
      neu_o[l]+=neu_i[l];
    num++;
  }
      not_null=0; //This variable counts the input weights which are not
                    null
      for(k=0;k<no_angles/2;k++)
  if(neu_o[k]==num && num>0)
  {
    neu_o[k]=1;
    not_null++;
  }
  else if(neu_o[k]==-num && num>0)
  {
    neu_o[k]=-1;
    not_null++;
  }
  else
    neu_o[k]=0;
```

```
    if(not_null)
    {
  for(k=0;k<no_regions;k++)
    if(j==k)
      out_file<<"1 ";
    else
      out_file<<"0 ";
  for(k=0;k<no_angles/2;k++)
    out_file<<neu_o[k]<<" ";
  out_file<<(-not_null-0.5)<<"\n";
  there_is_neu[i*no_regions+j]=1;
      }
      else
  there_is_neu[i*no_regions+j]=0;
    }
  delete neu_o;
  delete neu_i;
}
void write_second_layer(ofstream& out_file)
{
  int i,j,k,ones,no_inputs;
  int corelation[3][3]={{5,0,1},{1,2,3},{3,4,5}};
  for(i=0;i<3;i++)
  {
    no_inputs=0;
    for(j=0;j<7*no_regions;j++)
    {
      ones=0;
      for(k=0;k<3;k++)
  if(corelation[i][k] == j/no_regions || j/no_regions == 6)
  {
    ones=1;
    break;
  }
      if(there_is_neu[j])
      {
  if(ones)
  {
    out_file<<"1 ";
    no_inputs++;
  }
  else
    out_file<<"0 ";
      }
    }
    if(no_inputs)
    out_file<<(no_inputs-0.5)<<"\n";
  }
}
void main(int no_par, char** par)
{
  double* w;
```

```
  ifstream in_file;
  ofstream out_file;
  if(no_par<5)
  {
    cout<<"\n\aToo few parameters";
    exit(1);
  }
  no_angles=atoi(par[4]);
  if(no_angles%2==1)
  {
    cout<<"\a\nError:The number of angles cannot be odd";
    exit(1);
  }
  in_file.open(par[1],ios::in);

  if (!in_file.good())
  {
      cout<<"\n\aError: The input file could not be opened";
      exit(1);
  }
  w=read_matrix(w,in_file);
  there_is_neu=alloc_char(7*no_regions);
  in_file.close();
  out_file.open(par[2],ios::out);
  if (!out_file.good())
  {
      cout<<"\n\aError: The output file could not be opened";
      exit(1);
  }
  write_first_layer(w, out_file);
  out_file.close();
  out_file.open(par[3],ios::out);
  if (!out_file.good())
  {
      cout<<"\n\aError: The output file could not be opened";
      exit(1);
  }
  write_second_layer(out_file);
  out_file.close();
  delete w;
  delete there_is_neu;
}
```

## 11.3 Appendix C – Subnetworks VHDL models

### 11.3.1 The angle subnetwork

```
LIBRARY ieee;
USE ieee.std_logic_1164.all;
ENTITY network2 IS
  PORT(d_in : IN std_logic_vector(9 DOWNTO 0);
      d_out: OUT std_logic_vector(17 DOWNTO 0));
```

```
END network2;
ARCHITECTURE arch_network2 OF network2 IS
    SIGNAL n11,n12,n13,n14,n15,n16,n17,n18,n19,
           n20,n21,n22,n29,n30,n31,n32,n33,n34,n35,
           n36,n37,n38,n39,n40,n41,n43,n44,n46,n47,
           n48,n49,n57,n58,n59,n60,n62,n63,n64,n65,
           n69,n70,n76,n77,n78,n79,n80,n81,n82,n83,
           n84,n85,n89,n90,n91,n92,n93,n100,n101,n103,
           n104,n105,n106,n110,n111,n112,n113,n114,n115,n116,
           n117,n118,n120,n121,n122,n123,n124,n125,n127,n128,
           n129,n130,n131,n132,n133,n135,n137,n138,n139,n140,
           n147,n148,n150,n151,n152,n157,n158,n159,n160,n173,
           n174,n175,n176,n205,n206,n215,n216,n217,n218,n219,
           n222,n223,n225,n226,n230,n231,n232,n233,n235,n236,
           n238,n239,n240,n241,n242,n243,n244,n245,n246,n250,
           n251,n252,n253,n260,n261,n263,n264,n265,n266,n267,
           n270,n271,n272,n273,n274,n275,n276,n277,n278,n280,
           n281,n282,n283,n287,n288,n290,n291,n292,n293,n294,
           n302,n305,n306,n310,n311,n312,n313,n314,n315,n316,
           n317,n318,n319,n320,n321,n322,n330,n331,n332,n333,
           n334,n335,n336,n337,n343,n344,n345,n346,n347,n348,
           n349,n350,n351,n352,n353,n361,n362,n365,n366,n367,
           n368,n369,n370,n372,n373,n375,n376,n377,n378,n379,
           n380,n381,n382,n383,n384,n385,n395,n396,n397,n398,
           n399,n400,n401,n405,n406,n407,n408,n409,n410,n411,
           n412,n413,n414,n415,n419,n420,n421,n422,n423,n424,
           n426,n427,n437,n438,n439,n440,n441,n443,n444,n446,
           n447,n448,n450,n452,n453,n454,n455,n456,n457,n459,
           n460,n461,n462,n466,n467,n468,n469,n472,n473,n474,
           n475,n476,n477,n478,n479,n490,n491,n496,n497,n505,
           n506,n507,n508,n509,n529,n530,n542,n544,n545,n548,
           n551,n552,n553,n554,n555,n556,n557,n558,n559,n561,
           n562,n563,n564,n565,n566,n570,n571,n572,n573,n576,
           n577,n579,n580,n581,n582,n593,n595,n596,n597,n598,
           n602,n603,n604,n605,n606,n607,n608,n610,n611,n612,
           n616,n617,n618,n619,n620,n621,n622,n623,n624,n632,
           n633,n637,n638,n639,n640,n641,n643,n644,n646,n647,
           n648,n649,n650,n651,n652,n653,n661,n662,n667,n668,
           n669: std_logic;
BEGIN
  n11<= NOT d_in(4);
  n12<= NOT d_in(8);
  n13<= NOT d_in(7);
  n18<= NOT d_in(6);
  n19<= NOT d_in(5);
  n343<= NOT d_in(3);
  n344<= NOT d_in(2);
  n347<= NOT d_in(1);
  n348<= NOT d_in(0);
  n15<= d_in(0) AND n14;
  n17<= n11 AND n16;
  n20<= n18 AND n19;
```

```
n30<= d_in(2) AND n29;
n31<= d_in(0) AND n13;
n33<= n12 AND n32;
n36<= n11 AND n35;
n38<= n13 AND n37;
n41<= n12 AND n40;
n44<= n13 AND n43;
n47<= d_in(1) AND n46;
n58<= d_in(1) AND n57;
n60<= d_in(0) AND n59;
n63<= n13 AND n62;
n65<= d_in(2) AND n64;
n70<= d_in(1) AND n69;
n77<= n12 AND n76;
n79<= n11 AND n78;
n80<= d_in(0) AND n18;
n82<= d_in(1) AND n81;
n85<= d_in(2) AND n84;
n91<= n12 AND n90;
n101<= d_in(1) AND n100;
n104<= d_in(3) AND n103;
n106<= d_in(2) AND n105;
n111<= n13 AND n110;
n114<= n12 AND n113;
n118<= n11 AND n117;
n120<= n18 AND n115;
n121<= d_in(0) AND n19;
n123<= n12 AND n122;
n125<= n13 AND n124;
n128<= d_in(1) AND n127;
n130<= d_in(2) AND n129;
n133<= d_in(3) AND n132;
n135<= d_in(1) AND n37;
n140<= d_in(9) AND n139;
n148<= n18 AND n147;
n150<= d_in(1) AND n115;
n152<= n13 AND n151;
n158<= d_in(2) AND n157;
n160<= n12 AND n159;
n174<= d_in(3) AND n173;
n176<= d_in(9) AND n175;
n206<= n11 AND n205;
n215<= d_in(1) AND n19;
n217<= n12 AND n216;
n219<= n13 AND n218;
n223<= d_in(2) AND n222;
n226<= d_in(3) AND n225;
n233<= d_in(9) AND n232;
n236<= d_in(3) AND n235;
n238<= d_in(2) AND n124;
n240<= n18 AND n239;
n243<= n13 AND n242;
```

```
n246<= n12 AND n245;
n253<= n11 AND n252;
n261<= n18 AND n260;
n263<= d_in(2) AND n239;
n265<= n13 AND n264;
n267<= n18 AND n266;
n271<= d_in(3) AND n270;
n273<= d_in(9) AND n272;
n274<= d_in(1) AND d_in(0);
n276<= n18 AND n275;
n278<= n19 AND n277;
n281<= d_in(2) AND n280;
n283<= n13 AND n282;
n288<= n18 AND n287;
n292<= d_in(3) AND n291;
n302<= d_in(2) AND n147;
n306<= d_in(3) AND n305;
n311<= n18 AND n310;
n313<= d_in(9) AND n312;
n315<= d_in(3) AND n314;
n317<= n18 AND n316;
n319<= n13 AND n318;
n330<= d_in(2) AND n277;
n332<= d_in(9) AND n331;
n334<= n19 AND n333;
n335<= d_in(3) AND d_in(2);
n346<= d_in(9) AND n345;
n349<= n347 AND n348;
n351<= n343 AND n350;
n362<= n343 AND n361;
n366<= n19 AND n365;
n368<= n18 AND n367;
n370<= d_in(9) AND n369;
n373<= n343 AND n372;
n376<= n18 AND n375;
n379<= n13 AND n378;
n382<= d_in(4) AND n381;
n385<= n12 AND n384;
n396<= n344 AND n395;
n399<= n13 AND n398;
n401<= n18 AND n400;
n406<= n343 AND n405;
n409<= d_in(9) AND n408;
n411<= n18 AND n410;
n413<= n344 AND n412;
n415<= n13 AND n414;
n420<= n344 AND n419;
n422<= n343 AND n421;
n424<= d_in(4) AND n423;
n427<= n12 AND n426;
n438<= n347 AND n437;
n439<= n19 AND n348;
```

```
n441<= n12 AND n440;
n444<= n344 AND n443;
n448<= n343 AND n447;
n450<= n18 AND n412;
n455<= d_in(9) AND n454;
n457<= n343 AND n456;
n459<= n18 AND n395;
n460<= n19 AND n347;
n462<= n13 AND n461;
n469<= n12 AND n468;
n473<= n344 AND n472;
n478<= d_in(4) AND n477;
n491<= n13 AND n490;
n497<= n12 AND n496;
n506<= n343 AND n505;
n509<= d_in(9) AND n508;
n530<= d_in(4) AND n529;
n542<= n18 AND n437;
n545<= n343 AND n544;
n548<= n13 AND n443;
n552<= n12 AND n551;
n554<= n347 AND n553;
n559<= d_in(4) AND n558;
n561<= n12 AND n456;
n563<= n347 AND n562;
n564<= n18 AND n348;
n566<= n344 AND n565;
n573<= n343 AND n572;
n577<= n13 AND n576;
n582<= d_in(9) AND n581;
n593<= n13 AND n562;
n596<= n344 AND n595;
n598<= n347 AND n597;
n603<= n12 AND n602;
n606<= d_in(4) AND n605;
n608<= n347 AND n607;
n610<= n13 AND n553;
n612<= n344 AND n611;
n617<= n13 AND n616;
n619<= n12 AND n618;
n621<= d_in(9) AND n620;
n624<= n343 AND n623;
n633<= n12 AND n632;
n637<= n59 AND n348;
n639<= n347 AND n638;
n641<= d_in(4) AND n640;
n644<= n12 AND n643;
n647<= n347 AND n646;
n650<= n344 AND n649;
n662<= d_in(4) AND n661;
n667<= n12 AND n21;
n34<= d_in(1) AND d_in(0) AND n13;
```

```
n39<= d_in(0) AND n18 AND n19;
n83<= d_in(0) AND n13 AND n18;
n93<= d_in(9) AND d_in(3) AND n92;
n241<= d_in(1) AND d_in(0) AND n19;
n290<= d_in(2) AND n19 AND n277;
n294<= n11 AND n12 AND n293;
n322<= n11 AND n12 AND n321;
n397<= n19 AND n347 AND n348;
n446<= n18 AND n347 AND n437;
n453<= n13 AND n344 AND n452;
n467<= n18 AND n344 AND n466;
n476<= n13 AND n343 AND n475;
n555<= n18 AND n19 AND n348;
n557<= n13 AND n344 AND n556;
n571<= n13 AND n347 AND n570;
n580<= n12 AND n344 AND n579;
n653<= d_in(9) AND n343 AND n652;
n49<= d_in(9) AND d_in(3) AND d_in(2) AND n48;
n89<= d_in(0) AND n13 AND n18 AND n19;
n112<= d_in(1) AND d_in(0) AND n18 AND n19;
n138<= d_in(2) AND n12 AND n13 AND n137;
n231<= d_in(2) AND n13 AND n18 AND n230;
n244<= d_in(2) AND d_in(1) AND n18 AND n19;
n251<= d_in(3) AND d_in(2) AND n13 AND n250;
n377<= n19 AND n344 AND n347 AND n348;
n474<= n18 AND n19 AND n347 AND n348;
n648<= n13 AND n18 AND n19 AND n348;
n116<= d_in(2) AND d_in(1) AND n13 AND n18 AND n115;
n131<= d_in(1) AND d_in(0) AND n13 AND n18 AND n19;
n337<= n11 AND n12 AND n13 AND n18 AND n336;
n353<= d_in(4) AND n12 AND n13 AND n18 AND n352;
n380<= n18 AND n19 AND n343 AND n344 AND n347;
n407<= n18 AND n19 AND n344 AND n347 AND n348;
n604<= n13 AND n18 AND n19 AND n347 AND n348;
n651<= n12 AND n13 AND n18 AND n347 AND n348;
n669<= d_in(9) AND n343 AND n344 AND n347 AND n668;
n320<= d_in(3) AND d_in(2) AND d_in(1) AND d_in(0) AND n18 AND n19;
n507<= n13 AND n18 AND n19 AND n344 AND n347 AND n348;
n22<= d_in(9) AND d_in(3) AND d_in(2) AND d_in(1) AND d_in(0) AND n12
      AND n21;
n383<= n13 AND n18 AND n19 AND n343 AND n344 AND n347 AND n348;
n622<= n12 AND n13 AND n18 AND n19 AND n344 AND n347 AND n348;
n479<= n12 AND n13 AND n18 AND n19 AND n343 AND n344 AND n347 AND n348;
n14<= n12 OR n13;
n21<= n13 OR n20;
d_out(17)<= n17 OR n22;
n32<= d_in(1) OR n31;
n37<= d_in(0) OR n18;
n43<= d_in(0) OR n20;
n46<= n39 OR n44;
n48<= n41 OR n47;
d_out(16)<= n36 OR n49;
```

```
n59<= n18 OR n19;
n62<= n20 OR n60;
n76<= n63 OR n70;
n81<= n13 OR n80;
n90<= n82 OR n89;
n92<= n85 OR n91;
d_out(15)<= n79 OR n93;
n110<= n80 OR n101;
n115<= d_in(0) OR n19;
n127<= n18 OR n121;
n129<= n125 OR n128;
n137<= n39 OR n135;
n139<= n133 OR n138;
d_out(14)<= n118 OR n140;
n157<= n148 OR n150;
n173<= n152 OR n158;
n205<= n160 OR n174;
d_out(13)<= n176 OR n206;
n222<= n18 OR n150;
n230<= n121 OR n150;
n239<= d_in(1) OR n19;
n250<= n148 OR n241;
n252<= n246 OR n251;
d_out(12)<= n233 OR n253;
n270<= n263 OR n267;
n277<= d_in(1) OR d_in(0);
n280<= n274 OR n278;
n291<= n288 OR n290;
n293<= n283 OR n292;
d_out(11)<= n273 OR n294;
n310<= n215 OR n302;
n316<= d_in(2) OR n19;
n318<= n315 OR n317;
n321<= n319 OR n320;
d_out(10)<= n313 OR n322;
n336<= n334 OR n335;
d_out(9)<= n332 OR n337;
n350<= n344 OR n349;
n352<= n19 OR n351;
d_out(8)<= n346 OR n353;
n365<= n347 OR n348;
n367<= n344 OR n366;
n381<= n379 OR n380;
n384<= n382 OR n383;
d_out(7)<= n370 OR n385;
n412<= n19 OR n347;
n419<= n19 OR n349;
n421<= n411 OR n420;
n423<= n415 OR n422;
n426<= n383 OR n424;
d_out(6)<= n409 OR n427;
n437<= n19 OR n348;
```

```
n452<= n397 OR n450;
n466<= n438 OR n439;
n472<= n18 OR n438;
n475<= n473 OR n474;
n477<= n469 OR n476;
n553<= n18 OR n348;
n556<= n554 OR n555;
n570<= n439 OR n542;
n576<= n347 OR n542;
n579<= n474 OR n577;
n581<= n573 OR n580;
d_out(3)<= n559 OR n582;
n616<= n20 OR n348;
n618<= n608 OR n617;
n620<= n612 OR n619;
n623<= n621 OR n622;
d_out(2)<= n606 OR n624;
n638<= n13 OR n637;
n652<= n650 OR n651;
d_out(1)<= n641 OR n653;
n668<= n348 OR n667;
d_out(0)<= n662 OR n669;
n40<= d_in(1) OR n38 OR n39;
n57<= d_in(0) OR n13 OR n20;
n64<= n12 OR n58 OR n63;
n69<= n13 OR n20 OR n60;
n84<= n12 OR n82 OR n83;
n100<= d_in(0) OR n18 OR n19;
n113<= n106 OR n111 OR n112;
n132<= n123 OR n130 OR n131;
n147<= d_in(1) OR d_in(0) OR n19;
n151<= d_in(2) OR n148 OR n150;
n159<= d_in(3) OR n152 OR n158;
n175<= n11 OR n160 OR n174;
n225<= n112 OR n219 OR n223;
n242<= n238 OR n240 OR n241;
n245<= n236 OR n243 OR n244;
n264<= d_in(3) OR n261 OR n263;
n266<= d_in(2) OR d_in(1) OR n19;
n275<= d_in(2) OR n19 OR n274;
n282<= d_in(3) OR n276 OR n281;
n287<= d_in(2) OR n274 OR n278;
n305<= n18 OR n278 OR n302;
n314<= d_in(2) OR n18 OR n19;
n361<= n18 OR n19 OR n344;
n375<= n19 OR n344 OR n349;
n378<= n373 OR n376 OR n377;
n395<= n19 OR n347 OR n348;
n405<= n396 OR n397 OR n401;
n410<= n19 OR n344 OR n347;
n414<= n343 OR n411 OR n413;
n443<= n18 OR n347 OR n439;
```

```
n447<= n13 OR n444 OR n446;
n461<= n344 OR n459 OR n460;
n468<= n457 OR n462 OR n467;
d_out(5)<= n455 OR n478 OR n479;
n505<= n444 OR n474 OR n491;
n529<= n497 OR n506 OR n507;
d_out(4)<= n479 OR n509 OR n530;
n551<= n344 OR n446 OR n548;
n562<= n18 OR n19 OR n348;
n565<= n13 OR n563 OR n564;
n572<= n561 OR n566 OR n571;
n602<= n555 OR n593 OR n598;
n607<= n13 OR n18 OR n348;
n611<= n12 OR n608 OR n610;
n632<= n13 OR n347 OR n348;
n646<= n13 OR n20 OR n348;
n649<= n644 OR n647 OR n648;
n78<= d_in(9) OR d_in(3) OR n65 OR n77;
n105<= d_in(1) OR d_in(0) OR n13 OR n18;
n117<= d_in(9) OR n104 OR n114 OR n116;
n124<= d_in(1) OR d_in(0) OR n18 OR n19;
n232<= n11 OR n217 OR n226 OR n231;
n260<= d_in(2) OR d_in(1) OR d_in(0) OR n19;
n272<= n11 OR n12 OR n265 OR n271;
n333<= d_in(3) OR d_in(2) OR d_in(1) OR d_in(0);
n372<= n18 OR n19 OR n344 OR n349;
n398<= n18 OR n343 OR n396 OR n397;
n400<= n19 OR n344 OR n347 OR n348;
n454<= d_in(4) OR n441 OR n448 OR n453;
n490<= n18 OR n344 OR n347 OR n439;
n496<= n343 OR n444 OR n474 OR n491;
n508<= d_in(4) OR n497 OR n506 OR n507;
n558<= d_in(9) OR n545 OR n552 OR n557;
n595<= n12 OR n347 OR n555 OR n593;
n597<= n13 OR n18 OR n19 OR n348;
n643<= n13 OR n20 OR n347 OR n348;
n16<= d_in(9) OR d_in(3) OR d_in(2) OR d_in(1) OR n15;
n35<= d_in(9) OR d_in(3) OR n30 OR n33 OR n34;
n103<= d_in(2) OR n12 OR n13 OR n39 OR n101;
n122<= d_in(2) OR d_in(1) OR n13 OR n120 OR n121;
n216<= d_in(3) OR d_in(2) OR n13 OR n148 OR n215;
n218<= d_in(2) OR d_in(1) OR d_in(0) OR n18 OR n19;
n235<= d_in(2) OR d_in(1) OR n13 OR n18 OR n121;
n312<= n11 OR n12 OR n13 OR n306 OR n311;
n369<= d_in(4) OR n12 OR n13 OR n362 OR n368;
n408<= d_in(4) OR n12 OR n399 OR n406 OR n407;
n605<= d_in(9) OR n343 OR n596 OR n603 OR n604;
n640<= d_in(9) OR n343 OR n344 OR n633 OR n639;
n29<= d_in(1) OR d_in(0) OR n12 OR n13 OR n18 OR n19;
n440<= n13 OR n18 OR n343 OR n344 OR n438 OR n439;
n456<= n13 OR n18 OR n19 OR n344 OR n347 OR n348;
n544<= n12 OR n13 OR n344 OR n347 OR n439 OR n542;
```

```
  n331<= d_in(3) OR n11 OR n12 OR n13 OR n18 OR n19 OR n330;
  n345<= d_in(4) OR n12 OR n13 OR n18 OR n19 OR n343 OR n344;
  n661<= d_in(9) OR n12 OR n13 OR n343 OR n344 OR n347 OR n348;
END arch_network2;
CONFIGURATION conf_network2 OF network2 IS
  FOR arch_network2
  END FOR;
END conf_network2;
```

### 11.3.2 The position subnetwork

```
LIBRARY ieee;
USE ieee.std_logic_1164.all;
ENTITY network1 IS
  PORT(d_in : IN std_logic_vector(9 DOWNTO 0);
       d_out: OUT std_logic_vector(53 DOWNTO 0));
END network1;
ARCHITECTURE arch_network1 OF network1 IS
    SIGNAL   n11,n12,n13,n14,n16,n17,n18,n21,n22,
             n24,n25,n26,n28,n29,n30,n31,n32,n33,n34,
             n35,n36,n37,n38,n39,n40,n41,n42,n43,n44,
             n45,n46,n47,n48,n49,n50,n51,n52,n53,n54,
             n55,n57,n58,n59,n60,n61,n62,n63,n64,n65,
             n66,n67,n68,n74,n75,n80,n81,n86,n87,n93,
             n94,n95,n96,n97,n98,n99,n100,n108,n109,n110,
             n116,n117,n133,n134,n149,n150,n151,n152,n153,n154,
             n163,n164,n165,n166,n173,n174,n183,n184,n185,n197,
             n198,n199,n217,n218,n220,n221,n239,n240,n241,n242,
             n243,n245,n246,n247,n248,n249,n250,n251,n253,n254,
             n258,n259,n260,n261,n264,n265,n266,n267,n268,n269,
             n270,n271,n272,n273,n274,n275,n276,n277,n278,n279,
             n280,n281,n282,n283,n284,n285,n288,n289,n293,n294,
             n296,n297,n298,n304,n305,n314,n315,n316,n317,n324,
             n325,n326,n329,n330,n334,n335,n348,n349,n367,n368,
             n369,n372,n373,n379,n380,n390,n391,n399,n400,n401,
             n411,n412,n450,n451,n452,n453,n455,n457,n460,n462,
             n465,n469,n470,n487,n491,n492,n514,n521,n547: std_logic;
BEGIN
  n11<= NOT d_in(4);
  n28<= NOT d_in(8);
  n29<= NOT d_in(7);
  n30<= NOT d_in(6);
  n31<= NOT d_in(5);
  n243<= NOT d_in(9);
  n452<= NOT n154;
  n453<= NOT n285;
  n455<= NOT n14;
  n457<= NOT n199;
  n460<= NOT n18;
  n462<= NOT n242;
  n465<= NOT n11;
```

```
n469<= NOT n110;
n470<= NOT n326;
n487<= NOT n22;
n491<= NOT n68;
n492<= NOT n369;
n514<= NOT n26;
n521<= NOT n412;
n547<= NOT n451;
n13<= d_in(3) AND n12;
n22<= n11 AND n21;
n26<= n11 AND n25;
n34<= d_in(9) AND n33;
n36<= d_in(2) AND n35;
n38<= n30 AND n37;
n41<= d_in(3) AND n40;
n43<= d_in(2) AND n42;
n45<= n30 AND n44;
n47<= n29 AND n46;
n50<= n11 AND n49;
n52<= d_in(1) AND n51;
n53<= d_in(0) AND n31;
n55<= d_in(2) AND n54;
n59<= d_in(3) AND n58;
n60<= d_in(1) AND n31;
n62<= d_in(2) AND n61;
n65<= n29 AND n64;
n67<= n28 AND n66;
n75<= n29 AND n74;
n81<= d_in(9) AND n80;
n87<= n28 AND n86;
n94<= n29 AND n93;
n97<= d_in(3) AND n96;
n100<= n11 AND n99;
n117<= n28 AND n116;
n134<= d_in(9) AND n133;
n150<= n28 AND n149;
n153<= n11 AND n152;
n164<= n30 AND n163;
n166<= n11 AND n165;
n174<= n28 AND n173;
n185<= d_in(9) AND n184;
n218<= n29 AND n217;
n221<= n11 AND n220;
n240<= d_in(9) AND n239;
n246<= d_in(1) AND n245;
n247<= d_in(5) AND d_in(0);
n249<= n243 AND n248;
n251<= d_in(8) AND n250;
n254<= d_in(2) AND n253;
n259<= d_in(6) AND n258;
n261<= d_in(3) AND n260;
n265<= d_in(2) AND n264;
```

```
n268<= d_in(7) AND n267;
n270<= n11 AND n269;
n272<= d_in(7) AND n271;
n274<= d_in(6) AND n273;
n277<= d_in(2) AND n276;
n280<= d_in(3) AND n279;
n284<= d_in(8) AND n283;
n289<= d_in(3) AND n288;
n294<= d_in(7) AND n293;
n298<= n243 AND n297;
n305<= d_in(8) AND n304;
n317<= n11 AND n316;
n330<= n11 AND n329;
n335<= d_in(8) AND n334;
n349<= n243 AND n348;
n373<= d_in(7) AND n372;
n380<= n11 AND n379;
n391<= d_in(8) AND n390;
n401<= n243 AND n400;
n451<= n243 AND n450;
d_out(53)<= n452 AND n453;
d_out(52)<= n154 AND n455;
d_out(50)<= n199 AND n460;
d_out(48)<= n242 AND n465;
d_out(47)<= n11 AND n453;
d_out(46)<= n469 AND n470;
d_out(38)<= n22 AND n470;
d_out(37)<= n491 AND n492;
d_out(27)<= n26 AND n492;
d_out(26)<= n369 AND n455;
d_out(16)<= n242 AND n369;
d_out(15)<= n412 AND n460;
d_out(7)<= n199 AND n412;
d_out(6)<= n451 AND n465;
d_out(5)<= n11 AND n491;
d_out(3)<= n22 AND n469;
d_out(1)<= n26 AND n452;
d_out(0)<= n154 AND n451;
n17<= d_in(3) AND d_in(2) AND n16;
n24<= d_in(2) AND d_in(1) AND d_in(0);
n39<= d_in(1) AND d_in(0) AND n31;
n57<= d_in(1) AND n30 AND n51;
n183<= d_in(3) AND n29 AND n108;
n198<= n11 AND n28 AND n197;
n266<= d_in(6) AND d_in(5) AND d_in(1);
n275<= d_in(5) AND d_in(1) AND d_in(0);
n296<= d_in(6) AND d_in(2) AND n281;
n315<= d_in(7) AND d_in(3) AND n314;
n368<= d_in(8) AND n11 AND n367;
n399<= d_in(7) AND d_in(3) AND n324;
d_out(51)<= n14 AND n453 AND n457;
d_out(49)<= n18 AND n453 AND n462;
```

```
d_out(45)<= n110 AND n285 AND n455;
d_out(44)<= n14 AND n452 AND n470;
d_out(43)<= n154 AND n285 AND n460;
d_out(42)<= n18 AND n457 AND n470;
d_out(41)<= n199 AND n285 AND n465;
d_out(40)<= n11 AND n462 AND n470;
d_out(39)<= n242 AND n285 AND n487;
d_out(36)<= n68 AND n326 AND n455;
d_out(35)<= n14 AND n469 AND n492;
d_out(34)<= n110 AND n326 AND n460;
d_out(33)<= n18 AND n452 AND n492;
d_out(32)<= n154 AND n326 AND n465;
d_out(31)<= n11 AND n457 AND n492;
d_out(30)<= n199 AND n326 AND n487;
d_out(29)<= n22 AND n462 AND n492;
d_out(28)<= n242 AND n326 AND n514;
d_out(25)<= n14 AND n491 AND n521;
d_out(24)<= n68 AND n369 AND n460;
d_out(23)<= n18 AND n469 AND n521;
d_out(22)<= n110 AND n369 AND n465;
d_out(21)<= n11 AND n452 AND n521;
d_out(20)<= n154 AND n369 AND n487;
d_out(19)<= n22 AND n457 AND n521;
d_out(18)<= n199 AND n369 AND n514;
d_out(17)<= n26 AND n462 AND n521;
d_out(14)<= n18 AND n491 AND n547;
d_out(13)<= n68 AND n412 AND n465;
d_out(12)<= n11 AND n469 AND n547;
d_out(11)<= n110 AND n412 AND n487;
d_out(10)<= n22 AND n452 AND n547;
d_out(9)<= n154 AND n412 AND n514;
d_out(8)<= n26 AND n457 AND n547;
d_out(4)<= n68 AND n451 AND n487;
d_out(2)<= n110 AND n451 AND n514;
n63<= d_in(1) AND d_in(0) AND n30 AND n31;
n95<= d_in(2) AND d_in(1) AND n30 AND n31;
n109<= d_in(3) AND n28 AND n29 AND n108;
n278<= d_in(6) AND d_in(5) AND d_in(1) AND d_in(0);
n282<= d_in(7) AND d_in(6) AND d_in(2) AND n281;
n325<= d_in(8) AND d_in(7) AND d_in(3) AND n324;
n48<= d_in(2) AND d_in(1) AND d_in(0) AND n30 AND n31;
n411<= d_in(8) AND d_in(7) AND d_in(3) AND n11 AND n314;
n98<= d_in(2) AND d_in(1) AND d_in(0) AND n29 AND n30 AND n31;
n151<= d_in(3) AND d_in(2) AND d_in(1) AND n29 AND n30 AND n31;
n241<= d_in(3) AND d_in(2) AND d_in(1) AND n11 AND n28 AND n29 AND n30
      AND n31;
n12<= d_in(2) OR d_in(1);
n14<= n11 OR n13;
n16<= d_in(1) OR d_in(0);
n18<= n11 OR n17;
n21<= d_in(3) OR d_in(2);
n25<= d_in(3) OR n24;
```

```
n33<= d_in(0) OR n32;
n44<= d_in(1) OR n31;
n46<= n43 OR n45;
n51<= d_in(0) OR n31;
n61<= n30 OR n60;
n64<= n62 OR n63;
n66<= n59 OR n65;
n96<= n94 OR n95;
n108<= n55 OR n57;
n152<= n150 OR n151;
n154<= n134 OR n153;
n163<= d_in(1) OR n53;
n197<= n97 OR n98;
n199<= n185 OR n198;
n217<= n43 OR n164;
n239<= n67 OR n221;
n242<= n240 OR n241;
n245<= d_in(5) OR d_in(0);
n258<= n246 OR n247;
n264<= d_in(6) OR n246;
n267<= n265 OR n266;
n276<= n274 OR n275;
n281<= d_in(5) OR d_in(1);
n283<= n280 OR n282;
n314<= n254 OR n259;
n316<= n305 OR n315;
n324<= n277 OR n278;
n367<= n261 OR n268;
n369<= n349 OR n368;
n412<= n401 OR n411;
n450<= n270 OR n284;
n37<= d_in(1) OR d_in(0) OR n31;
n42<= d_in(1) OR n30 OR n31;
n54<= n30 OR n52 OR n53;
n58<= n29 OR n55 OR n57;
n68<= n34 OR n50 OR n67;
n93<= n36 OR n38 OR n39;
n99<= n87 OR n97 OR n98;
n110<= n81 OR n100 OR n109;
n149<= n41 OR n47 OR n48;
n184<= n166 OR n174 OR n183;
n253<= d_in(6) OR d_in(1) OR n247;
n260<= d_in(7) OR n254 OR n259;
n269<= n251 OR n261 OR n268;
n273<= d_in(5) OR d_in(1) OR d_in(0);
n279<= n272 OR n277 OR n278;
n285<= n249 OR n270 OR n284;
n326<= n298 OR n317 OR n325;
n390<= n289 OR n294 OR n296;
n400<= n380 OR n391 OR n399;
n35<= d_in(1) OR d_in(0) OR n30 OR n31;
n40<= n29 OR n36 OR n38 OR n39;
```

```
  n49<= n28 OR n41 OR n47 OR n48;
  n86<= d_in(3) OR n29 OR n43 OR n45;
  n133<= n11 OR n59 OR n65 OR n117;
  n173<= d_in(3) OR n62 OR n63 OR n75;
  n220<= n28 OR n41 OR n48 OR n218;
  n271<= d_in(6) OR d_in(5) OR d_in(2) OR d_in(1);
  n293<= d_in(6) OR d_in(2) OR n246 OR n247;
  n304<= d_in(7) OR d_in(3) OR n265 OR n266;
  n348<= n280 OR n282 OR n330 OR n335;
  n74<= d_in(2) OR d_in(1) OR d_in(0) OR n30 OR n31;
  n165<= d_in(3) OR n28 OR n29 OR n43 OR n164;
  n288<= d_in(7) OR d_in(6) OR d_in(5) OR d_in(2) OR d_in(1);
  n297<= d_in(8) OR n11 OR n289 OR n294 OR n296;
  n372<= d_in(6) OR d_in(5) OR d_in(2) OR d_in(1) OR d_in(0);
  n379<= d_in(8) OR d_in(3) OR n265 OR n266 OR n373;
  n80<= d_in(3) OR n11 OR n28 OR n62 OR n63 OR n75;
  n334<= d_in(7) OR d_in(6) OR d_in(3) OR d_in(2) OR n246 OR n247;
  n116<= d_in(3) OR d_in(2) OR d_in(1) OR d_in(0) OR n29 OR n30 OR n31;
  n250<= d_in(7) OR d_in(6) OR d_in(5) OR d_in(3) OR d_in(2) OR d_in(1)
        OR d_in(0);
  n32<= d_in(3) OR d_in(2) OR d_in(1) OR n11 OR n28 OR n29 OR n30 OR n31;
  n248<= d_in(8) OR d_in(7) OR d_in(6) OR d_in(3) OR d_in(2) OR n11 OR
        n246 OR n247;
  n329<= d_in(8) OR d_in(7) OR d_in(6) OR d_in(5) OR d_in(3) OR d_in(2)
        OR d_in(1) OR d_in(0);
END arch_network1;
CONFIGURATION conf_network1 OF network1 IS
  FOR arch_network1
  END FOR;
END conf_network1;
```

## 11.4 Appendix D – VHDL model of sine wave ROM

### 11.4.1 SIN_ROM.CPP

```
/* This program generates the VHDL model of the internal look-up table
used by tier1.
*/
#include <iostream.h>
#include <fstream.h>
#include <math.h>
#include <process.h>
#include <string.h>
#include <conio.h>
#define AMPL 255
#define N_STEPS 64
#define FileName "c:\\andrei\\sin_rom.vhd"
const int upper_index=((int)floor(log(N_STEPS-1)/log(2)));
void write_header(ofstream& f)
{
  f<<"LIBRARY IEEE;"<<endl;
```

```
  f<<"USE IEEE.std_logic_1164.ALL;"<<endl;
  f<<"USE IEEE.std_logic_unsigned.ALL;"<<endl<<endl;
  f<<"ENTITY sin_rom IS"<<endl;
  f<<" PORT("<<endl;
  f<<"   A: IN std_logic_vector("<<upper_index;
  f<<" DOWNTO 0);"<<endl;
  f<<"   DO: OUT std_logic_vector(2 DOWNTO 0));"<<endl;
  f<<"END sin_rom;"<<endl<<endl;
  f<<"ARCHITECTURE sin_rom_arch OF sin_rom IS"<<endl;
  f<<"   TYPE mem_data IS ARRAY (0 TO "<<(pow(2,upper_index+1)-1);
  f<<") OF std_logic_vector(2 downto 0);"<<endl;
  f<<"   constant VD: mem_data :="<<endl<<" (";
}
void write_end(ofstream& f)
{
  f<<"BEGIN"<<endl;
  f<<"  PROCESS(A)"<<endl;
  f<<"  begin"<<endl;
  f<<"     DO<=VD(conv_integer(A));"<<endl;
  f<<" END PROCESS;"<<endl;
  f<<"END sin_rom_arch;";
}
void main(void)
{
  clrscr();
  ofstream f;
  int sample;
  double step=M_PI/2.0/N_STEPS;
  int sum=-AMPL,max=0;
  f.open(FileName,ios::out);
  if(f.fail())
  {
    cout<<"Error:The file could not be opened"<<endl;
    exit(1);
  }
  write_header(f);
  for(int i=0;i<N_STEPS;i++)
  {
    sample=floor(AMPL*sin(-M_PI_2+(i+1)*step)-sum+0.5);
    if(max<sample)
      max=sample;
    sum+=sample;
    cout<<sample<<endl;
    switch(sample)
    {
      case 0: f<<"    ('0','0','0')";
        if (i<N_STEPS-1)
               f<<","<<endl;
        break;
      case 1: f<<"    ('0','0','1')";
        if (i<N_STEPS-1)
               f<<","<<endl;
```

```
          break;
        case 2: f<<"    ('0','1','0')";
          if (i<N_STEPS-1)
                 f<<","<<endl;
          break;
        case 3: f<<"    ('0','1','1')";
          if (i<N_STEPS-1)
                 f<<","<<endl;
          break;
        case 4: f<<"    ('1','0','0')";
          if (i<N_STEPS-1)
                 f<<","<<endl;
          break;
        case 5: f<<"    ('1','0','1')";
          if (i<N_STEPS-1)
                 f<<","<<endl;
          break;
        case 6: f<<"    ('1','1','0')";
          if (i<N_STEPS-1)
                 f<<","<<endl;
          break;
        default: f<<"   ('1','1','1')";
          if (i<N_STEPS-1)
                 f<<","<<endl;
      }
    }
    f<<");"<<endl;
    write_end(f);
    f.close();
}
```

# 11.5 Appendix E – VHDL code for simulation

## Plant models for simulation

### 11.5.1 Generator and rectifier model

```
--   File:  Genrect.vhd
--   Model of Synchronous Generator-Rectifier system.
--   Steady state model of generator-rectifier system
--   is derived from the equivalent circuit model of synchronous generators.
library ieee;
use ieee.std_logic_1164.all;
use ieee.math_real.all;
entity Genrect is
    port (
-- inputs
        n                : in REAL;
       Ifield            : in REAL;
       Idc               : in REAL;
       theta             : in REAL;
```

```
        CLK            : in STD_LOGIC;
-- outputs
       Vph             : out REAL;
       Vdc             : out REAL;
       torque          : out REAL);
end Genrect;
architecture Genrect_arc of Genrect is
constant PI             :REAL:=3.1416;
constant KG_CONST       :REAL:=4.7997;
constant R_CONST        :REAL:=0.0015;
constant X_CONST        :REAL:=1.0000;
constant RECT1_CONST    :REAL:=0.8165;
constant RECT2_CONST    :REAL:=1.3500;
-- Star Configuration
constant YD_CURR        :REAL:=1.0000;
constant YD_VOLT        :REAL:=1.7320;
begin
-- REM: Architecture is Synchronous and Sequential
GENREC_PROCESS:
process(CLK)
  variable Ei_VAR : REAL;
  variable Iph_VAR, Vph_VAR : REAL;
  variable torque_VAR : REAL;
  variable delta_VAR : REAL;
begin
    if (CLK'event and CLK='1') then
-- Avoid division by zero when n=0.0
if (n=0.0) then
  -- Assign output signals
  -- No speed -> No voltage
  Vph <= 0.0;
  Vdc <= 0.0;
  torque<=50.0;
else
  Ei_VAR := KG_CONST * n * Ifield;
  Iph_VAR  := RECT1_CONST * YD_CURR * Idc;
  delta_VAR  :=  arcsin(  (Iph_VAR/Ei_VAR)  *  (X_CONST*cos(theta)  +
R_CONST*sin(theta)) );
  Vph_VAR := Ei_VAR*cos(delta_VAR) + Iph_VAR*(X_CONST*sin(theta) -
                        R_CONST*cos(theta) );
-- Assign output signals
  Vph <= Vph_VAR;
  Vdc <= YD_VOLT * RECT2_CONST * Vph_VAR;
  torque<=(6.0*PI*Vph_VAR*Iph_VAR*cos(theta))/n;
end if;
end if;
end process;
end Genrect_arc;
configuration Genrect_conf1 of Genrect is
  for Genrect_arc
  end for;
end Genrect_conf1;
```

## 11.5.2 Diesel engine model

```
-- File: Engine.vhd
-- Diesel Engine Model Based on linearised torque-speed characteristics.
library ieee;
use ieee.MATH_REAL.all;
use ieee.std_logic_1164.all;
entity Engine is
    port (
     -- inputs
     TL           : in REAL;
     Q            : in REAL;
     Period       : in REAL;
     CLK          : in STD_LOGIC;
     -- output
     Te           : out REAL;
     Nout         : out REAL);
end Engine;
architecture Engine_arc of Engine is
  signal Nz_SIG   : REAL:=0.0;
  constant A_CONST :REAL:= 0.9;
  constant B_CONST :REAL:= 50.53;
  -- Total inertia
  constant J_CONST :REAL:= 10.0;
begin
-- REM Architecture is Synchronous and Sequential
ENGINE_PROCESS:
  process(CLK)
    variable Te_VAR   :REAL:=0.0;
    variable N_VAR    :REAL:=0.0;
    variable TL_VAR   :REAL:=0.0;
  begin
    if (CLK'event and CLK='1') then
    TL_VAR := TL;
    Te_VAR := (B_CONST*Q)-(A_CONST*Nz_SIG);
    N_VAR  := (((Te_VAR-TL_VAR)/J_CONST)*Period)+Nz_SIG;
-- Assign output/signal with MAX/MIN limit
    if (N_VAR > -5000.0 or N_VAR < 5000.0) then
      Nz_SIG <= N_VAR;
      Nout <= N_VAR;
  elsif (N_VAR < -5000.0) then
      Nz_SIG <= -5000.0;
      Nout <= -5000.0;
  elsif (N_VAR > 5000.0) then
      Nz_SIG <= 5000.0;
      Nout <= 5000.0;
  end if;
  if (Te_VAR > -5000.0 or Te_VAR < 5000.0) then
            Te <= Te_VAR;
  elsif (Te_VAR < -5000.0) then
            Te <= -5000.0;
  elsif (Te_VAR > 5000.0) then
            Te <= 5000.0;
```

```
    end if;
    end if;
  end process;
end Engine_arc;
configuration Engine_conf1 of Engine is
    for Engine_arc
    end for;
end Engine_conf1;
```

## Fuzzy logic controller

### 11.5.3 Input interface

```
-- File: Interface1.vhd
-- Part of Fuzzy Logic Controller
-- Simulation Version
library ieee;
use ieee.std_logic_1164.all;
entity Interface1 is
    port (
        CLK      :in STD_LOGIC;
        Vdc      :in INTEGER;
        Vref     :in INTEGER;
        x1       :out INTEGER;
        x2       :out INTEGER);
end Interface1;
architecture Interface1_arc of Interface1 is
begin
    LATCH_PROCESS:
    process(CLK)
    variable NOW_VAR:INTEGER:=0;
    variable PAST_VAR:INTEGER:=0;
    variable error:INTEGER:=0;
    variable DIFF :INTEGER:=0;
    begin
    if CLK'event and CLK='1' then
    -- error
    error:= (Vdc-Vref)/3;
    -- x1 -> (x1, x2)
    PAST_VAR:=NOW_VAR;
    NOW_VAR:=error;
    -- Output Assignment
    if (NOW_VAR<=(-127)) then
              x1<=(-127);
    elsif (NOW_VAR>=128) then
              x1<=128;
    else
              x1<=NOW_VAR;
    end if;
    DIFF:= NOW_VAR-PAST_VAR;
    if (DIFF>=30) then
```

```
                x2<=100;
    elsif (DIFF<=-30) then
                x2<=-100;
    else
                x2<=DIFF*3;
  end if;
                x2<=0;
  end if;
  end process;
end Interface1_arc;
configuration Interface1_conf1 of Interface1 is
  for Interface1_arc
  end for;
end Interface1_conf1;
```

### 11.5.4 Fuzzifier

```
--   File:  Fuzzify.vhd
--   Fuzzifier
--   Part of Fuzzy Logic Controller
--   Simulation Version
library ieee;
use ieee.std_logic_1164.all;
use ieee.numeric_std.all;
entity Fuzzify is
    port (
        -- inputs
        x1: in INTEGER range -127 to 128;
        x2: in INTEGER range -127 to 128;
        -- fuzzy sets for x1
        B1_1: out INTEGER range 0 to 128;
        B1_2: out INTEGER range 0 to 128;
        B1_3: out INTEGER range 0 to 128;
        B1_4: out INTEGER range 0 to 128;
        B1_5: out INTEGER range 0 to 128;
        -- fuzzy sets for x2
        B2_1: out INTEGER range 0 to 128;
        B2_2: out INTEGER range 0 to 128;
        B2_3: out INTEGER range 0 to 128;
        B2_4: out INTEGER range 0 to 128;
        B2_5: out INTEGER range 0 to 128);
end Fuzzify;
architecture Fuzzify_arc of Fuzzify is
    constant a1:INTEGER:=-60;
    constant b1:INTEGER:=-10;
    constant a2:INTEGER:=-60;
    constant b2:INTEGER:=-10;
    constant c2:INTEGER:=0;
    constant a3:INTEGER:=-10;
    constant b3:INTEGER:=0;
    constant c3:INTEGER:=10;
```

```
    constant a4:INTEGER:=0;
    constant b4:INTEGER:=10;
    constant c4:INTEGER:=60;
    constant a5:INTEGER:=10;
    constant b5:INTEGER:=60;
begin
  -- Concurrent Architechture
--| Fuzzify input x1 |--
-- 1. Very Small
B1_1 <= 100 when x1<=a1 else(100*(x1-b1))/(a1-b1)when(x1>a1 and x1<=b1)else
     0;
-- 2. Small
B1_2 <= (100*(x1-a2))/(b2-a2) when (x1>=a2 and x1<=b2) else
          (100*(x1-c2))/(b2-c2) when (x1>b2 and x1<=c2) else 0;
-- 3. Optimum
B1_3 <= (100*(x1-a3))/(b3-a3) when (x1>=a3 and x1<=b3) else
          (100*(x1-c3))/(b3-c3) when (x1>b3 and x1<=c3) else 0;
-- 4. Big
B1_4 <= (100*(x1-a4))/(b4-a4) when (x1>=a4 and x1<=b4) else
          (100*(x1-c4))/(b4-c4) when (x1>b4 and x1<=c4) else 0;
-- 5. Very Big
B1_5 <= 0 when x1<=a5 else
    (100*(x1-b5))/(a5-b5) when (x1>a5 and x1<b5) else 100;
--| Fuzzify input x2 |--
-- 1. Very Small
B2_1 <= 100 when x2<=a1 else
       (100*(x2-b1))/(a1-b1) when (x2>a1 and x2<=b1) else 0;
-- 2. Small
B2_2 <= (100*(x2-a2))/(b2-a2) when (x2>=a2 and x2<=b2) else
          (100*(x2-c2))/(b2-c2) when (x2>b2 and x2<=c2) else 0;
-- 3. Optimum
B2_3 <= (100*(x2-a3))/(b3-a3) when (x2>=a3 and x2<=b3) else
        (100*(x2-c3))/(b3-c3) when (x2>b3 and x2<=c3) else 0;
-- 4. Big
B2_4 <= (100*(x2-a4))/(b4-a4) when (x2>=a4 and x2<=b4) else
        (100*(x2-c4))/(b4-c4) when (x2>b4 and x2<=c4) else 0;
-- 5. Very Big
  B2_5 <= 0 when x2<=a5 else
          (100*(x2-b5))/(a5-b5) when (x2>a5 and x2<b5) else 100;
end Fuzzify_arc;
configuration Fuzzify_conf1 of Fuzzify is
  for Fuzzify_arc
  end for;
end Fuzzify_conf1;
```

### 11.5.5 Rule base and inference engine

```
--   File:  Infer.vhd
--   Rule base and Inference Engine
--   Part of Fuzzy Logic Controller
--   Simulation Version
```

```
library ieee;
use ieee.std_logic_1164.all;
entity Infer is
    port (
      CLK: in STD_LOGIC;
     -- Inputs
        B1_1: in INTEGER range 0 to 128;
        B1_2: in INTEGER range 0 to 128;
        B1_3: in INTEGER range 0 to 128;
        B1_4: in INTEGER range 0 to 128;
        B1_5: in INTEGER range 0 to 128;
        B2_1: in INTEGER range 0 to 128;
        B2_2: in INTEGER range 0 to 128;
        B2_3: in INTEGER range 0 to 128;
        B2_4: in INTEGER range 0 to 128;
        B2_5: in INTEGER range 0 to 128;
    --  Outputs
        D1: out INTEGER range 0 to 128;
        D2: out INTEGER range 0 to 128;
        D3: out INTEGER range 0 to 128;
        D4: out INTEGER range 0 to 128;
        D5: out INTEGER range 0 to 128;
        D6: out INTEGER range 0 to 128;
        D7: out INTEGER range 0 to 128;
        D8: out INTEGER range 0 to 128;
        D9: out INTEGER range 0 to 128);
end Infer;
architecture Infer_arc of Infer is
begin
Sequential:
process(CLK)
    variable c1,c2,c3,c4,c5: INTEGER range 0 to 128;
    variable c6,c7,c8,c9,c10: INTEGER range 0 to 128;
    variable c11,c12,c13,c14,c15: INTEGER range 0 to 128;
    variable c16,c17,c18,c19,c20: INTEGER range 0 to 128;
    variable c21,c22,c23,c24,c25: INTEGER range 0 to 128;
begin
    -- Fuzzy Inference Engine -- Ci = min(U1_x,U2_y)
    if B1_1 < B2_1 then c1:=B1_1;
    else c1:=B2_1;
    end if;
  if B1_1 < B2_2 then c2:=B1_1;
    else c2:=B2_2;
    end if;
  if B1_1 < B2_3 then c3:=B1_1;
    else c3:=B2_3;
    end if;
  if B1_1 < B2_4 then c4:=B1_1;
    else c4:=B2_4;
    end if;
  if B1_1 < B2_5 then c5:=B1_1;
    else c5:=B2_5;
```

```
    end if;
  if B1_2 < B2_1 then c6:=B1_2;
    else c6:=B2_1;
    end if;
  if B1_2 < B2_2 then c7:=B1_2;
    else c7:=B2_2;
    end if;
  if B1_2 < B2_3 then c8:=B1_2;
    else c8:=B2_3;
    end if;
  if B1_2 < B2_4 then c9:=B1_2;
    else c9:=B2_4;
    end if;
  if B1_2 < B2_5 then c10:=B1_2;
    else c10:=B2_5;
    end if;
  if B1_3 < B2_1 then c11:=B1_3;
    else c11:=B2_1;
    end if;
  if B1_3 < B2_2 then c12:=B1_3;
    else c12:=B2_2;
    end if;
  if B1_3 < B2_3 then c13:=B1_3;
    else c13:=B2_3;
    end if;
  if B1_3 < B2_4 then c14:=B1_3;
    else c14:=B2_4;
    end if;
  if B1_3 < B2_5 then c15:=B1_3;
    else c15:=B2_5;
    end if;
  if B1_4 < B2_1 then c16:=B1_4;
    else c16:=B2_1;
    end if;
  if B1_4 < B2_2 then c17:=B1_4;
    else c17:=B2_2;
    end if;
  if B1_4 < B2_3 then c18:=B1_4;
    else c18:=B2_3;
    end if;
  if B1_4 < B2_4 then c19:=B1_4;
    else c19:=B2_4;
    end if;
    if B1_4 < B2_5 then c20:=B1_4;
    else c20:=B2_5;
    end if;
    if B1_5 < B2_1 then c21:=B1_5;
    else c21:=B2_1;
    end if;
    if B1_5 < B2_2 then c22:=B1_5;
    else c22:=B2_2;
    end if;
```

```
   if B1_5 < B2_3 then c23:=B1_5;
   else c23:=B2_3;
   end if;
   if B1_5 < B2_4 then c24:=B1_5;
   else c24:=B2_4;
   end if;
   if B1_5 < B2_5 then c25:=B1_5;
   else c25:=B2_5;
   end if;
-- 25 Fuzzy Rules -> 9 Fuzzy Sets (Get MAX) --
   -- Negative Very Big
   D1 <= c25;
   -- Negative Big
   if ( c20=0 and c24=0 ) then
     D2<= 0;
   elsif (c20>=c24) then
     D2<=c20;
   else
     D2<=c24;
   end if;
   -- Negative
   if (c15=0 and c19=0 and c23=0) then D3<=0;
   elsif (c15>=c19 and c15>=c23) then D3<=c15;
   elsif (c19>=c15 and c19>=c23) then D3<=c19;
   else
   D3<=c23;
   end if;
   -- Negative Small
   if (c10=0 and c14=0 and c18=0 and c22=0) then D4<=0;
   elsif (c10>=c14 and c10>=c18 and c10>=c22)then D4<=c10;
   elsif (c14>=c10 and c14>=c18 and c14>=c22)then D4<=c14;
   elsif (c18>=c10 and c18>=c14 and c18>=c22) then D4<=c18;
   else  D4<=c22;
   end if;
   -- Zero
   if (c5 =0 and c9 =0 and c13=0 and c17=0 and c21=0)then D5<=0;
   elsif (c5>=c9  and c5>=c13 and c5>=c17 and c5>=c21)then D5<=c5;
   elsif (c9>=c5  and c9>=c13 and c9>=c17 and c9>=c21)then D5<=c9;
   elsif (c13>=c5 and c13>=c9 and c13>=c17 and c13>=c21)then D5<=c13;
   elsif (c17>=c5 and c17>=c9 and c17>=c13 and c17>=c21)then D5<=c17;
   else D5<=c21;
   end if;
   -- Positive Small
   if (c4 =0 and c8 =0 and c12=0 and c16=0) then D6<=0;
   elsif (c4 >=c8 and c4 >=c12 and c4 >=c16) then D6<=c4;
   elsif (c8 >=c4 and c8 >=c12 and c8 >=c16) then D6<=c8;
   elsif (c12>=c4 and c12>=c8  and c12>=c16) then D6<=c12;
   else D6<=c16;
   end if;
-- Positive
   if (c3 =0 and c7 =0 and c11=0) then D7<=0;
   elsif (c3 >=c7 and c3 >=c11) then D7<=c3;
```

```
    elsif (c7 >=c3 and c7 >=c11) then D7<=c7;
    else    D7<=c11;
    end if;
-- Positive Big
    if (c2=0 and c6=0) then D8<=0;
    elsif (c2>=c6) then D8<=c2;
    else  D8<=c6;
    end if;
-- Positive Very Big
  D9 <= c1;
end process;
end Infer_arc;
configuration Infer_conf1 of Infer is
    for Infer_arc
    end for;
end Infer_conf1;
```

### 11.5.6 Defuzzifier and output interface

```
--  File: Defuzz.vhd
--  Defuzzifier and Output Interface
--  Part of Fuzzy Logic Controller
--  Simulation Version
library ieee;
use ieee.std_logic_1164.all;
entity Defuzz is
    port (
    CLK: in STD_LOGIC;
        D1: in INTEGER range 0 to 128;
        D2: in INTEGER range 0 to 128;
        D3: in INTEGER range 0 to 128;
        D4: in INTEGER range 0 to 128;
        D5: in INTEGER range 0 to 128;
        D6: in INTEGER range 0 to 128;
        D7: in INTEGER range 0 to 128;
        D8: in INTEGER range 0 to 128;
        D9: in INTEGER range 0 to 128;
    --  Crisp control signal
        U: out INTEGER);
end Defuzz;
architecture Defuzz_arc of Defuzz is
      constant E1: INTEGER:=-4;
      constant E2: INTEGER:=-3;
      constant E3: INTEGER:=-2;
      constant E4: INTEGER:=-1;
      constant E5: INTEGER:=0;
      constant E6: INTEGER:=1;
      constant E7: INTEGER:=2;
      constant E8: INTEGER:=3;
      constant E9: INTEGER:=4;
begin
```

```
-- Defuzz Engine : Weighted average method
  DEFUZZ_PROCESS:
  process(CLK)
    variable Dividend :INTEGER:=0;
    variable Divisor :INTEGER:=1;
    variable Y :INTEGER;
    variable Uz :INTEGER:=20;
    variable U_var :INTEGER;
  begin
    if CLK'event and CLK='1' then
    Dividend:=(E1*D1)+(E2*D2)+(E3*D3)+(E4*D4)+(E5*D5)+(E6*D6)
              +(E7*D7)+(E8*D8)+(E9*D9);
    Divisor:=(D1+D2+D3+D4+D5+D6+D7+D8+D9);
  -- To avoid division by zero
        if Divisor=0 then
                Y:=0;
        else
  -- Definition of crisp output
               Y := (Dividend/Divisor);
  end if;
Ys<=Y;
-- Output interface
    U_var  :=Uz+Y;
    Uz     :=U_var;
if (U_VAR<=(-254)) then
          U<=(-254);
  elsif (U_VAR>=255) then
          U<=255;
  else
          U<=U_var;
end if;
end if;
end process;
end Defuzz_arc;
configuration Defuzz_conf1 of Defuzz is
  for Defuzz_arc
  end for;
end Defuzz_conf1;
```

### 11.5.7 Top hierarchy of FLC

```
--  File: Controller.vhd
--  Binds all the components together
--  Simulation Version
library ieee;
use ieee.std_logic_1164.all;
entity Controller is
    port (
        CLK      :in STD_LOGIC;
        Vdc      :in INTEGER;
        Vref     :in INTEGER;
```

```
        U        :out INTEGER);
end Controller;
architecture Controller_arc of Controller is
component Interface1
    port (
    CLK            :in STD_LOGIC;
    Vdc            :in INTEGER;
    Vref           :in INTEGER;
    x1             :out INTEGER;
    x2             :out INTEGER);
  end component Interface1;
component Fuzzify
    port (
        -- inputs
        x1: in INTEGER range -127 to 128;
        x2: in INTEGER range -127 to 128;
        -- fuzzy sets for x1
        B1_1: out INTEGER range 0 to 128;
        B1_2: out INTEGER range 0 to 128;
        B1_3: out INTEGER range 0 to 128;
        B1_4: out INTEGER range 0 to 128;
        B1_5: out INTEGER range 0 to 128;
        -- fuzzy sets for x2
        B2_1: out INTEGER range 0 to 128;
        B2_2: out INTEGER range 0 to 128;
        B2_3: out INTEGER range 0 to 128;
        B2_4: out INTEGER range 0 to 128;
        B2_5: out INTEGER range 0 to 128);
  end component Fuzzify;
  component Infer
    port (
      CLK: in STD_LOGIC;
      -- Inputs
        B1_1: in INTEGER range 0 to 128;
        B1_2: in INTEGER range 0 to 128;
        B1_3: in INTEGER range 0 to 128;
        B1_4: in INTEGER range 0 to 128;
        B1_5: in INTEGER range 0 to 128;
        B2_1: in INTEGER range 0 to 128;
        B2_2: in INTEGER range 0 to 128;
        B2_3: in INTEGER range 0 to 128;
        B2_4: in INTEGER range 0 to 128;
        B2_5: in INTEGER range 0 to 128;
      -- Outputs
        D1: out INTEGER range -127 to 128;
        D2: out INTEGER range -127 to 128;
        D3: out INTEGER range -127 to 128;
        D4: out INTEGER range -127 to 128;
        D5: out INTEGER range -127 to 128;
        D6: out INTEGER range -127 to 128;
        D7: out INTEGER range -127 to 128;
        D8: out INTEGER range -127 to 128;
```

```
        D9: out INTEGER range -127 to 128
        );
  end component Infer;
  component Defuzz
    port (
        CLK: in STD_LOGIC;
        D1: in INTEGER range -127 to 128;
        D2: in INTEGER range -127 to 128;
        D3: in INTEGER range -127 to 128;
        D4: in INTEGER range -127 to 128;
        D5: in INTEGER range -127 to 128;
        D6: in INTEGER range -127 to 128;
        D7: in INTEGER range -127 to 128;
        D8: in INTEGER range -127 to 128;
        D9: in INTEGER range -127 to 128;
      -- Crisp control signal
        U: out INTEGER);
  end component Defuzz;
  signal x1_SIG, x2_SIG : INTEGER;
  signal SB1_1,SB1_2,SB1_3,SB1_4,SB1_5 : INTEGER range 0 to 128;
  signal SB2_1,SB2_2,SB2_3,SB2_4,SB2_5 : INTEGER range 0 to 128;
  signal D1_SIG,D2_SIG,D3_SIG : INTEGER range -127 to 128;
  signal D4_SIG,D5_SIG,D6_SIG : INTEGER range -127 to 128;
  signal D7_SIG,D8_SIG,D9_SIG : INTEGER range -127 to 128;
begin
Interface1_U: Interface1 port map(CLK=>CLK, Vdc=>Vdc, Vref=>Vref,
              x1=>x1_SIG, x2=>x2_SIG);
Fuzzify_U: Fuzzify port map(x1=>x1_SIG, x2=>x2_SIG,
  B1_1=>SB1_1,B1_2=>SB1_2, B1_3=>SB1_3, B1_4=>SB1_4,
  B1_5=>SB1_5,B2_1=>SB2_1, B2_2=>SB2_2, B2_3=>SB2_3, B2_4=>SB2_4,
  B2_5=>SB2_5);
Infer_U: Infer port map(CLK=>CLK, B1_1=>SB1_1, B1_2=>SB1_2, B1_3=>SB1_3,
  B1_4=>SB1_4, B1_5=>SB1_5, B2_1=>SB2_1, B2_2=>SB2_2, B2_3=>SB2_3,
  B2_4=>SB2_4, B2_5=>SB2_5,
  -- Outputs D1=>D1_SIG,D2=>D2_SIG,D3=>D3_SIG, D4=>D4_SIG, D5=>D5_SIG,
  D6=>D6_SIG, D7=>D7_SIG,D8=>D8_SIG,D9=>D9_SIG);
Defuzz_U:Defuzz port map(
  -- Inputs
  CLK=>CLK, D1=>D1_SIG, D2=>D2_SIG, D3=>D3_SIG, D4=>D4_SIG, D5=>D5_SIG,
  D6=>D6_SIG, D7=>D7_SIG,D8=>D8_SIG,D9=>D9_SIG,
  -- Output
  U=>u);
end Controller_arc;
configuration Controller_conf1 of Controller is
    for Controller_arc
    for Interface1_U: Interface1 use configuration work.Interface1 conf1;
    end for;
    for Fuzzify_U:Fuzzify use configuration work.Fuzzify_conf1;
    end for;
    for Infer_U:Infer use configuration work.Infer_conf1;
    end for;
    for Defuzz_U:Defuzz use configuration work.Defuzz_conf1;
```

```
    end for;
    end for;
end Controller_conf1;
```

## Test-benches and simulations

### 11.5.8 Delay

```
--  File: Delay.vhd
--  Part of Simulation component
library ieee;
use ieee.std_logic_1164.all;
entity delay is
    port (
        Tin: in REAL;
        Uin: in REAL;
        MUX: in STD_LOGIC;
    Uout:   out REAL:=2.0;
    Tout:   out REAL:=10.0);
end delay;
architecture delay_arc of delay is
  begin
    Tout <= Tin when MUX='1';
    Uout <= Uin when MUX='1';
end delay_arc;
configuration delay_conf1 of delay is
  for delay_arc
  end for;
end delay_conf1;
```

### 11.5.9 Simulator

```
--  File:  Sim.vhd
--  Simulation component
--  This test unit comprises:
--  FLC, Engine, Genrect and a delay component
library ieee;
use ieee.math_real.all;
use ieee.std_logic_1164.all;
entity Sim is
port (
  -- inputs
      CLK1          : in STD_LOGIC;
      CLK2          : in STD_LOGIC;
      Period        : in REAL;
      Idc           : in REAL;
      theta         : in REAL;
      Ifield        : in REAL;
      Vref          : in INTEGER;
    -- outputs
```

```
  TE          : out REAL;
      Vph    : out REAL;
      Vdc    : out REAL);
end Sim;
architecture Sim_arc of Sim is
  component Controller
  port(      CLK :in std_logic;
             Vdc :in INTEGER;
             Vref:in INTEGER;
             U :out INTEGER);
  end component;
  component Engine
  port(
      TL : in REAL;
      U : in REAL;
      CLK : in STD_LOGIC;
      Period : in REAL;
      TE : out REAL;
      N : out REAL);
  end component;
  component Genrect
  port(
        N : in REAL;
        Ifield : in REAL;
        Idc : in REAL;
        theta : in REAL;
        CLK : in STD_LOGIC;
        Vph : out REAL;
        Vdc : out REAL;
        TG : out REAL);
    end component;
    component delay
    port(
      Tin  : in REAL;
      Uin  : in REAL;
      MUX  : in STD_LOGIC;
      Uout : out REAL;
      Tout : out REAL);
    end component;
--  Controller input flow
    signal Vdc_SIG :REAL;
    signal Vdc_rSIG :REAL;
    signal Vdc_iSIG :INTEGER;
--  Controller output flow
    signal U_iSIG :INTEGER;
    signal U_rSIG, Uo_SIG :REAL;
--  Torque & Speed signals
    signal TL_SIG :REAL :=50.0;
    signal TG_SIG :REAL;
    signal N_SIG :REAL;
--  Delay element
    signal MUX :STD_LOGIC:='0';
```

```
    signal UE_SIG :REAL :=2.0;
begin
Controller_U: Controller
    port map(
    -- Inputs
    CLK=>CLK1, Vdc =>Vdc_iSIG, Vref=>Vref,
    -- Outputs
    U=>U_iSIG);
Engine_U: Engine port map(TL=>TL_SIG,U=>UE_SIG,CLK=>CLK2,Period =>Period,
  TE=>TE,N=>N_SIG);
  Genrect_U: Genrect
  port map(N=>N_SIG, Ifield=>Ifield,Idc=>Idc, theta=>theta, CLK=>CLK2,
  Vph=>Vph,Vdc=>Vdc_SIG,TG=>TG_SIG);
  delay_U: delay
  port map(Tin=>TG_SIG,Uin=>Uo_SIG,MUX=>MUX,Uout=>UE_SIG,Tout
=>TL_SIG);
-- Torque in/out delay to overcome problem
-- caused by propagation of unknown values during the starting transient.
  MUX <='1' after 30ns;
  -- error: Normalise
  Vdc_rSIG <= Vdc_SIG * 1.0;
  -- error: Real-Integer Conversion
  Vdc_iSIG <= 2000 when (Vdc_rSIG>2000.0) else
  0 when (Vdc_rSIG<0.0) else
  INTEGER(Vdc_rSIG);
  -- U: Integer-Real Conversion
  U_rSIG <= REAL(U_iSIG);
  -- UnNormalise U (from -127/128 to 12.7/12.8)
  OutputU:
  process(CLK1)
  variable COUNT_VAR : STD_LOGIC := '0';
  variable U_VAR :REAL;
  begin
  if (CLK1'event and CLK1='1') then
  if (COUNT_VAR='1') then
                U_VAR := (U_rSIG/10.0);
  if (U_VAR>25.5) then
                Uo_SIG <= 25.5;
  elsif (U_VAR<-25.4) then
                Uo_SIG <= -25.4;
                else
                Uo_SIG <= U_VAR;
  end if;
  else
  -- Initial condition
                COUNT_VAR:='1';
                Uo_SIG <= 2.0;
  end if;
  end if;
  end process;
-- Assign output Vdc
  Vdc <= Vdc_SIG;
```

```
end Sim_arc;
configuration Sim_conf1 of Sim is
  for Sim_arc
  for Controller_U: Controller use configuration work.Controller_conf1;
  end for;
  for Engine_U: Engine use configuration work.Engine_conf1;
  end for;
  for Genrect_U: Genrect use configuration work.Genrect_conf1;
  end for;
  for delay_U: delay use configuration work.delay_conf1;
  end for;
end for;
end Sim_conf1;
```

### 11.5.10 Test-bench for 'Sim'

```
--   File: TB_Sim.vhd
--   Testbench for Sim
--   Provides the appropriate stimuli and observes the simulated signals
--   Writes the observed values into a text file
library ieee;
use ieee.math_real.all;
use ieee.std_logic_1164.all;
use std.textio.all;
entity TB_Sim is
end TB_Sim;
architecture TB_Sim_arc of TB_Sim is
  -- Component declaration of the tested unit
  component Sim
  port(
  --Inputs
  CLK1      :in std_logic;
  CLK2      :in std_logic;
  Period in REAL;
  Idc       :in REAL;
  theta     :in REAL;
  Ifield    :in REAL;
  Vref      :in INTEGER;
  --Outputs
  TE        :out REAL;
  Vph       :out REAL;
  Vdc       :out REAL );
  end component;
  -- Stimulus signals - signals mapped to the input and inout ports of
tested entity
  signal CLK1        : std_logic:='1';
  signal CLK2        : std_logic:='1';
  signal Period      : REAL;
  signal Idc         : REAL;
  signal theta       : REAL;
  signal Ifield      : REAL;
```

```
  signal Vref       : INTEGER;
  -- Observed signals - signals mapped to the output ports of tested
entity
  signal TE  : REAL;
  signal Vph : REAL;
  signal Vdc : REAL;
  -- Signals for simulation purposes
begin
  -- Unit Under Test port map
  UUT : Sim
    port map(
  --Inputs
  CLK1 => CLK1,CLK2 => CLK2, Period => Period, Idc => Idc, theta => theta,
  Ifield => Ifield, Vref => Vref,
  --Outputs
  TE => TE, Vph => Vph, Vdc => Vdc );
-- ***Stimulus***--
  CLK1 <= not CLK1 after 10ns;
  CLK2 <= not CLK2 after 1ns;
  -- Period:
  Period   <= 3.0;
  theta    <= 0.0;
  Ifield   <= 2.5;
  Vref     <= 1000;
  Idc      <= 25.0,1.0 after 4100ns;
--Write results into file
process (CLK1)
  file outfile : text is out
  "C:\My Designs\Simulation\src\Results\Further25.txt";
  variable out_line                                              : line;
begin
  write(out_line, Vdc);
--write(out_line, " "); write(out_line, Idc);
  writeline(outfile, out_line);
end process;
end TB_Sim_arc;
configuration TB_Sim_conf1 of TB_Sim is
  for TB_Sim_arc
  for UUT : Sim
  use entity work.Sim(Sim_ARC);
  end for;
  end for;
end TB_Sim_conf1;
```

## 11.6 Appendix F – VHDL code for synthesis

Fuzzy logic controller

### 11.6.1 Input interface

```
--   File: Deriv.vhd
```

```
library ieee;
use ieee.std_logic_1164.all;
use ieee.std_logic_signed.all;
use ieee.std_logic_arith.all;
entity Deriv is
    port (
        CLK      : in STD_LOGIC;
        RST      : in STD_LOGIC;
        Vdc      : in std_logic_vector(7 downto 0);
        Vref     : in std_logic_vector(7 downto 0);
        x1: out std_logic_vector(8 downto 0);
        x2: out std_logic_vector(8 downto 0));
end Deriv;
architecture Deriv_arc of Deriv is
begin
LATCH_PROCESS:
process(CLK,RST)
    variable x: std_logic_vector(8 downto 0);
    variable NOW_VAR: std_logic_vector(8 downto 0);
    variable PAST_VAR: std_logic_vector(8 downto 0);
    variable DIFF: std_logic_vector(8 downto 0);
    variable Vdc_var: std_logic_vector(8 downto 0);
    variable Vref_var: std_logic_vector(8 downto 0);
    variable x_temp: std_logic_vector(8 downto 0);
    variable error: std_logic_vector(8 downto 0);
begin --process
if RST='1' then
    NOW_VAR :="000000000";
    PAST_VAR :="000000000";
    DIFF:= "000000000";
    x1<= "000000000";
    x2<= "000000000";
elsif CLK'event and CLK='1' then
    --Convert from unsigned to signed
    Vdc_var(8):='0';
    Vdc_var(7 downto 0):=Vdc;
    Vref_var(8):='0';
    Vref_var(7 downto 0):=Vref;
    --Get Error
    error:=Vdc_var-Vref_var;
    --x is error*(Gain=3)
    x:=error+shl(error,"1");
    --Overflow check
    if (error(8) XOR x(8))='1' then
    if error(8)='1' then
          x:="110011100";
    else
          x:="001100100";
    end if;
    end if;
    --Block: x -> x1, x2--
    -- Effect immediately
```

```
        PAST_VAR:=NOW_VAR;
        NOW_VAR:=x;
        DIFF:= NOW_VAR-PAST_VAR;
        --x1 is NOW_VAR
        x1<=NOW_VAR;
        --x2--
        x2<="000000000";
end if; -- Clock,Reset
end process;
end Deriv_arc;
-- Configuration
configuration Deriv_conf1 of Deriv is
        for Deriv_arc
        end for;
end Deriv_conf1;
```

### 11.6.2 Fuzzifier

```
--  File: Fuzzify2.vhd
library ieee;
use ieee.std_logic_1164.all;
use ieee.std_logic_signed.all;
use ieee.std_logic_arith.all;
entity Fuzzify2 is
        port (
            -- inputs
            CLK       : in STD_LOGIC;
            RST       : in STD_LOGIC;
            x1        : in std_logic_vector(8 downto 0);
            x2        : in std_logic_vector(8 downto 0);
            -- address
            ADR1      : out std_logic_vector(1 downto 0);
            ADR2      : out std_logic_vector(1 downto 0);
            -- fuzzy sets for x1 (data)
            B1_A      : out std_logic_vector(8 downto 0);
            B1_B      : out std_logic_vector(8 downto 0);
            -- fuzzy sets for x2 (data)
            B2_A      : out std_logic_vector(8 downto 0);
            B2_B      : out std_logic_vector(8 downto 0);
    READY             : out std_logic);
end Fuzzify2;
architecture Fuzzify2_arc of Fuzzify2 is
    signal temp    : std_logic_vector(1 downto 0);
    signal temp_A  : std_logic_vector(8 downto 0);
    signal temp_B  : std_logic_vector(8 downto 0);
    signal R_sig   : std_logic;
begin
-- Sequential Architecture (Synchronous)
process(CLK,RST)
constant AA_const:std_logic_vector(8 downto 0): =conv_std_logic_vector
                                                        (-60,9);
```

```
constant BB_const:std_logic_vector(8 downto 0): =conv_std_logic_vector
                                                   (-10,9);
constant CC_const:std_logic_vector(8 downto 0):="000000000";
constant  DD_const:std_logic_vector(8  downto  0):  =conv_std_logic_
  vector(10,9);
constant  EE_const:std_logic_vector(8  downto  0):  =conv_std_logic_
  vector(60,9);
constant hundred: std_logic_vector(8 downto 0):="001100100";
constant zero: std_logic_vector(8 downto 0):="000000000";
  variable dumbs: std_logic_vector(8 downto 0);
  variable x: std_logic_vector(8 downto 0);
  variable ADR: std_logic_vector(1 downto 0);
  variable B_A: std_logic_vector(8 downto 0);
  variable B_B: std_logic_vector(8 downto 0);
begin
if RST='1' then
        ADR1<="00";
        ADR2<="00";
        -- fuzzy sets for x1 (data)
        B1_A<= zero;
        B1_B<= zero;
        -- fuzzy sets for x2 (data)
        B2_A<= zero;
        B2_B<= zero;
      R_sig<='1';
      READY<='0';
elsif CLK'event and CLK='1' then
--MUX
if R_sig='1' then
    x:=x1;
else
    x:=x2;
end if; --select bit
--| Fuzzify input x (Sequential) |--
    if x<=AA_const then
                  --Zone 0a
                  ADR:="00";
                  --Very Small(B_1)
                  B_A:=hundred;
                  --Small(B_2)
                  B_B:=zero;
    elsif (x>AA_const and x<=BB_const) then
                  --Zone 0b
                  ADR:="00";
                  --Very Small(B_1)=2(-x-10)
                  dumbs:=not(x-"01")-"01010";
                  B_A :=shl(dumbs,"1");
                  --Small(B_2)=100-2(-x-10)
                  B_B:=hundred-B_A;
        elsif (x>BB_const and x<=CC_const) then
                  --Zone1
                  ADR:="01";
```

```
                    --Small(B_2)=x*(-10)=pos_x*10
                    dumbs:=not(x-"01");
                    B_A:=shl(dumbs,"1")+shl(dumbs,"11");
                    --Optimum(B_3)
                    B_B:=hundred-B_A;
          elsif (x>CC_const and x<=DD_const) then
                    --Zone2
                    ADR:="10";
                    --Optimum(B_3)
                    dumbs:=shl(x,"1")+shl(x,"11");
                    B_A:=hundred-dumbs;
                    --Big(B_4)
                    B_B:=dumbs;
          elsif (x>DD_const and x<=EE_const) then
                    --Zone3a
                    ADR:="11";
                    --Big(B_4)=100-2(x-10)
                    dumbs:=x-"01010";
                    --Very Big(B_5)
                    B_B:=shl(dumbs,"1");
                    B_A:=hundred-B_B;
          else --(x>EE_const)
                    --Zone3b
                    ADR:="11";
                    --Big(B_4)
                    B_A:=zero;
                    --Very Big(B_5)
                    B_B:=hundred;
          end if; -- fuzzy sets for x
      --Storing of process results
if R_sig='1' then
      temp<=ADR;
      temp_A<=B_A;
      temp_B<=B_B;
        READY<='0';
        R_sig<='0';
elsif R_sig='0' then
      ADR1<=temp;
      B1_A<=temp_A;
      B1_B<=temp_B;
      ADR2<=ADR;
      B2_A<=B_A;
      B2_B<=B_B;
      READY<='1';
      R_sig<='1';
end if; --R_sig
end if; --CLK,RESET
end process; -- main
end Fuzzify2_arc;
-- Configuration
configuration Fuzzify2_conf1 of Fuzzify2 is
  for Fuzzify2_arc
```

```
  end for;
end Fuzzify2_conf1;
```

### 11.6.3 Inference engine

```
--  File: Infer.vhd
library ieee;
use ieee.std_logic_1164.all;
use ieee.std_logic_signed.all;
use ieee.std_logic_arith.all;
entity Infer is
      port (
      -- Standard Input
      CLK  : in std_logic;
      LOAD: in std_logic;
      RST  : in std_logic;
      -- Inputs
      ADR1: in std_logic_vector(1 downto 0);
      B1_Ad: in std_logic_vector(8 downto 0);
      B1_Bd: in std_logic_vector(8 downto 0);

      ADR2: in std_logic_vector(1 downto 0);
      B2_Ad: in std_logic_vector(8 downto 0);
      B2_Bd: in std_logic_vector(8 downto 0);
      -- Outputs
      win: out std_logic_vector(3 downto 0);
      c1: out std_logic_vector(8 downto 0);
      c2: out std_logic_vector(8 downto 0);
      c3: out std_logic_vector(8 downto 0);
      c4: out std_logic_vector(8 downto 0));
end Infer;
architecture Infer_arc of Infer is
begin
MAIN_PROCESS:
process(CLK,RST)
begin --process
if RST='1' then
    win<="0000";
    c1<="000000000";
    c2<="000000000";
    c3<="000000000";
    c4<="000000000";
elsif CLK'event and CLK='1' then
    if LOAD='1' then
--| Fuzzy Inference Engine | Ci=min(B1_a,B2_b)|--
win<=("00"& ADR1)+ADR2;
--| Mini Fuzzy Inference Engine |--
--:B->C (min operation)
-- c1:=min(B1_Ad,B2_Ad)
    if (B1_Ad<B2_Ad) then
                    c1<=B1_Ad;
  else
```

```
                c1<=B2_Ad;
  end if;
-- c2:=min(B1_Bd,B2_Ad)
  if (B1_Bd<B2_Ad) then
             c2<=B1_Bd;
  else
             c2<=B2_Ad;
  end if;
-- c3:=min(B1_Ad,B2_Bd)
  if (B1_Ad<B2_Bd) then
             c3<=B1_Ad;
  else
             c3<=B2_Bd;
  end if;
- - c4:=min(B1_Bd,B2_Bd)
  if (B1_Bd<B2_Bd) then
             c4<=B1_Bd;
  else
             c4<=B2_Bd;
  end if;
  end if; --LOAD
end if; --RST,CLK
end process;
end Infer_arc;
-- Configuration
configuration Infer_conf1 of Infer is
  for Infer_arc
  end for;
end Infer_conf1;
```

### 11.6.4 Defuzzifier

```
--  File: Defuzz.vhd
library ieee;
use ieee.std_logic_1164.all;
use ieee.std_logic_signed.all;
use ieee.std_logic_arith.all;
entity Defuzz is
    port (
      -- Standard Input
      CLK: in std_logic;
      RST: in std_logic;
      LOAD: in std_logic;
      WIN: in std_logic_vector(3 downto 0);
      c1: in std_logic_vector(8 downto 0);
      c2: in std_logic_vector(8 downto 0);
      c3: in std_logic_vector(8 downto 0);
      c4: in std_logic_vector(8 downto 0);
    -- Outputs
      READY: out std_logic;
      divA: out std_logic_vector(13 downto 0);
```

```
      divB: out std_logic_vector(8 downto 0));
end Defuzz;
architecture Defuzz_arc of Defuzz is
  signal READY_sig: std_logic;
begin
MAIN_PROCESS:
process(CLK,RST)
    variable COUNT: std_logic_vector(3 downto 0);
    variable DA: std_logic_vector(8 downto 0);
    variable DB: std_logic_vector(8 downto 0);
    variable DC: std_logic_vector(8 downto 0);
    variable VA1: std_logic_vector(13 downto 0);
    variable VB1: std_logic_vector(13 downto 0);
    variable VC1: std_logic_vector(13 downto 0);
    variable VA: std_logic_vector(13 downto 0);
    variable VB: std_logic_vector(13 downto 0);
    variable VC: std_logic_vector(13 downto 0);
    variable SA: std_logic_vector(13 downto 0);
    variable SB: std_logic_vector(13 downto 0);
    variable SC: std_logic_vector(13 downto 0);
begin --process
if RST='1' then
  divA<="00000000000000";
  divB<="000000001";
  READY<='1';
  READY_sig<='1';
  COUNT:="0000";
  DA:="000000000";
  DB:="000000000";
  DC:="000000000";
  VA:="00000000000000";
  VB:="00000000000000";
  VC:="00000000000000";
elsif CLK'event and CLK='1' then
  if LOAD='1' and (READY_sig='1' or (COUNT>WIN)) then
  COUNT:="0000";
  --Sample
  --DA:=c1
  DA:=c1;
  -- DB:=max(c2,c3)
  if c2>c3 then DB:=c2;
  else DB:=c3;
  end if;
  --DC:=c3
  DC:=c4;
  --Defuzz: Multiplication
  --VA:=DA*POS_40;
    VA1:="00000"&DA;
  --SA:=shl(VA1,"1")+shl(VA1,"11"); --VA1*10
    SA:=VA1;
    VA:=shl(SA,[10"); --SAv*4
  --VB:=DD*POS_30;
```

```
    VB1:="00000"&DB;
--SB:=shl(VB1,"1")+shl(VB1,"11"); --VB1*10
    SB:=VB1;
    VB:=shl(SB,"1")+SB; --SAc*3
--  VC:=DC*POS_20;
    VC1:="00000"&DC;
--  SC:=shl(VC1,"1")+shl(VC1,"11"); --VC1*10
    SC:=VC1;
    VC:=shl(SC,"1"); --SCv*2
elsif READY_sig='0' then
  COUNT:=COUNT+1;
  VA:=VA-SA;
  VB:=VB-SB;
  VC:=VC-SC;
else
-- LOAD=0, READY_sig=1
-- Do nothing
end if;
--READY,LOAD
if COUNT=WIN then
  --Inputs to the divider
  divA<=VA+VB+VC;
  divB<=DA+DB+DC;
  READY_sig<='1';
  READY<='1';
else
  READY_sig<='0';
  READY<='0';
end if; --WIN=COUNT
end if; --RST,CLK
end process;
end Defuzz_arc;
-- Configuration
configuration Defuzz_conf1 of Defuzz is
    for Defuzz_arc
    end for;
end Defuzz_conf1;
```

### 11.6.5 Divider

```
--  File: Divide.vhd
--  Remarks:
--     Part of Fuzzy Logic Controller
--     Implementation Version
--     Includes output interface
--     Output of divider is Y
--     Output of interface is U
--     Overflow limit included
library ieee;
use ieee.std_logic_1164.all;
use ieee.std_logic_unsigned.all;
```

```
use ieee.std_logic_arith.all;
entity Divider is
     port (
       -- Standard Input
     CLK: in std_logic;
     RST: in std_logic;
     LOAD: in std_logic;
     -- Inputs
         divA: in std_logic_vector(13 downto 0);
         divB: in std_logic_vector(8 downto 0);
     -- Outputs
     READY: out std_logic;
     U: out std_logic_vector(7 downto 0));
end Divider;
architecture Divider_arc of Divider is
     signal READY_sig: std_logic;
begin
MAIN_PROCESS:
process(CLK,RST)
     --Variables for divider
     variable SIGN: std_logic;
     variable divBp: std_logic_vector(13 downto 0);
     variable divBn: std_logic_vector(13 downto 0);
     variable A: std_logic_vector(13 downto 0);
     variable divA_var: std_logic_vector(12 downto 0);
     variable Y1_var: std_logic_vector(12 downto 0);
     variable Y: std_logic_vector(7 downto 0);
     variable U_past: std_logic_vector(7 downto 0);
     variable U_var: std_logic_vector(7 downto 0);
     variable COUNT: integer range -1 to 11;
begin
if RST='1' then
     U<="01111111";
     U_past:="01111111";
     READY_sig<='1';
     READY<='0';
     COUNT:=11;
elsif CLK'event and CLK='1' then
--Loading new input values and perform first division sequence
--Condition: Load=1 & Ready=1
if LOAD='1' and READY_sig='1' then
--Division by zero check
if divB="00000000" then
     --Avoid division by zero: assign Y=0
     Y1_var:="0000000000000";
     READY<='1';
     READY_sig<='1';
else --divB
     --Assign D(+ve) and D(-ve)
     divBp:="00000"&divB;
     divBn:=not(divBp)+"00000000000001";
     --Convert SIGNED into UNSIGNED
```

```
    SIGN:=divA(13);
    if SIGN='1' then
            divA_var:=not(divA(12 downto 0))+"01";
  else
            divA_var:=divA(12 downto 0);
  end if; --SIGN type conversion
  --First division sequence after loading
  A(13 downto 1):="0000000000000";
  A(0):=divA_var(12);
  divA_var:=shl(divA_var,"1");
  A:=A+divBn;
  Y1_var(12):=not(A(13));
  COUNT:=11; --END First division sequence
end if;  --divB
--Subsequent Division sequence.
--Condition: Load=[don't care] & Ready=0
elsif (READY_sig='0' and COUNT>0) then
```

---

```
-- DIVIDER   --
-- divA: in std_logic_vector(13 downto 0);          --
-- divB: in std_logic_vector(7 downto 0);           --
-- Y: out std_logic_vector(7 downto 0);             --
-- U: out std_logic_vector(7 downto 0);             --
    COUNT:=COUNT-1;
    A:=shl(A,"1");
    A(0):=divA_var(12);
    divA_var:=shl(divA_var,"1");
    if Y1_var(COUNT+1)='0' then
            A:=A+divBp;
  else
            A:=A+divBn;
  end if;
  Y1_var(COUNT):=not(A(13));
--LOAD=0, READY=1
--No operation
end if; --LOAD, READY
if COUNT=0 then
    --Ready to spit out the answer
    --Converts back into SIGNED value (2's complement)
    --Assume that Y2_var's value does not exceed 8bits(signed)
  Y:=Y1_var(7 downto 0);
  if SIGN='1' then
--  Y2_var:=not(('0'&Y1_var)-"01");
--  Y<=Y2_var(7 downto 0);
--Negative
    U_var:=U_Past-Y;
--Set lower limit "00000000"
    if U_var>U_past then
            U_past:="00000000";
            U<="00000000";
    else
            U_past:=U_var;
```

```
              U<=U_var;
              end if;
  else
--Positive
              U_var:=U_Past+Y;
--Set upper limit "11111111"
     if U_var<U_past then
              U_past:="11111111";
              U<="11111111";
  else
              U_past:=U_var;
              U<=U_var;
  end if;
  end if; --SIGN
  READY<='1';
  READY_sig<='1';
else --COUNT
    --Not ready: condition: COUNT != 0
    READY<='0';
    READY_sig<='0';
end if; --COUNT
end if; --CLK,RST
end process;
end Divider_arc;
-- Configuration
configuration Divider_conf1 of Divider is
    for Divider_arc
    end for;
end Divider_conf1;
```

### 11.6.6 Top hierarchy component of FLC

```
--  File: Control.vhd
--  Remarks: Fuzzy Logic Controller (Top hierarchy component)
--  Contains Deriv, Fuzzify2, Infer, Defuzz, Divider
--  Modified to fit 9bits
library ieee;
use ieee.std_logic_1164.all;
use ieee.std_logic_signed.all;
use ieee.std_logic_arith.all;
entity Control is
    port (
        -- inputs
    CLK1: in std_logic;
        CLK2: in std_logic;
        RST: in std_logic;
        Vdc: in std_logic_vector(7 downto 0);
        Vref: in std_logic_vector(7 downto 0);
        --Test Probes
        --x1: out std_logic_vector(8 downto 0);
        --x2: out std_logic_vector(8 downto 0);
```

```
        -- Outputs
        U: out std_logic_vector(7 downto 0);
        READY: out std_logic);
end Control;
architecture Control_arc of Control is
component Deriv
  port(
        CLK: in STD_LOGIC;
        RST: in STD_LOGIC;
        Vdc: in std_logic_vector(7 downto 0);
        Vref: in std_logic_vector(7 downto 0);

        x1: out std_logic_vector(8 downto 0);
        x2: out std_logic_vector(8 downto 0));
end component;
component Fuzzify2
    port(
        -- inputs
        CLK: in STD_LOGIC;
        RST: in STD_LOGIC;
        x1: in std_logic_vector(8 downto 0);
        x2: in std_logic_vector(8 downto 0);
        -- address
        ADR1: out std_logic_vector(1 downto 0);
        ADR2: out std_logic_vector(1 downto 0);
        -- fuzzy sets for x1 (data)
        B1_A: out std_logic_vector(8 downto 0);
        B1_B: out std_logic_vector(8 downto 0);
        -- fuzzy sets for x2 (data)
        B2_A: out std_logic_vector(8 downto 0);
        B2_B: out std_logic_vector(8 downto 0);
        READY: out std_logic);
end component;
component Infer
    port(
    -- Standard Input
    CLK: in std_logic;
    LOAD: in std_logic;
    RST: in std_logic;
    -- Inputs
        ADR1: in std_logic_vector(1 downto 0);
        B1_Ad: in std_logic_vector(8 downto 0);
        B1_Bd: in std_logic_vector(8 downto 0);
        ADR2: in std_logic_vector(1 downto 0);
        B2_Ad: in std_logic_vector(8 downto 0);
        B2_Bd: in std_logic_vector(8 downto 0);
        -- Outputs
        win: out std_logic_vector(3 downto 0);
        c1: out std_logic_vector(8 downto 0);
        c2: out std_logic_vector(8 downto 0);
        c3: out std_logic_vector(8 downto 0);
        c4: out std_logic_vector(8 downto 0));
end component;
```

```
component Defuzz
    port (
    -- Standard Input
    CLK: in std_logic;
    RST: in std_logic;
    LOAD: in std_logic;
    WIN: in std_logic_vector(3 downto 0);
    c1: in std_logic_vector(8 downto 0);
    c2: in std_logic_vector(8 downto 0);
    c3: in std_logic_vector(8 downto 0);
    c4: in std_logic_vector(8 downto 0);
    -- Outputs
    READY: out std_logic;
    divA: out std_logic_vector(13 downto 0);
    divB: out std_logic_vector(8 downto 0));
end component;
component Divider
    port (
      -- Standard Input
      CLK: in std_logic;
      RST: in std_logic;
      LOAD: in std_logic;
      -- Inputs
        divA: in std_logic_vector(13 downto 0);
        divB: in std_logic_vector(8 downto 0);
      -- Outputs
      READY: out std_logic;
      U: out std_logic_vector(7 downto 0));
end component;
--Signal Declaration
signal SIG2_4, SIG4_5: std_logic;
signal x1_sig, x2_sig: std_logic_vector(8 downto 0);
signal adr1_sig, adr2_sig: std_logic_vector(1 downto 0);
signal B1a_sig, B1b_sig, B2a_sig, B2b_sig: std_logic_vector(8 downto 0);
signal c1_sig, c2_sig, c3_sig, c4_sig: std_logic_vector(8 downto 0);
signal win: std_logic_vector(3 downto 0);
signal DA_sig: std_logic_vector(13 downto 0);
signal DB_sig: std_logic_vector(8 downto 0);
signal READY2, READY4: std_logic;
begin
Deriv_U: Deriv
    port map(
    --in
    CLK=>CLK1,
        RST=>RST,
        Vdc=>Vdc,
        Vref=>Vref,
        --out
        x1=>x1_sig,
        x2=>x2_sig);
Fuzzify2_U: Fuzzify2
    port map(
```

```
    --in
    CLK=>CLK2,
        RST=>RST,
        x1=>x1_sig,
        x2=>x2_sig,
        --out
        ADR1=>adr1_sig,
        ADR2=>adr2_sig,
        B1_A=>B1a_sig,
        B1_B=>B1b_sig,
        B2_A=>B2a_sig,
        B2_B=>B2b_sig,
        READY=>READY2);
Infer_U: Infer
      port map(
      --in
    CLK=>CLK2,
    LOAD=>READY2,
    RST=>RST,
    ADR1=>adr1_sig,
        B1_Ad=>B1a_sig,
        B1_Bd=>B1b_sig,
        ADR2=>adr2_sig,
        B2_Ad=>B2a_sig,
        B2_Bd=>B2b_sig,
        -- Outputs
        win=>win,
        c1=>c1_sig,
      c2=>c2_sig,
      c3=>c3_sig,
      c4=>c4_sig);
Defuzz_U: Defuzz
      port map(
      -- Standard Input
      CLK=>CLK2,
      RST=>RST,
      LOAD=>SIG2_4,
      WIN=>win,
      c1=>c1_sig,
      c2=>c2_sig,
      c3=>c3_sig,
      c4=>c4_sig,
      READY=>READY4,
      divA=>DA_sig,
      divB=>DB_sig);
Divider_U: Divider
      port map(
      CLK=>CLK2,
      RST=>RST,
      LOAD=>SIG4_5,
      divA=>DA_sig,
      divB=>DB_sig,
      READY=>READY,
```

```
        U=>U);
--Infer-Defuzz
process(CLK2,RST)
begin
if RST='1' then
      SIG2_4<='0';
elsif CLK2'event and CLK2='1' then
   if READY2='1' then
      SIG2_4<='1';
   else
      SIG2_4<='0';
   end if;
end if;
end process;
--Defuzz-Divider
process(CLK2,RST)
begin
if RST='1' then
      SIG4_5<='0';
elsif CLK2'event and CLK2='1' then
   if READY4='1' then
      SIG4_5<='1';
   else
      SIG4_5<='0';
   end if;
end if;
end process;
--Probe
--x1<=x1_sig;
--x2<=x2_sig;
end Control_arc;
--configuration Control_conf1 of Control is
--for Control_arc
--    for Deriv_U: Deriv use configuration work.Deriv_conf1;
--    end for;
--    for Fuzzify2_U: Fuzzify2 use configuration work.Fuzzify2_conf1;
--    end for;
--    for Infer_U: Infer use configuration work.Infer_conf1;
--    end for;
--    for Defuzz_U: Defuzz use configuration work.Defuzz_conf1;
--    end for;
--    for Divider_U: Divider use configuration work.Divider_conf1;
--    end for;
--    end for;
--    end Control_conf1;
```

## 11.7 Appendix G – PWM controllers

### 11.7.1 C++ program for PWM waveform generation

```
// This program generates the PWM waveforms based on the desired
parameters.
```

```
#include <iostream.h>
#include <fstream.h>
#include <string.h>
#include <math.h>
#include <stdio.h>
// Function prototype
int Sign(float);
int main(void)
  {
  fstream out_file;
  //define stream object
  char filename[30] ;
  char T;
  int Choice;
  float n,NT,M,N;
  //initialise values
  N=4096; //EPROM memory spaces, Sinewave period
  int A[4096],B[4096],C[4096],NA[4096],Out[4096];
  float Tri[4096];
  cout <<"\n Pulse Width Modulation Pattern Generation Program";
  // Request for parameters
  cout <<"\n";
  cout <<"\n1. Three Phase.";
  cout <<"\n2. Single Phase, bipolar voltage switching.";
  cout <<"\n3. Single Phase, half controlled switching.";
  cout <<"\n\nEnter selection (1-3): ";
  cin >>Choice;
  cout <<"\nEnter triwave period, NT (N=4096) : ";
  cin >>NT;
  cout <<"Enter amplitude modulation factor : ";
  cin >>M;
  //Triangular wave generation
  Tri[0]=0;
  for(n=1;n<N;n++)
          {
          Tri[n]=Tri[n-1]+(Sign(sin(2*M_PI*(n/NT)+M_PI)-
sin(2*M_PI*((n-1)/NT)+M_PI))*(4.0/NT));
          }
  //Comparator - Phase A only
  for(n=0;n<N;n++)
          {
          // Phase A
          if((M*sin(2*M_PI*(n/N))) >= Tri[n])
          {
          A[n]=2;
          NA[n]=1;
          }
        else
          {
          A[n]=1;
          NA[n]=2;
          }
        }
```

```
// Defining Phases B & C - shift by 120deg (1365)
for (n=0;n<N;n++)
        {
        // Phase B
        if((M*sin((2*M_PI*(n/N))+2.094395)) >= Tri[n])
        {
        B[n]=2;
        }
        else
        {
        B[n]=1;
        }

        // Phase C
        if((M*sin((2*M_PI*(n/N))-2.094395)) >= Tri[n])
        {
        C[n]=2;
        }
        else
        {
        C[n]=1;
        }
        }
// Assigning values for Out[n]
switch(Choice) {
        case 1:
        for (n=0;n<N;n++)
        {
        Out[n]= A[n]+(B[n]*4)+(C[n]*16);
        }
        break;
        case 2:
        for (n=0;n<N;n++)
        {
        Out[n]= A[n]+(NA[n]*4);
        }
        break;
        case 3:
        for (n=0;n<N;n++)
        {
        if (n<(N/2))
        {
        Out[n]=A[n]+(2*4);
        }
        else
        {
        Out[n]=A[n]+(1*4);
        }
        }
        }
/***** Filing Operation *****/
cout <<"\nEnter name of input file : ";
gets(filename);
```

```
  cout <<"\nInclude test parameters (y/n) ? ";
  cin >>T;
  out_file.open(filename, ios::out);
  // Filing Error
  if(! out_file)
          {
          cout << "\nUnable to open file ";
          return 1;
          }
  cout << "\nWriting data into file : " << filename;
  // Writing into file
  if(T=='y'|| T=='Y')
          {
          out_file <<"<Testing Parameters>\n";
          switch(Choice)
          {
          case 1:
          out_file <<"Three Phase\n";
          break;
          case 2:
          out_file <<"Single Phase (full)\n";
          break;
          case 3:
          out_file <<"Single Phase (half)\n";
          break;
          }
  out_file <<"N=" <<N <<" ; NT=" <<NT <<" ; Ma="<<M <<"\n\n";
          }
  for(n=0;n<N;n++)
          {
          out_file << hex << Out[n] <<" ";
          }
    out_file.close();
    cout <<"\n\nOk.";
    return 0;
    }
/***** End of Main() *****/
//FUNCTION : Sign
//PURPOSE  : To return the sign of the float a.
int Sign(float a)
    {
    if(a<0)
          {
          return -1;
          }
    else
          {
          return 1;
          }
    }
```

### 11.7.2 PIC assembly code to control SA828

```
;This program initialises the SA828 pulse width modulation chip
LIST C=80, P=16C84, F=INHX8M
         include "c:\pic\p16cxx.inc"
         list
;*********************** EQUATES *************************
;*******************PIN/BIT DEFINITIONS*********************
        __FUSES _CP_OFF&_PWRTE_OFF&_WDT_OFF&_XT_OSC
;        __CONFIG 11H

#DEFINE ALE   0 ;PORT A
#DEFINE WRTE  1 ;PORT A
#DEFINE RST   2 ;PORT A
#DEFINE LIGHT 3 ;PORT A CONFIDENCE LIGHT
#DEFINE SWTCH 4 ;PORT A INPUT
;PORTB IS ALL ADDRESS LINES AND DATA LINES
ADDRSS        EQU          11H
DAT           EQU          12H
COUNTER       EQU          13H
COUNTER2      EQU          14H
ORG     000H
RESET  GOTO   START
;**********************MAIN PROGRAM*************************
START    CLRF      STATUS            ;Initialise port b as outputs
         MOVLW     0X00
         MOVWF     PORTB
         MOVLW     0X00
         TRIS      PORTB
         ; port a arranged as all outputs
         CLRF      PORTA
         MOVLW     0X10
         TRIS      PORTA
         BCF       EEADR,7           ;Clear EE top addresses to minimise
                                      power
         BCF       EEADR,6
;SORT OUT ALL THE INTERRUPTS
         BCF       INTCON,GIE                ;GLOBAL INTERRUPT
                                              DISABLE
         BCF       INTCON,EEIE               ;NO EEPROM INTERRUPT
         BCF       INTCON,T0IE               ;NO TIMER INTERRUPT
         BSF       INTCON,INTE               ;PORT B PIN 6 INTERRUPT
                                              ENABLED
         BCF       INTCON,RBIE               ;CHANGE ON PORT B
                                              INTERRUPT BISABLED
         BSF       STATUS,RP0
         BCF       0X1,INTEDC                ;INTERRUPT ON FALLING
                                              EDGE
                                             ;FOR PORT B PIN 6
                                              INTERRUPT
         BCF       STATUS,RP0
;****FIRST MAKE SURE DEVICE IS RESET*************
```

```
        BCF         PORTA,RST
MAIN
                                        ;*******SENDS D2 TO ADDRESS 0
        MOVLW       0
        MOVWF       ADDRSS
        MOVLW       0XD2
        MOVWF       DAT
        CALL        SENDIT
                                        ;*******SENDS 0 TO ADDRESS 1
        MOVLW       1
        MOVWF       ADDRSS
        MOVLW       0X00
        MOVWF       DAT
        CALL        SENDIT
                                        ;*******SENDS 7F TO ADDRESS 2
        MOVLW       2
        MOVWF       ADDRSS
        MOVLW       0X7F
        MOVWF       DAT
        CALL        SENDIT
                                        ;*******SENDS FF TO ADDRESS 4
        MOVLW       4
        MOVWF       ADDRSS
        MOVLW       0XFF
        MOVWF       DAT
        CALL        SENDIT
                                        ;*******SENDS CD TO ADDRESS 0
        MOVLW       0
        MOVWF       ADDRSS
        MOVLW       0XCD
        MOVWF       DAT
        CALL        SENDIT
                                        ;*******SENDS 0C TO ADDRESS 1
        MOVLW       1
        MOVWF       ADDRSS
        MOVLW       0X0C
        MOVWF       DAT
        CALL        SENDIT
                                        ;*******SENDS CC TO ADDRESS 2
        MOVLW       2
        MOVWF       ADDRSS
        MOVLW       0XCC
        MOVWF       DAT
        CALL        SENDIT
                                        ;*******SENDS FF TO ADDRESS 3
        MOVLW       3
        MOVWF       ADDRSS
        MOVLW       0XFF
        MOVWF       DAT
        CALL        SENDIT
;Enable PWM Output
        BSF         PORTA,RST
```

```
                                        ;*******SENDS CD TO ADDRESS 0
        MOVLW       0
        MOVWF       ADDRSS
        MOVLW       0XCD
        MOVWF       DAT
        CALL        SENDIT
                                        ;*******SENDS 2C TO ADDRESS 1
        MOVLW       1
        MOVWF       ADDRSS
        MOVLW       0X2C
        MOVWF       DAT
        CALL        SENDIT
                                        ;*******SENDS CC TO ADDRESS 2
        MOVLW       2
        MOVWF       ADDRSS
        MOVLW       0XCC
        MOVWF       DAT
        CALL        SENDIT
                                        ;*******SENDS FF TO ADDRESS 3
        MOVLW       3
        MOVWF       ADDRSS
        MOVLW       0XFF
        MOVWF       DAT
        CALL        SENDIT
WAIT1   MOVLW       0XFF
        MOVWF       COUNTER
        MOVWF       COUNTER2
WAIT2   NOP
        NOP
        DECFSZ      COUNTER,F
        GOTO        WAIT2
        DECFSZ      COUNTER2,F
        GOTO        WAIT2
        BSF         PORTA,LIGHT
        MOVLW       0XFF
        MOVWF       COUNTER
        MOVWF       COUNTER2
WAIT3   NOP
        NOP
        DECFSZ      COUNTER,F
        GOTO        WAIT3
        DECFSZ      COUNTER2,F
        GOTO        WAIT3
        BCF         PORTA,LIGHT
        GOTO        WAIT2
        GOTO        MAIN

;THIS SUBROUTINE IS USED TO DRIVE ALE AND WRTE HIGH AND LOW
;AS PER THE INTEL TIMING SPECIFICATIONS
SENDIT
        BSF         PORTA,ALE           ;take ale high
        MOVF        ADDR33,W            ;send the address
```

```
        MOVWF       PORTB
        BCF         PORTA,ALE           ;take ale low
        BCF         PORTA,WRTE          ;take write low
        MOVF        DAT,W
        MOVWF       PORTB               ;send the data
        BSF         PORTA,WRTE          ;take write high
        RETURN
END
```

# Index